Suppose and Tell

T0347631

What does 'if' mean?

This book argues for a new approach to understanding conditionals, based on the idea that in assessing them we are guided by psychological heuristics, fast and frugal methods, mostly but not always reliable, although the user may have no advance warning from the inside of their fallibility. As a result, philosophers and linguists have been led astray in theorizing about conditionals, because they have taken erroneous judgements about examples as data; simple theories have been too quickly dismissed. The main heuristic involves making a supposition and exploring its consequences, often in the imagination. Although powerful, this method can lead to paradoxes. A secondary heuristic is to accept conditionals on the word of someone you trust. The two heuristics are sometimes in tension. However, formal results help show why the simple 'material' semantics makes the best sense of the two heuristics, even though they generate apparent counter-examples to that semantics. The second half of the book presents a new theory of the meaning of counterfactuals about what *would* be *if* things were different; it is explained in terms of the separate meanings of 'if' and 'would'. This account is applied to the role of counterfactuals in thought experiments. *Suppose and Tell: The Semantics and Heuristics of Conditionals* analyses a wide variety of linguistic data, and discusses the cognitive value of conditionals in our lives, from everyday decision-making to mathematical proofs. Wider implications are drawn for the nature of meaning and its non-transparency to native speakers, vagueness in thought and language, and the need for semantics to attend to the unreliable heuristics underlying our judgements.

Timothy Williamson is Wykeham Professor of Logic at Oxford University and Whitney Griswold Visiting Professor at Yale University. He has also taught at MIT, Princeton, Michigan, Edinburgh, Trinity College Dublin, Chinese University of Hong Kong, University of Canterbury (New Zealand), and elsewhere. He works on logic, philosophy of language, epistemology, metaphysics, and metaphilosophy. He is a fellow of the British Academy and the Royal Society of Edinburgh, foreign honorary member of the American Academy of Arts and Sciences, member of Academia Europaea, honorary member of the Royal Irish Academy, foreign member of the Norwegian Academy of Science and Letters, and member of the Institut International de Philosophie.

Suppose and Tell

The Semantics and Heuristics of Conditionals

Timothy Williamson

OXFORD
UNIVERSITY PRESS

OXFORD
UNIVERSITY PRESS

Great Clarendon Street, Oxford, OX2 6DP,
United Kingdom

Oxford University Press is a department of the University of Oxford.
It furthers the University's objective of excellence in research, scholarship,
and education by publishing worldwide. Oxford is a registered trade mark of
Oxford University Press in the UK and in certain other countries

First published 2020
First published in paperback 2022

Published in the United States of America by Oxford University Press
198 Madison Avenue, New York, NY 10016, United States of America

British Library Cataloguing in Publication Data
Data available

Library of Congress Cataloging in Publication Data
Data available

ISBN 978–0–19–886066–2 (Hbk.)
ISBN 978–0–19–287104–6 (Pbk.)

Preface

Conditionals are hard to avoid in reasoning; reasoning about conditionals is correspondingly hard to avoid in reasoning about reasoning. For many years, I kept bumping into them in connection with other philosophical issues: counterfactual conditionals as alleged counterexamples to the principle of bivalence (Williamson 1988), indicative conditionals in non-bivalent semantic theories of vagueness (Williamson 1994), non-standard 'consequential' conditionals defined within modal logic (Pizzi and Williamson 1997, 2005), differences between indicative and counterfactual conditionals as they interact with the modal operator 'actually' (Williamson 2006, 2009), the logic of counterfactual conditionals underlying thought experiments and generating principles of modal logic (Williamson 2007a, 2013). At the interface of epistemology and psychology, humans' cognitive use of supposition and imagination to assess counterfactual conditionals, often quite reliably, is a template for our capacity to learn about objective possibility and necessity through thought experiments (Williamson 2007, 2016d). That line of thought drew me into a debate over the orthodox view that counterfactual conditionals with impossible antecedents are vacuously true—an academic question if ever there was one. However, it has significant ramifications. The putative counterexamples to orthodoxy *feel* compelling; they rely on the same form of supposition-based thinking as we use in assessing other counterfactual conditionals. But that form of thinking can be shown to be subtly inconsistent, and so less than perfectly reliable, in particular for conditionals whose antecedents are themselves inconsistent. The objections to orthodoxy tacitly rely on a usually but not universally reliable heuristic in just the sort of case where it can be predicted to fail (Williamson 2016c, 2017a).

Despite my inability to leave conditionals alone, until spring 2018 I expected never to write a book on them. I thought that I had too little to add but details to the extant literature. On counterfactual conditionals, I feared that David Lewis's theory of them might be about as good as we can get. On indicative conditionals, my unfashionable sympathies were with the plain truth-functional account, on which many plausible logical considerations converge, but no adequate counter seemed available to many of the apparent counterexamples. However, deeper analysis of a suppositional heuristic for indicative conditionals showed that all the data which look so bad for the truth-functional view are predicted by the hypothesis that we rely on that heuristic—even though it cannot be fully reliable. I could thereby see my way to a full defence of the truth-functional view, the simplest semantics for a conditional. Given that plain 'if' is just the truth-functional conditional, it also becomes much easier to see the counterfactual conditional, not as another *sui generis* conditional, but as generated within a compositional semantics by the interaction of plain 'if' with an autonomous modal operator, such as 'would', resulting in a contextually restricted strict conditional. Along those lines, I wrote the first draft of this book in a year, pretty much from scratch.

This book does not contain extensive discussion of alternative theories of conditionals. I indicate what I take to be their key faults, but I do not survey all their strengths and weaknesses. Such discussion is available elsewhere, and this book is long enough as it is. My central aim has been to explain the idea that conditional thinking relies on a fallible heuristic, how it fits the proposed simple semantics, and how the combined view makes sense of a wide range of complex and recalcitrant-looking data. My hope is that this presentation will encourage theorists of conditionals to take a more critical attitude to the data, and enable the positive view to be judged on its own merits.

Readers may recognize a common theme to this book and my work on vagueness: the danger of giving semantic solutions to epistemic problems. In both cases, the trap is sprung by over-hasty articulations of the data in semantic terms. Alas, good practice offers no ritual observance to avoid all such traps: as in the rest of science, whatever 'observation language' we speak, no statement of the evidence is in principle beyond question. Instead, we must introduce local safeguards after bitter experience shows them to be needed. For both vagueness and conditionals, by now the experience has been bitter enough. Some sections of this book tighten the analogy, by explaining how the 'tolerance principles' which induce paradoxes for vague terms when treated semantically are better understood as convenient psychological heuristics for applying such terms, reliable but not infallible.

Obviously, this book could not have been written without the long tradition of previous work on conditionals and related topics. Indeed, the bare bones of its semantic treatments of indicative and counterfactual conditionals were already familiar to the ancient Greeks. Even my more personal debts go back several decades, and can be traced through the acknowledgments in my previous publications, cited above, and of course the co-authorship with Claudio Pizzi. I will not repeat them here. Since the present project took shape, I have been the beneficiary of excellent feedback from audiences at presentations of various parts of it. The venues included: Yale University, the University of Oxford, the 2018 joint conference of the Australasian Association of Philosophy and the New Zealand Association at the Victoria University of Wellington, the 2019 Eastern Division meeting of the American Philosophical Association in New York, where Michael Caie was the commentator, and conferences, workshops, and colloquia at the universities of Belgrade, Bochum, Cologne, Edinburgh, Tübingen, Warsaw, King's College London, University College London, and (by Skype) the Humboldt University of Berlin. Outside those events, I have gained from specific comments on the material from many individuals, including John Bigelow, Will Davies, Christina Dietz, Igor Douven, Dorothy Edgington, John Hawthorne, Matthew Hewson, Laurence Horn, James Kirkpatrick, Angelika Kratzer, Matthew Mandelkern, Sarah Moss, Jennifer Nagel, Daniel Rothschild, Jason Stanley, Zoltán Szabó, and three anonymous readers for Oxford University Press, who provided thoughtful and constructive comments. Apologies to all who have contributed to the book in ways which no longer come to mind. Peter Momtchiloff has been, as always, a supportive and encouraging editor.

For no special reason, I strongly associate this book with the places where it was written, especially a room on the island of Korcula with a view across the channel to Sveti Ilija mountain, a darker room in a Milan hotel, and my usual rooms in New College, Oxford, Connecticut Hall, Yale, and at home in Iffley (= Ifandonlyifley). My wife Ana finds conditional speeches irritating, and sometimes complains: "Too many 'if' clauses!" This book has *far* too many 'if' clauses. But, with all their faults, conditionals can be understood.

Contents

PART I

If

1

The Value of Conditionals

1.1 Introduction

What if...? Hypothetical thinking is central to human cognitive life, from the naïve to the super-sophisticated. A sign of its importance is how often we use the word 'if', or its equivalents in other languages, in discourse from everyday conversation to abstract science. If a creature cannot think hypothetically, how intelligently can it live? That is itself a hypothetical question.

We rely on hypothetical thinking in deciding what to do. Choosing between two alternative courses of action, you compare what will happen if you take one course with what will happen if you take the other. 'If I turn right, I'll come to the river. If I turn left, I'll come to the forest.' What will happen depends too on unknown circumstances beyond your control. 'If it rains, I'll get wet.' You make a plan and form the intention to put it into effect. That can include intentions conditional on the circumstances. 'If the river is low, I'll wade across. If it is high, I'll walk all the way down to the bridge.' Such thinking is no mere quirk of human psychology. It simply reflects the general predicament of decision-making under uncertainty, a central feature of intelligent life.

Many conditionals are not one-off devices, to be discarded after immediate use. One can retain a conditional intention for years. 'If I ever meet him, I'll tell him what I think of him.' One can also retain conditional knowledge for later use. 'If I go by the forest path, it will take me four hours.' One can communicate such conditional knowledge in words to someone else. 'If you look west from the top of the hill, you can see the river.' Conditionals can survive robustly from one time to another and from one person to another.

We can communicate our intentions, including our conditional intentions, to other people. Some of those intentions can be *for* other people, indeed, for those we are speaking to. We may issue requests, instructions, or orders, in the form of imperatives, including conditional imperatives. 'If the river is low, wade across.' Conditional commands are central to computer programming languages. 'If condition C holds do X, else do Y.' Both humans and computers need conditional instructions to deal with the unpredictable variety of situations they may encounter.

Imperatives are of course closely related to declaratives (even when understood as predictions): 'If the river is low, wade across' corresponds to 'If the river is low, you will wade across.' You can make sure that you comply with both, either cognitively, by observing that the river is not low, or conatively, by wading across. Such alternatives may be available even when you know neither whether the antecedent

Suppose and Tell: The Semantics and Heuristics of Conditionals. Timothy Williamson, Oxford University Press (2020).
© Timothy Williamson.
DOI: 10.1093/oso/9780198860662.001.0001

holds nor whether the consequent does. At a party where you have no idea who belongs to which family, you can make sure that you comply with both the instruction 'If you speak to a Montague, don't speak to a Capulet' and the prediction 'If you speak to a Montague, you won't speak to a Capulet', either cognitively, by being reliably informed that no party is mixed, or conatively, by speaking to only one person. Uses of 'if' in declaratives and imperatives are part of the same overall practice. Section 7.2 has more discussion of conditional commands.

Hypothetical thinking comes in more than one flavour. A distinction is often drawn between *indicative* conditionals, such as 'If those mushrooms were poisonous, he is dead' and so-called *subjunctive* or *counterfactual* conditionals, such as 'If those mushrooms were poisonous, he would be dead'. Both correspond to hypothetical thinking, but of different sorts: the hypothesis is treated differently. Chapters 9–11 discuss the difference.

The most plausible initial hypothesis is that 'if' means the same in the two types of sentence: the source of the difference is an extra modal operator in subjunctive conditionals, introduced by 'would' or by some other means. Chapter 10 vindicates that hypothesis: it derives the semantics of 'subjunctive conditionals' compositionally, by combining the semantics of 'if' with the independent semantics of 'would'. Chapters 1–8 concentrate on indicative conditionals, as crudely distinguished by their superficial form.

Unsurprisingly, 'if' is one of the most commonly used words of English. On one count, it comes in fortieth place.[1] It is never far away when we are uncertain. One may wonder what 'if' means, to have this cognitive power. But the question has the tail wagging the dog. 'If' matters because we use it to articulate the results of hypothetical thinking. Until we have properly understood the practice of hypothetical thinking, we shall not appreciate what the meaning of 'if' has to fit. And once we have properly understood that rich practice, we shall see why that strict and literal meaning has to be utterly impoverished by comparison. Conditionals are too important to be left to semantics alone.

To use a conditional in practice, we need more than bare knowledge of its strict and literal meaning. If we cannot tell whether it is applicable to the situation we find ourselves in, it is to that extent less useful. We must be able to combine conditionals productively with our background knowledge or belief. We need to *reason* with conditionals—sometimes to reason *from* them, when they are premises, sometimes to reason *to* them, when they are conclusions. But deductive reasoning is only one aspect of our use of conditionals. Often, when we address a 'What if?' question, our background information does not entail an answer. Instead, we have to assess candidate answers by less formal means. We may have no feasible alternative but to use what psychologists call 'heuristics': 'fast and frugal' or, in less positive terms, 'quick and dirty' methods, which tend to give the right answer under normal conditions, but the wrong one in unfavourable cases.[2] For practical purposes, a

---

[1] See http://www.bckelk.ukfsn.org/words/uk1000n.html.

[2] The term 'heuristic' is intended to recall both the heuristics and biases programme in psychology (going back to works such as Kahneman, Slovic, and Tversky 1982) and the partly contrasting tradition of work on adaptive rationality (see for example Gigerenzer, Hertwig, and Pachur 2011). The heuristics

small decrease in the reliability of the answer may be a reasonable price to pay for a large increase in the willingness to give a prompt answer in the first place.

Of course, it is not for a semantic theory of conditionals to determine which specific conditionals 'If A, C' are true or credible: it depends on the specific cognitive properties and relations of the antecedent A and the consequent C, which do not concern that theory.[3] But it *is* a task for a general theory of conditionals to explain how agents get from cognitive capacities to apply sentences A and C to a cognitive capacity to apply the sentence 'If A, C.' As just noted, such cognitive capacities have both deductive and non-deductive aspects. Formal semantic theories of conditionals are typically informative on the deductive side of the task. They predict a deductive logic for conditionals, usually by determining which argument patterns with them are truth-preserving in all models, although they may not say much about how far and how speakers implement that logic in practice. What such theories are much less informative about is the more usual case: the non-deductive cognitive assessment of conditionals. For example, they do not explain how speakers get the practical capacity to apply the sentence 'If the guests all left within three hours, the party was a failure' from the practical capacities to apply the sentences 'The guests all left within three hours' and 'The party was a failure.' The semantic theory may specify which proposition the conditional expresses, as a function of the propositions its antecedent and consequent express, but that does not explain how speakers recognize the truth-value of the proposition in practice. Arguably, it is not the job of semantics to provide such an explanation. That is fine, but then whose job is it? An informative account of speakers' general cognitive capacity to apply conditionals is needed from *somewhere*.

Pragmatics is the usual first resort for filling the gaps between semantics and linguistic practice. However, it is not what is wanted here. In analysing conversational phenomena, pragmatics (legitimately) takes for granted speakers' capacities to make the very cognitive assessments we are now seeking to understand. At the basic cognitive level, what we seek is a matter of psychology rather than linguistics. But it is also hooked up to natural language, by words like 'if'.

Of course, the semantic and cognitive-psychological tasks are not mutually independent. The cognitive methods are for shooting at a target set by the semantics. If a cognitive theory specifies methods quite inappropriate for shooting at the target specified by a semantic theory, presumably something is wrong with at least one of the two theories. In the long run, what need to be compared are theoretical packages combining both semantic and cognitive components. At present, the trouble is that semantic theories of conditionals are often proposed with no accompanying cognitive account: half the theoretical package is missing. Without the full package, they cannot properly allay the suspicion that semantic benefits have been bought at hidden cognitive costs.

---

postulated in this book may be adaptively rational, despite importing biases. Exactly how the connections go with those traditions is left as an open question.

[3] Here and elsewhere in the book, the reader is asked to imagine schematic letters replaced by expressions of the relevant grammatical category, with quotation marks where appropriate.

From the cognitive side, we may expect constraints of *feasibility*, both computational feasibility and feasibility of other cognitive kinds. The semantic theory should not set a target invisible to humans, or out of our range. More subtly, even if many visible potential targets are within range, the semantics should not specify the actual target by a formula too complex to be computed in real time. For, within reasonable limits, normal human beings *do* very often apply conditionals successfully.

Just as in archery, the relevant computations need not be conscious or reflective. Some may be hardwired into our brains, or acquired through long practice with good feedback. However, conditionals are all-purpose devices, at home in virtually any context, including those in which we are reasoning consciously and reflectively. The semantics must reflect our ability to use conditionals in those contexts too. Of course, even when speakers are reasoning consciously and reflectively, they almost never invoke an explicit semantic theory, correct or incorrect. In principle, the complexity of a semantic theory does not prevent its outputs from matching those of a humanly feasible capacity in relevant respects. For a start, the complexity might be eliminable, by formulating an equivalent semantic theory in simpler terms. But if the complexity is ineliminable, a serious question arises as to how humans manage to be so good at the cognitive task of hitting the targets set by such a theory.

Complex semantic theories also tend to validate weaker logics, because their complexities generate more counterexamples to putative logical principles. That in itself is no objection, since human cognition deals easily with many familiar terms which fail to satisfy any non-trivial logical principles. But weakening the deductive logic of conditionals puts still more weight on the non-deductive side, and makes the need for a proper cognitive account of it even more urgent.

In the light of such considerations, other things equal, we might expect 'if' to work quite simply, in both thought and language. This is not just the very general point that simple operations can be characterized by a simple hypothesis, and we should give simple hypotheses priority here, as elsewhere in science. The more complicated the operation of 'if', the harder conditionals will probably be to use—harder to assess, harder to act on. The more moving parts we postulate, the more likely we make the device to break down, or to catch out its users. The density of conditionals in our cognitive life is evidence that 'if' is robust and comparatively easy to handle. Since it is all-purpose equipment, it also needs to be flexible: tailoring its meaning too closely to one task would make it unserviceable for another. In a construction set, the basic components in most frequent use are typically plain, simple ones, without special features to get in the way. 'If' is just such a basic, commonplace component of our thought and talk.

Of course, other things are often not equal; we cannot exclude in advance the possibility that conditionals work in surprisingly complex ways. But every complication has a non-trivial cost, for practitioners as well as theorists. The onus is on those who postulate the complication to make the case that it is really indispensable.[4] This book shows how a very simple account of the semantics and heuristics of

---

[4] There sometimes seems to be an element of macho competition over who can argue for the most complicated semantics for a simple expression. Of course, it would be quite unfair to implicate most semanticists in that tendency.

conditionals may be able to do the explanatory work. In principle, it should be compared against other combined semantic-cognitive accounts, although this is made difficult by the shortage of adequately developed cognitive alternatives. A purely semantic theory of conditionals is only half an alternative.

The rest of this chapter further illustrates the cognitive value of 'if', which an overcomplicated conditional operator would risk breaking. The next chapters provide a very general account of the simple heuristics which guide our use of 'if'. These chapters rely on an ordinary pre-theoretic practice of using 'if' as a speaker of English, without assuming any special semantic theory of the conditional. I focus on 'if', not on conditionals in other languages, because English is the language I know best. Nevertheless, on the view developed in this book, the key to understanding 'if' is its connection to a fundamental cognitive capacity for hypothetical thinking shared by virtually all humans. One would expect all human languages to have a conditional device similarly connected to that capacity, though the details of its implementation may well vary from language to language. As far as I know, that claim of universality is true, but it will be left as a plausible conjecture, for others to test.

1.2 Conditionals and generalizations

A child is going off to pick strawberries. An adult gives her two simple rules:

If a strawberry is red, pick it.
If a strawberry is green, leave it.

Those are conditional imperatives. One might wonder whether that form is really needed. The adult could just as well have said:

Pick the red strawberries.
Leave the green strawberries.

Those are general imperatives, with no 'if'. Even the conditional imperatives had a general element: the indefinite 'a strawberry'. But when the child recognizes a berry in the patch before her, she must instantiate the general rules to the case at hand. She might mutter to herself:

If this is a red strawberry, pick it.
If this is a green strawberry, leave it.

Here the perceptual demonstrative 'this' refers to the particular berry she is looking at. A conditional is the natural form in which to instantiate a general rule to a particular case, the obvious way to articulate what the rule says about that case. More specifically, an indicative conditional is the natural form in which to instantiate a general rule to a particular case, treated as actual rather than merely as possible.

The same connection holds between declarative conditionals and declarative generalizations. The adult might explain the point of the rules to the child by saying:

If a strawberry is red, it is ripe.
If a strawberry is green, it is unripe.

One might again wonder whether the conditional form is really needed. The adult could just as well have said:

Red strawberries are ripe.
Green strawberries are unripe.

Those are general declaratives, with no 'if'. More specifically, they are *generic* statements, not falsified by the occasional anomalous unripe red strawberry or ripe green one. Thus they offer only defeasible support to their instances. But, as before, the natural form for those instances is a conditional:

If this is a red strawberry, it is ripe.
If this is a green strawberry, it is unripe.

The same conditionals are also the natural instances of the corresponding universal generalizations, which do entail them deductively, as when we add 'all' to the generics or, to slightly different effect, say:

Every red strawberry is ripe.
Every green strawberry is unripe.

Unlike the generic generalizations, the universal ones are falsified by anomalous cases, except in contexts where the domain is restricted to exclude such cases.

Instantiating universal generalizations by indicative conditionals is not just an ordinary conversational move. It also plays a central role in scientific theory-testing. In a simplified schema, one moves from the universal generalization 'Every F is a G' to the prediction 'If a is an F, a is a G', where a is the thing to be observed. Less formally, one goes from 'Salt dissolves in water' to 'If this is salt, this dissolves in water.' Scientific practice relies on such inferences, even when the generalization is put forward only as a hypothesis, not an assertion—we may be testing it precisely because it is only a conjecture, so far. The practice relies on the assumption that the argument from a universal generalization to a specific instance in conditional form is truth-preserving, whatever the status of the premise. In scientific contexts, native speakers still fluently speak natural language, using words in their normal senses, including the word 'if', despite the added technical scientific vocabulary and mathematical notation.

For a natural language to lack a simple, easy means of capturing what a universal generalization entails and a generic generalization defeasibly implies about a given particular case would be surprisingly dysfunctional. A conditional like 'if' is the obvious candidate—if it does not do the job, what does?

As alternative candidates for what 'Every F is a G' says about a, someone might suggest the disjunction 'a is not an F or a is a G' or the negated conjunction 'It is not the case that a is an F and a is not a G.' However, each of those is a much less natural candidate than the plain 'If a is an F, a is a G.' They each introduce a negative element that seems extraneous to both the original generalization and the plain conditional. Such negative elements tend to cause computational difficulties for normal speakers, so their role as instances of a generalization would incur significant cognitive costs. To make the point vivid, consider the blatantly trivial generalization 'Every F is an F' (the special case where $F = G$). What it says about a should also be blatantly trivial, given that *a* belongs to the contextually relevant domain for 'every'. The blatantly trivial conditional 'If *a* is an *F*, *a* is an *F*' meets that condition. By contrast, the proposed disjunction '*a* is not an *F* or *a* is an *F*' is much less blatantly trivial; it is indeed a classical logical truth, but it is a typical instance of the law of excluded middle, disputed by some non-classical logicians. The proposed negated conjunction 'It is not the case that *a* is an *F* and *a* is not an *F*' is also less blatantly trivial than the corresponding conditional; it too is indeed a classical logical truth, but it is a typical instance of the law of non-contradiction, disputed by some non-classical logicians (though fewer than for excluded middle). Other proposed alternatives do even worse.

Thus the price of denying that what a universal generalization says about a member of its domain is the corresponding plain conditional may be that one ends up with *no* good candidate in natural language for what a universal generalization says about a member of its domain. It would be quite surprising, though not impossible, for natural language to suffer from such a basic expressive deficiency. Thus we may regard it as at least a defeasible desideratum on a theory of conditionals to explain how they follow, in the relevant sense, from the corresponding generalizations.

The challenge is not just to explain why, when one is in a position to *assert* a generalization, one is also in a position to *assert* its instances in conditional form. As noted, we also draw instances in conditional form from a merely hypothesized generalization, to test its predictions. The form of argument from 'Every *F* is a *G*' to 'If *a* is an *F*, *a* is a *G*' had better be *valid*, in an appropriate sense. This tightly constrains the meaning of 'if'. If a theory imposes a requirement on the truth-conditions of indicative conditionals with no corresponding requirement on the truth-conditions of generalizations, it will invalidate the argument form. For example, since the mere truth of a universal generalization does not in general impose any epistemic requirement (on the speaker's knowledge, for instance), the mere truth of an instance of it should not in general impose any epistemic requirement either (on the speaker's knowledge, for instance).

The truth-conditions of indicative conditionals must not be too strong. There is also a Scylla to that Charybdis. The truth-conditions of indicative conditionals must not be too weak. Not only must the conditional avoid saying *more* about the particular case than the generalization says, it must also avoid saying *less* about the particular case than the generalization says. Otherwise, even though the argument form may be validated, the conditional fails to capture fully what the generalization says about the particular case. But what 'Every *F* is a *G*' says about a particular case *a* is in general quite simple and straightforward—apart from any complexities generated by '*F*' and '*G*', rather than by the universally quantified form itself. We should therefore expect 'If *a* is an *F*, *a* is a *G*' to say something quite simple and straightforward—apart from any complexities generated by '*F*', '*G*', and '*a*', rather than by the conditional form itself (sections 8.1-2 discuss the connection in more detail).

None of this means that there are no serious challenges to the entailment from a universal generalization to the corresponding conditionals. For example, you have massive inductive evidence that all swans are white. Now you glimpse what looks like a black swan. It is clearly black, but you reasonably (though mistakenly) suspect that when properly examined it will turn out not to be a swan. On your evidence, we may suppose, it is still 80 per cent probable that all swans are white. But how probable on your evidence is it that if this bird is a swan, it is white? A tempting answer is that the probability of the conditional is much closer to 0 per cent than to 80 per cent. But, by the laws of probability, a logical consequence of a hypothesis is at least as probable as the hypothesis itself. One might therefore conclude that 'If this bird is a swan, it is white' is *not* a genuine logical consequence of 'All swans are white'.

Chapters 3 and 6 discuss the application of probability to conditionals in detail. They explain how a good but fallible heuristic for assessing conditionals can sometimes lead us drastically to underestimate their probabilities. For now, the reader is not asked to take that explanation on trust. Rather, the point is just that if the explanation works, and we can hold onto the equation of what a universal generalization says about a member of its domain with the corresponding plain conditional, we thereby avoid imputing a strange expressive dysfunctionality to natural language. Of course, opposing theorists of conditionals may claim that natural language *is* dysfunctional in this respect, and perhaps even argue that this dysfunctionality is the obverse of functionality in some other respect. But the onus is on them to provide such an argument, in detail. The admission of dysfunctionality is a significant cost to their view.

The case of the universal quantifier also serves as an apt reminder that we should not expect to read off the full cognitive story about a logical constant from a bare semantic account of its truth-conditional meaning. Very roughly, the truth-conditional semantics for 'every' says that 'Every F is a G' is true if and only if the extension of 'F' is a subset of the extension of 'G'. It does not specify how a speaker is supposed to go about trying to verify 'Every F is a G'—whether by an exhaustive check on each member of the contextually relevant domain, one by one, or by an induction from checks on a sample of members of the domain, or by a deduction from the open formula 'x is an F' to the open formula 'x is a G'. Of course, the truth-conditional meaning should somehow make sense of the cognitive practice, but that does not require the truth-conditional meaning to spell out how the practice works. The connection may be more distant than that. Such latitude will be significant when we come to the semantics of conditionals in chapter 6. Indeed, given the close connection between the universal quantifier and the conditional, we should not expect the cognitive story to be more legible in the bare semantics for 'if' than it is in the bare semantics for 'every'.

1.3 Conditionals across contexts

If the close link between 'if' and 'every' holds, a further point may follow: we should not expect the truth-conditional contribution of 'if' to vary across contexts of utterance in any way in which the truth-conditional contribution of 'every' does not. For if what 'If a is an F, a is a G' says varies between a context C_1 and a context C_2, while what 'Every F is a G' says about a is the same in C_1 as in C_2, then, in at least one of those two contexts, what 'If a is an F, a is a G' says is not what 'Every F is a G' says about a. Of course, the quantifier is itself an independent source of contextual variation, since two contexts may impose different restrictions on the domain of quantification. But if C1 and C2 have the same domain, to which a belongs, and 'F', 'G', and 'a' do not vary in reference between C1 and C2, then 'Every F is a G' should say the same about a in C1 as in C2, so 'If a is an F, a is a G' should also say the same in C1 as in C2. That suggests that 'if' is not an independent source of contextual variation.

Independently of the link with quantifiers, other evidence also tells against 'if' as a source of contextual variation. As seen in section 1.1, conditionals can be communicated from one context to another, without change of wording. Through memory, they can be passed on from one time to another. Through testimony, they can be passed on from one agent to another.

Two clarifications are needed straight away:

First, when the antecedent A or the consequent C is context-dependent, the conditional 'If A, C' will typically inherit that context-dependence. For example, what the sentences 'He is hungry' and 'He is grumpy' express depends on the contextually determined referent of the pronoun 'he'; unsurprisingly, the same goes for the conditional 'If he is hungry, he is grumpy'. The key question is whether 'if' is a *further* source of context-dependence.

Second, the assertibility of any declarative sentence whatsoever depends on the epistemic circumstances of the speaker. However constant the content it expresses across different contexts, that by itself does not prevent it from being assertible in one context and not in another, owing to differences in the available evidence. Such variability does not imply that its content varies with context. In particular, for conditional sentences, we must not confuse the two kinds of epistemic sensitivity.

Indicative conditionals are often held to be context-dependent in ways not reducible to those trivialities. Apart from anything in the antecedent and consequent, 'if' is claimed to introduce a further dimension of context-dependence in which proposition the conditional expresses—not just in whether the speaker is in an epistemic position to assert that proposition. On the most fashionable view of indicative conditionals, the context-dependence does indeed concern the speaker's epistemic position, but as a determinant of which proposition the conditional expresses. Very roughly, it is held to express a proposition about an epistemic connection between its antecedent and its consequent, with respect to the relevant stock of evidence.[5]

The usual analogy for the alleged epistemic context-dependence of indicative conditionals is with the behaviour of epistemic modals such as 'must' and 'may'. For example, as a detective gathers more evidence, the sentence 'Smith may be innocent' can go from true to false, and 'Smith must be guilty' from false to true. Smith's innocence was epistemically possible, then ceased to be so. The referent of 'Smith' and the crime in question remained constant, as did the truth-values of the plain sentences 'Smith is innocent' and 'Smith is guilty.' When the detective rightly asserts 'Smith must be guilty', she does not regard her earlier assertion of 'Smith may be innocent' as in any way wrong. It was not a *mistake* made on the basis of incomplete but misleading evidence. That was the point of saying 'Smith may be innocent' rather than 'Smith is innocent.' If she had initially been deceived by Smith's

---

[5] Stalnaker's work (1968, 1975, 1984) has been seminal in the development of the view.

false but highly plausible alibi, she might have asserted 'Smith is innocent' and would later have regarded *that* assertion as false, a mistake. Instead, she kept an open mind from the beginning. She only ever went as far as asserting 'Smith may be innocent', and in retrospect can still regard what she then said as true. Of course, now she knows that Smith is guilty, she should no longer assert 'Smith may be innocent', but that is unproblematic, since it no longer expresses the same proposition, given the change in her evidence. Thus epistemically modal sentences are non-trivially context-dependent.

The same point can be made synchronically rather than diachronically. Two detectives are simultaneously conducting independent inquiries into the same crime, each hiding their results from the other. Smith's innocence is compatible with Detective One's evidence, but not with Detective Two's. In Detective One's mouth, 'Smith may be innocent' is true and 'Smith must be guilty' false. In Detective Two's mouth, 'Smith may be innocent' is false and 'Smith must be guilty' true. Of course, if they really wanted, they could both use 'may' and 'must' with respect to the combination of both bodies of evidence, in which case Detective One would not know the truth-value of 'Smith may be innocent.' But that pooled use of 'may' and 'must' would be of less use to either detective, given their ignorance of the pooled body of evidence. For each of them, the contextually relevant use of epistemic modals is with respect to their own body of evidence.

Do such phenomena have analogues for indicative conditionals? That is not obvious. Of course, we can easily imagine that a conditional such as 'If Smith was in Berlin, he is innocent' was highly probable on the original evidence but is highly improbable on the present evidence, or that it is highly probable on Detective One's evidence but highly improbable on Detective Two's evidence. But that does not mean that which proposition the conditional expresses varies over time or between the two detectives. Even if it expresses the same proposition throughout, its probability on the evidence can vary as the evidence varies.

The dependence of epistemic modals on a relevant body of evidence is easily accessible to speakers; the epistemic role of 'may' and 'must' is more or less overt. For example, imagine a school for detectives, where they are taught to discriminate what is compatible with their evidence from what is not. They are trained on real-life cases, but are given only the evidence available at an early stage of the actual investigation. The instructor, who knows the full story about the cases, can recognize that the trainee speaks truly when he says 'Smith may be innocent', even though she knows that Smith is guilty. She adjusts her assessments in the light of her awareness of differences between her evidence and his. Any competent teacher would do likewise. By contrast, it is much less obvious whether she should treat a conditional like 'If Smith was in Berlin, he was smuggling drugs' as false in the trainee's mouth, because he lacks the required evidence, but true in her own mouth (outside the training context), because she has the required evidence.

We can probe some differences between epistemic modals and indicative conditionals by considering a sentence such as (1):

(1)   Smith may be innocent and he may be guilty.

Things like (1) are often natural, correct things to say. Sentence (1) is neutral between the non-epistemic hypotheses 'Smith is innocent' and 'Smith is guilty', and yields the answer 'No' to the epistemic question whether the evidence decides between them. If indicative conditionals work like epistemic modals, we may expect there to be a sentence involving indicative conditionals (but not epistemic modals) which is neutral between the non-epistemic hypotheses 'Smith was in Berlin and he is innocent' and 'Smith was in Berlin and he is guilty', and yields the answer 'No' to the epistemic question whether the evidence decides between them. What is that sentence?

Suppose that the evidence is compatible with Smith's having been in Berlin. Then it decides in favour of the second conjunction over the first just in case it combines with 'Smith was in Berlin' to determine that Smith is guilty. Consequently, the evidence is compatible with Smith's innocence (does not determine his guilt) just in case it does *not* combine with 'Smith was in Berlin' to determine that Smith is guilty. By parallel reasoning, the evidence is compatible with Smith's guilt just in case it does *not* combine with 'Smith was in Berlin' to determine that Smith is innocent. Thus one might be tempted to propose (2) as the desired sentence:

(2)   It is not the case that if Smith was in Berlin, he is guilty, and it is not the case that if Smith was in Berlin, he is innocent.

However, compared to (1), (2) sounds unnatural. If someone uttered (2), we might conjecture that they intended to convey something analogous to (1), but it would be guesswork, quite unlike our fluent understanding of (1). More specifically, it is not obvious whether any similarity between (1) and (2) can somehow be calculated from their literal meanings, or instead concerns some shaky conversational implicatures.

One awkward feature of (2) is that it involves negated conditionals. As has long been observed, we tend to be more comfortable negating the consequent of a conditional than negating the whole conditional: we prefer 'If *A*, not(*C*)' to 'Not(if *A*, *C*)'.[6] If we work the negations in (2) through into the consequents, the result is (3):

(3)   If Smith was in Berlin, he is not guilty, and if Smith was in Berlin, he is not innocent.

But since the evidence is compatible with Smith's having been in Berlin, and he is either guilty or innocent, (3) seems hopeless.

To avoid negated conditionals, we can consider a natural contradictory of (1), like (4):

(4)   Either Smith must be guilty or he must be innocent.

Although (4) sounds less natural than (1), we could use it with reference to the detective's evidence rather than our own, when we see that she has received the decisive DNA report but we do not yet know what it says. Like (1), (4) is neutral between the non-epistemic hypotheses 'Smith is innocent' and 'Smith is guilty.' Unlike (1), (4) yields the opposite answer 'Yes' to the epistemic question whether the evidence decides between the two non-epistemic hypotheses. Again, if indicative conditionals work like epistemic modals, we may expect there to be a sentence

---

[6] The 'not(.)' stands in for any suitable way of forming the negation of a sentence, perhaps simply by prefixing it with 'it is not the case that'.

involving indicative conditionals (but not epistemic modals) which is neutral between the two non-epistemic hypotheses 'Smith was in Berlin and he is innocent' and 'Smith was in Berlin and he is guilty', but yields the answer 'Yes' to the question whether the evidence decides between the two non-epistemic hypotheses. What is that sentence?

Reasoning as before, the natural candidate is (5):

(5)   Either if Smith was in Berlin, he is guilty, or if Smith was in Berlin, he is innocent.

Compared to (4), (5) still sounds unnatural. If someone uttered (5), we might conjecture that they intended to convey something analogous to (4), but it would again be guesswork, quite unlike our understanding of (4). One might put the difference by saying that with (1) and (4) we hear the epistemic meaning in the words, whereas with the conditional constructions (2) and (5) we make an effort to reconstruct an epistemic meaning because it is unclear what else the speaker might be getting at. More specifically, it is not obvious whether any similarity between (4) and (5) can somehow be calculated from their literal meanings, or instead concerns some shaky conversational implicature.[7]

Of course, indicative conditionals could be epistemically context-dependent in some way which did not enable them to clarify epistemic matters like (1) and (4). But then one starts to question the *point* of their epistemic context-dependence. Why load conditionals with such complications if it does not enable them to clarify such elementary epistemic matters? The question is urgent because context-dependence can confuse and obstruct communication across contexts, especially when it is not salient to speakers. Of course, indicative conditionals *might* have a dysfunctional semantics, but it seems unlikely. At any rate, the comparison of indicative conditionals with epistemic modals seems to do more harm than good to the claim that indicative conditionals are epistemically context-dependent, absent a plausible story to explain either the added value of such an accretion to the meaning of our commonest conditional expression, or else the needlessness of such added value.

So far, we have only been scratching at the surface. It would be premature to draw any final conclusions without examining the practice of using conditionals in much more detail and depth. We have begun to see how resistant conditionals are to reflection from the standpoint of a native speaker. It is not only semantic theorists who find them difficult. Ordinary speakers too find conditionals hard to handle once embedded under logical operators such as negation, and correspondingly hard to assert or deny with confidence. That in turn often makes the ordinary language data for theorizing elusive or untrustworthy.

In chapter 2, we examine the basic ways in which an indicative conditional is assessed, and the problems they raise. That will eventually enable us to understand, amongst other things, the phenomena which have been misinterpreted as revealing epistemic context-dependence in indicative conditionals.

---

[7] See chapter 5 and section 7.3 for detailed discussion of some alleged connections between conditionals and epistemic modals.

2

The Suppositional Rule

2.1 Supposing and imagining

What if a strawberry is orange? What if you get lost? What if the government falls? A normal human way to answer hypothetical questions is by using one's imagination—not arbitrarily, of course, but constrained by one's background knowledge and expectations of what the world is like. One thinks through various scenarios, imagining and comparing their consequences. Imagining that *A* may prompt one to say something of the form 'If *A*, *C*'. Conversely, someone else saying something of the form 'If *A*, *C*' may prompt one to imagine that *A*.

Imagination is sometimes contrasted with *supposition*. Imagining is envisaged as a much richer and more dynamic activity, typically involving visual imagery, while supposing is a minimal mental or linguistic act, perhaps accomplished simply in the words 'Suppose that *A*'. But the contrast is overblown. For a start, sensory imagery is not essential to all imagining. Without employing any such imagery, someone might imagine the political effects of a Christian fundamentalist becoming President of the United States, or what it is like to be depressed. Conversely, the point of supposing that *A* is typically to initiate a process of answering a hypothetical question 'What if *A*?', by mentally exploring the relevant consequences of the supposition that *A*, often by imagining them.

The method of making suppositions and exploring their consequences can also be applied in more rigorous and systematic ways. The prime example is mathematics, which normally limits itself to strictly deductive consequences. Proofs by *reductio ad absurdum* establish a proposition by supposing that it does *not* hold and deducing a contradiction. An early example is the Pythagorean proof that $\sqrt{2}$ is irrational, that is, for any two integers p and q (with $q \neq 0$), $p^2 \neq 2q^2$. The proof starts by supposing that $\sqrt{2}$ is rational, and climaxes with a contradiction—of course, only *on the supposition* that $\sqrt{2}$ is rational.

Not all mathematical suppositions are made to be reduced to absurdity. Often they are needed simply to achieve the required level of generality. A typical example is the standard proof of Lagrange's Theorem, named after the eighteenth-century Piedmontese mathematician Joseph-Louis Lagrange, who proved a special case of it. In modern terms, it concerns the algebraic structures known as *groups*; it says that the order (number of members) of any subgroup of a finite group divides the order of the group. To prove it, one starts by supposing that *H* is any subgroup of a finite group *G*, and then establishes, on that supposition, that the order of *H* divides the order of *G*. Here the letters '*G*' and '*H*' are variables. Of course, one cannot universally generalize the statement 'the order of *H* divides the order of *G*', because for many values of the

Suppose and Tell: The Semantics and Heuristics of Conditionals. Timothy Williamson, Oxford University Press (2020).
© Timothy Williamson.
DOI: 10.1093/oso/9780198860662.001.0001

variables H is *not* a subgroup of G and the order of H does *not* divide the order of G. Instead, one must first discharge the assumption and form the conditional conclusion 'If H is a subgroup of a finite group G, then the order of H divides the order of G.' Since that conditional statement depends on no further supposition, it can be universally generalized, which yields the theorem. Thus the conditional form enables one to express the residue of an argument on a supposition from outside that supposition. Without such a form, we could not properly articulate what we have learned by making the supposition and want to generalize—even though, as one might expect from section 1.2, once we have the universal generalization we can paraphrase it without using 'if': the order of a subgroup of a finite group always divides the order of the group.

Significantly, mathematicians write and speak their proofs with ordinary logical words, including 'if', using them in unreflectively fluent, linguistically standard ways. The standard medium of mathematics is a natural language afforced with mathematical notation and a few diagrams, not a purely formal language. Even when a proof is fully formalized, mathematicians use extended natural language to explain what is going on in it. In particular, there is the normal interplay between conditionals and universal generalizations.

Of course, we often assess and apply conditionals without involving generalizations, at least not consciously. For example, going back to basics, one might ask oneself:

If I start eating that food, will the alpha male notice?

In effect, one is assessing two opposite conditionals:

If I start eating that food, he will notice.
If I start eating that food, he will not notice.

A natural way to do it is to imagine starting to eat the food, and imagine how the alpha male will react, taking into account his direction of gaze, his apparent degree of alertness, the sight lines, how noisy the food is to eat, how hungry he is, how well disposed, and so on. One's imagining is reality-oriented, but one is typically not conscious of applying any generalizations to the particular case.[1] Whether one is applying some unconsciously is another matter. One's initial supposition is that one starts eating the food. It initiates an imaginative effort. If that exercise robustly leads to the conclusion that he will notice, from outside the initial supposition one accepts the first conditional and rejects the second; one leaves the food well alone. But if the exercise robustly leads to the conclusion that he will not notice, from outside the initial supposition one rejects the first conditional and accepts the second; one starts eating the food. If the exercise is inconclusive—under the constraints one imagines either alternative with similar ease—from outside the supposition one accepts neither conditional; one may or may not start eating the food, depending on one's hunger and one's caution. The method of trial and error is too risky when the price of error is high. One must assess the relation between the antecedent and

---

[1] For more on this function of the imagination see Williamson 2016c.

consequent *before* deciding whether to make true the antecedent, 'I start eating the food', or to make it false.

Even in contemporary life, one often has no option but to use a similar method. For example, you ask yourself:

If I live in this apartment, will it suit me?

In other cases, one has no control over whether the antecedent holds:

If it rains overnight, will the stream be passable in the morning?

One may have no idea whether it will in fact rain overnight, and no idea whether the stream will in fact be passable in the morning, yet by realistically imagining the effects of overnight rain one may come to know that the stream will *not* be passable in the morning *if* it rains overnight. One need not consciously apply any generalization to the case. One's imagination enables one to access a connection that verifies the conditional, even when one is in no position to verify or falsify its components, its antecedent and consequent.

2.2 Ways of assessing conditionals

There are many contrasting but interrelated ways of assessing conditionals.

Sometimes we can test a conditional 'If *A*, *C*' by learning that *A* and whether *C*. When we test the conditional *experimentally*, we bring it about that *A*, then observe whether *C*. For example, the theory that salt dissolves in water predicts that if I drop this quantity of salt in my hand into the bowl of water, it will dissolve. I drop the salt into the water and see whether it dissolves. If not, the conditional fails the test. In such a case, one has grounds to reject the conditional, and with it the theory. But if the salt does dissolve in the water, the conditional passes the test. In such a case, one has grounds to accept the conditional, and to gain confidence in the theory. Schematically, we take '*A* and *C*' to verify 'If *A*, *C*' and '*A* and not(*C*)' to falsify it. This way of falsifying a conditional connects closely with the rule of *modus ponens*, by which one can deduce '*C*' from 'If *A*, *C*' and '*A*'. *Modus ponens* is also known as the *elimination* rule for the conditional, since in formal systems for natural deduction it is the basic way of deriving conclusions in which the conditional symbol does not occur from premises in which it does occur.

However, the experimental testing of conditionals is post hoc. Once we know that *A*, and whether *C*, we typically have no further use for the conditional 'If *A*, *C*'. Our thinking about the case need no longer be hypothetical. We can use '*A* and *C*' or '*A* and not(*C*)' instead, as a more informative replacement. We need 'If *A*, *C*' most when we do *not* know that *A*. Then we need to assess 'If *A*, *C*' *prospectively* rather than retrospectively. For instance, you may need to know whether, if you try to climb the cliff, you will fall. You do not test the conditional experimentally, because the cost of falsifying it experimentally is too high. Often, when we most need conditionals, we must test them prospectively.

The examples in section 2.1 suggest a schematic procedure for assessing a conditional 'If *A*, *C*' prospectively, the *Suppositional Procedure*. For clarity, the procedure will be characterized as operating on interpreted sentences rather than propositions, though one can regard the former as vehicles for the latter, and the procedure is typically implemented in thought rather than out loud (more detail on this later in the section). Thus supposing, accepting, and rejecting are here relations to interpreted sentences, rather than to propositions, and the variables 'A' and 'C' are to be understood accordingly. Informally, however, we sometimes describe examples in terms of attitudes to propositions, when it is more natural to do so. The agent is assumed to engage with the propositions as expressed by corresponding sentences.

The Suppositional Procedure for assessing 'If A, C' works as follows. First, suppose A. Then, on that supposition, develop its consequences by whatever appropriate means you have available: constrained imagination, background knowledge, deduction, ... If the development leads to accepting C conditionally, on the supposition A, then accept the conditional 'If A, C' unconditionally, from outside the supposition. If instead the development leads to *rejecting* C conditionally, on the supposition A, then *reject* 'If A, C' unconditionally, from outside the supposition. Naturally, the firmness of the unconditional acceptance or rejection of the conditional will correspond to the firmness of the conditional acceptance or rejection of the consequent on the antecedent. Of course, sometimes the exercise is inconclusive: the development leads neither to accepting nor to rejecting C conditionally, on the supposition A. In that case, neither accept nor reject 'If A, C' unconditionally, from outside the supposition.

An immediate upshot of the Suppositional Procedure is that conditionals of the form 'If A, A' are accepted, for trivially one accepts A on the supposition A.

Here is a less trivial example. You have to judge whether, if you cancel the meeting, Alex will be disappointed. In your imagination, you suppose that you cancel the meeting and consider Alex's reaction. If you judge, on that supposition, that Alex will be disappointed, you then non-hypothetically accept that, if you cancel the meeting, Alex will be disappointed. If instead you judge, on the supposition, that Alex will not be disappointed, you then non-hypothetically reject the idea that, if you cancel the meeting, Alex will be disappointed. But if you cannot make up your mind, on the supposition, whether Alex will be disappointed, you also fail to make up your mind non-hypothetically on whether, if you cancel the meeting, Alex will be disappointed.

To take another example, you are considering whether a mushroom is poisonous. You are confident that all mushrooms of type M are poisonous, and that no mushrooms of type N are. On the supposition that this mushroom is of type M, you remain confident that all mushrooms of type M are poisonous, so you are also confident that this mushroom is poisonous. Thus you are confident in unconditionally accepting the conditional 'If this mushroom is of type M, it is poisonous', and in unconditionally rejecting the opposite conditional 'If this mushroom is of type M, it is not poisonous.' Here, developing the supposition involves deducing 'This mushroom is poisonous' from 'This mushroom is of type M' and 'All mushrooms of type M are poisonous.' But on the supposition that this mushroom is of type N, you remain confident that no mushrooms of type N are poisonous, so you are also confident that this mushroom is not poisonous. Thus you are confident in unconditionally rejecting the conditional 'If this mushroom is of type N, it is poisonous', and in unconditionally accepting the opposite conditional 'If this mushroom is of type N, it is not poisonous.' Here developing the supposition involves deducing 'This mushroom is not poisonous' from 'No mushrooms of type N are poisonous' and

'This mushroom is of type N.' In such ways, the procedure upholds the role of conditionals as instances of generalizations, emphasized in section 1.2.

The Suppositional Procedure involves reaching an unconditional attitude to 'If A, C' by means of first reaching a conditional attitude to C on the supposition A. But it also involves something more specific. Once one has reached an attitude to C conditionally on A, the question is: *which* attitude to 'If A, C' should one take unconditionally? The natural answer was exemplified above for a variety of attitudes: acceptance and rejection of various degrees of firmness, and agnosticism: take the *same* attitude to 'If A, C' unconditionally that you take to C conditionally on A. In general, the Suppositional Procedure involves coming to take an attitude to 'If A, C' by first coming to take that very attitude to C conditionally on A.

The Suppositional Procedure is applicable to a wide range of attitudes, understood as cognitive appraisals. Of course, not everything counts as an attitude for these purposes. On the supposition that it is hot, you may judge the sentence 'It is not snowing' not to contain the word 'if'. From that, the Suppositional Procedure should not license you to conclude unconditionally that the sentence 'If it is hot, it is not snowing' does not contain the word 'if'. That is not a cognitive appraisal in the relevant sense, by contrast with acceptance, rejection, and agnosticism. Although the category of attitudes has not been delimited precisely, it is clear enough to work with.

What has just been described is a single application of the Suppositional Procedure. The Procedure can also be repeated several times, with the same antecedent and consequent. For example, a single application may lead one to *uncon*fident acceptance or rejection of the consequent on the supposition of the antecedent. In that case one may repeat the procedure, perhaps many times, varying how one imagines the antecedent realized, to see what difference it makes, in the hope of reaching a more robust verdict on the conditional. If each trial of the procedure gives the same result, one can be correspondingly more confident that the result is correct. In that case, the attitude one finally takes to the consequent conditionally on the antecedent becomes the attitude one finally takes to the conditional unconditionally.

One can also run the Suppositional Procedure in reverse. Sometimes, you may first take an attitude to 'If A, C' unconditionally: for example, you may rely on the word of an expert (see below for such cases). You may then come to take the very attitude to C conditionally on A that you already have to 'If A, C' unconditionally. The governing rule is that the attitudes should be the *same*, in whichever temporal or causal order you come to them:

Suppositional Rule   Take an attitude unconditionally to 'If A, C' just in case you take it conditionally to C on the supposition A.

The Suppositional Rule merely requires the attitudes to be the same, whereas the Suppositional Procedure also gives temporal and causal priority to taking the attitude to A on C over taking it to 'If A, C', in effect to processing the constituents over processing the whole conditional.

The Suppositional Procedure and Rule may also be applied under further background suppositions, held constant between input and output. Many proofs in mathematics do just that. Discussion of this feature is postponed to chapter 3; for now we concentrate on fully unconditional attitudes to conditionals.

So far, we have mainly considered *cognitive* attitudes, such as acceptance and rejection. However, the Suppositional Rule arguably applies to other attitudes too. Some of them are more relevant to conditional commands and questions. For example, to order 'If *A*, do *X*!' ('If the window is open, close it!') is in effect to order 'Do *X*!' ('Close it!') on the supposition *A* ('The window is open'). To query 'If *A*, *Q*?' ('If the window is open, who opened it?') is in effect to query '*Q*?' ('Who opened it?') on the same supposition (see also section 7.2). For now, however, the focus will remain on conditional statements and how we assess them cognitively.

As already suggested, we sometimes assess conditionals prospectively without carrying out the Suppositional Procedure. For example, we sometimes accept conditionals on someone else's testimony. When an expert on cricket tells you 'If that batsman stays in for another ten overs, Pakistan will win', you may simply take his word for it. One may also defer to one's own past judgements, and accept a conditional because it is so similar to many other conditionals one already accepts, without bothering to apply the Suppositional Procedure to the new conditional itself. In such cases, we pass the buck of assessment: in the first example to the cricket expert's assessment of the same conditional, in the second example to one's own previous assessments of other conditionals. Those earlier assessments may themselves have employed the Suppositional Procedure. The final assessments in the examples may be regarded as *non-basic*. All of that is compatible with all prospective assessments of conditionals ultimately depending on applications of the Suppositional Procedure, and in that sense all non-basic prospective assessments of conditionals tracing back to *basic* assessments by the Suppositional Procedure.

That is still not quite right. Perhaps the cricket expert was hoaxing you: he did not use the Suppositional Procedure to assess that conditional, and your acceptance of it did not ultimately depend on applications of the Procedure. Of course, you may still have relied on the false assumption that he did use the Suppositional Procedure, but in other cases not even that much is assumed.

For example, superstitious people may regard a certain coin as magic, and use it as an oracle on special occasions. They ask '*P*?', toss the coin, and accept '*P*' if it comes up heads, 'Not(*P*)' if it comes up tails. They explain away apparent falsifications of the oracle as punishments for tossing the coin in a disrespectful spirit. They may accept 'If we set sail today, a storm will blow up' because when they respectfully asked 'If we set sail today, will a storm blow up?' and tossed the coin, it came up heads. They do not imagine that the coin or its guiding spirit applied the Suppositional Procedure. Rather, they believe that the coin simply has a direct line to the truth. In that case, their acceptance of the conditional does not depend on real or imagined applications of the Suppositional Procedure. However, if mundane conditionals play anything like the ubiquitous cognitive role which they play in ours, they will presumably also need a more reliable way of assessing them: if not the Suppositional Procedure, what else? With respect to the normal human practice of using conditionals, we may provisionally treat prospective assessments not directly by the Suppositional Procedure as in a broad sense *secondary*.

To classify the testimony-based assessment of conditionals as secondary is not to dismiss it as unimportant. Indeed, taking it seriously will turn out, in chapter 5, to be crucial to understanding our practice of using conditionals, and their meaning.

Nevertheless, it obviously depends on some prospective way of assessing conditionals which does not itself depend on conditional testimony, most importantly on something like the Suppositional Procedure.

Assessments of conditionals may also count as secondary when the process is mediated by amateur or professional theorizing about conditionals, in logical, semantic, pragmatic, or other terms. It is not always obvious whether a judgement about a case is theory-laden in such ways. There is no principled limit to what theories, however crazy, may interfere in speakers' assessments of ordinary sentences in their native language. However, it is too simple to say that the Suppositional Procedure only concerns *unreflective* judgements of conditionals. For any amount of reflective theorizing may legitimately figure in the derivation of C from A even in straightforward applications of the Procedure. Thus, in the proof of Lagrange's Theorem, a complex mathematical argument is needed to get from the antecedent '*H* is a subgroup of a finite group *G*' to the consequent 'The order of *H* divides the order of *G*.' What is typically much less reflective, more automatic, is the mathematician's subsequent step *from* that derivation *to* the conditional 'If *H* is a subgroup of a finite group *G*, the order of *H* divides the order of *G*.' That step still constitutes an application of the Suppositional Procedure. By contrast, if a mathematical crank rejects the conditional on the grounds that the word 'if' has no place in serious mathematics, theoretical interference is second-guessing the Procedure.

Another kind of violation of the Suppositional Procedure includes utterances like:

If he's a qualified doctor, I'm the Pope.

(The speaker is not the Pope.) Obviously, the reality-oriented development of the supposition 'He's a qualified doctor' does *not* lead to the conclusion 'I'm the Pope.' The utterance has a similar effect to this:

If he's a qualified doctor, I'll eat my hat.

In both cases, the point of the utterance is to display one's supreme confidence that the antecedent is false, that he is not a qualified doctor, by gratuitously committing oneself to a humiliating penalty payable if the antecedent is true. In the second example, one is committed to eating one's hat. In the first example, one is committed to an absurd claim to be the Pope. Normally, it would be unreasonable to make such conditional commitments if the risk of the condition's being met were serious. The utterances would not sound so extravagant if the consequent were unsurprising on the supposition of the antecedent. The apparent disregard of the Suppositional Procedure signals that more is going on. These cases too can be classified as secondary assessments of the conditionals. On a deeper understanding of our practice of using conditionals, such examples will turn out to be less anomalous than they first look (see section 5.1 and chapter 6).

With these preliminary clarifications of the distinction between primary and secondary prospective assessments of conditionals, everything seen so far is compatible with this conjecture:

Suppositional Conjecture   The Suppositional Procedure is humans' primary way of prospectively assessing conditionals.

One main aim of this book is to explore the implications and status of the Suppositional Conjecture.

Admittedly, the Suppositional Conjecture is not super-precise. The remarks above about the distinction between primary and secondary means of assessment are surely not enough to eliminate all vagueness from the term 'primary' in the statement of the Conjecture. Think of it this way. The proposal is that we get a good *model* of the practice of using conditionals by treating it as based on the Suppositional Procedure. Like most models in science, that involves various simplifications or idealizations, often quite severe ones—like treating a planet as a point mass, or an environment as containing only two species. Thus we should not expect a perfect or even near-perfect fit with experimental data, since the latter concern complex, messy systems that grossly violate the model's simplifications and idealizations. Nevertheless, the track record of the natural and social sciences shows that a good simple model of a phenomenon can provide crucial insights into its nature and structure. In the long run, it may also lead to better models of the phenomenon that explain more of its features.[2] Indeed, specific intrinsic features of the Suppositional Procedure will turn out to point in such a direction.

The Suppositional Conjecture is a *psychological* hypothesis, which in the end must live or die by psychological evidence. However, this book does not contain much discussion of experimental data. Before we can sensibly test a model against such data, we must properly understand the model itself, and what it does or does not imply. Science needs well-developed theories to make sense of its almost intractably messy data, which often result from dozens of interacting variables. Without such theories to guide it, data-gathering risks becoming a parody of directionless Baconian inductive inquiry. Properly developing a theory is no easy task. It is enough for one book.[3]

The Suppositional Conjecture is not a *semantic* hypothesis. By itself, it says nothing about conditionals' truth-conditions and falsity-conditions, or their lack of them, or any other semantic features attributed to conditionals. However, the semantics and the psychology of conditionals should not run completely free of

---

[2] On models in natural science see Weisberg 2013. On models in philosophy and semantics see Williamson 2017b.

[3] In psychology, there are several contrasting traditions of work on conditionals, including the mental models approach (Johnson-Laird and Byrne 2002, Byrne and Johnson-Laird 2009, Khemlani, Byrne, and Johnson-Laird 2018), the suppositional approach (Evans and Over 2004), the Bayesian approach (Oaksford and Chater 2007), and more recently the erotetic approach (Koralus forthcoming, Koralus and Mascarenhas 2013). Given the present emphasis on the role of supposition in the primary heuristic for conditionals, the obvious comparison is with Evans and Over's work; however, the view to be defended in this book differs sharply from them on the semantics of conditionals, and makes a wider separation than they do between the semantics and the psychology. Most work on the semantics of conditionals has engaged very little with the psychology of conditionals, and indeed the two traditions are often hard to relate to each other. One might wonder whether a subsidiary reason has been that, within psychology, conditionals have mainly been studied within the psychology of reasoning, with the conditionals figuring as premises or conclusions (or components thereof), and subjects having to consider questions about what does or does not follow from what, which in turn raises questions about how they understand relations of logical or non-logical consequence or validity. One can generate a conditional such as 'If she doesn't come, I'll be disappointed' without addressing issues of validity. However, the shortage of interaction between semantics and psychology is a more general phenomenon.

each other, just as the semantics and psychology of natural kind terms should not run completely free of each other. The extension and intension of the noun 'horse' are somehow constrained by psychological facts about how speakers of English use the word, and in causal and perceptual relations to which features of the environment they do so, mediated by some principle of charity in interpretation. Other things being equal, speakers of English should come out knowing what they are talking about when they use the word 'horse'. We should not waste time on the hypothesis that 'horse' really refers to waterfalls, while speakers of English use it under the influence of a false dogma that it refers to a kind of animal. Similarly, by the principle of charity, other things being equal, we should come out knowing what we are talking about when we use the word 'if'. Thus, given the Suppositional Conjecture, we should expect the semantics of conditionals to make the Suppositional Procedure come out reliable, as far as possible. At the very least, the semantics should not gratuitously make the Procedure come out *un*reliable.

One might try to understand the Suppositional Conjecture in terms of a distinction between *competence* and *performance*. On this view, competence with conditionals involves the capacity to assess them by correctly applying the Suppositional Procedure. In actual performance, all sorts of errors and deviations can be expected. They need not all be mere random noise, which can be filtered out over large samples. They may include far more systematic effects, various forms of bias and interference. Since experimental data measure performance rather than competence, they may cast only very indirect light on the Suppositional Conjecture, which concerns competence rather than performance.

There is something to such a view of the Suppositional Conjecture. However, the term 'competence' could do with considerable clarification. In particular, we should not assume without argument that a competent attitude to a proposition is *ipso facto* correct, in whatever sense of 'correct' is appropriate to that type of attitude.

Compare the case of perceptual judgements. They often rely on various *heuristics*, fast and frugal ways of judging which are normally reliable, especially under ecologically realistic conditions, but are not 100 per cent reliable. For example, we use discontinuities in colour as a heuristic in judging the boundaries of three-dimensional objects, which is mostly reliable but can be exploited by experts in camouflage to make us misjudge. Such 'glitches' in our heuristics cause predictable illusions in special circumstances. They are not individual performance errors, failures to execute the heuristic properly, or interferences with its outputs. Rather, they result systematically from its proper execution. We can regard the ability to apply the colour-discontinuity heuristic as a competence. Thus, in unfavourable circumstances, competent judgements can be false.

Similarly, one might understand the Suppositional Procedure as a heuristic, a fast and frugal way of assessing conditionals which is normally reliable, especially under ecologically realistic conditions, but not 100 per cent reliable. It may have glitches, which cause predictable illusions in special circumstances. They too will not be individual performance errors, failures to execute the Procedure properly, or interferences with its outputs. Rather, they will result systematically from its proper execution. We might still regard the ability to apply the Suppositional Procedure as a competence, which, in unfavourable circumstances, can deliver competent but false assessments of conditionals. Chapters 3–8 will confirm this view of the Suppositional Procedure as an imperfectly reliable heuristic.

One might also wonder where to fit such heuristics for assessing sentences into a standard picture of linguistic architecture, with semantics built on syntax and pragmatics built on semantics. The heuristics must come above the semantics, for normally one is in no position to decide between accepting and rejecting a sentence until one knows what it means. But if the semantics takes the syntax of a sentence and a context as inputs and delivers the proposition expressed by the sentence in the context as output, then the semantic output may be too coarsely individuated to provide the heuristic with the information it needs to work on. For a start, the conditional *word*, such as 'if', may be needed to activate the heuristic. Moreover, if the proposition expressed by a conditional is a set of possible worlds, or something similarly coarse-grained, not even the propositions expressed by its antecedent and consequent will be recoverable from the proposition expressed by the conditional itself. For example, given the principle of contraposition, the logically equivalent conditionals 'If Jan is outside, it is sunny' and 'If it is not sunny, Jan is not outside' express the very same proposition (in a given context), even though their antecedents 'Jan is outside' and 'It is not sunny' express two different propositions, as do their consequents 'It is sunny' and 'Jan is not outside' (in that context). But the Suppositional Procedure needs to take the antecedent and the consequent as separate inputs.

The short answer is that the heuristic fits in wherever verbal reasoning fits in. Since verbal reasoning fits in *somewhere*, so does the heuristic. Some verbal reasoning may play a role in calculating pragmatic implicatures, but much of it goes far beyond both semantic and pragmatic comprehension. Mathematicians understood the statement of Fermat's Last Theorem centuries before it was eventually proved. Similarly, conversational participants have considerable discretion in how far they think through the consequences of what is said. Although for gross simplicity we may model them as logically omniscient, that is just a convenient first approximation. When we seek to understand cognitive dynamics in more detail, we must acknowledge that reasoning takes time and is costly in other resources too. For these purposes, a psychologically realistic theory of verbal reasoning will have to treat it as operating over structured mental representations such as interpreted sentences, not over bare coarse-grained propositions. For example, an inference by disjunctive syllogism from 'P or Q' and 'Not(P)' to 'Q' depends on the thinker's capacity for pattern recognition, and in particular for recognizing 'P' as a constituent common to the two premises. Some participants may notice that the conversation has supplied the premises for a step of disjunction syllogism, and draw the conclusion, while others do not. None of this is to deny that sentences express coarse-grained propositions; it is just to insist that the fine-grained dynamics can only be properly understood at a level where sentential structure has not been left behind.

A framework adequate for understanding the dynamics of verbal reasoning will also help us situate the Suppositional Procedure. The framework will deal with structured mental representations of some sort, just as the Procedure requires. Those representations may still express coarse-grained propositions. The framework will allow thinkers considerable discretion in how far they apply the Procedure:

exhaustively applying it is no requirement for ordinary linguistic comprehension. For example, let A logically entail C. If the entailment is obvious to native speakers, they will normally accept 'If A, C', by a special case of the Suppositional Procedure. But if the entailment is not obvious to them, they are not obliged to continue tracing out the consequences of A until (if ever) they eventually come to C. Similarly, let A be logically inconsistent with D. If the inconsistency is obvious to native speakers, they will normally reject 'If A, D', by another special case of the Suppositional Procedure. But if the entailment is not obvious to them, they are not obliged to continue tracing out the consequences of A until (if ever) they eventually come to 'Not(D)'.

Further support for this view comes from the need for theories of verbal reasoning to handle *generality*. In arguing for 'Every F is a G', one may argue from the supposition 'x is an F' to the conclusion 'x is a G', where 'x' is a free variable. In ordinary English, one might put it by arguing from 'Something is an F' to 'It is a G', where the pronoun 'it' is anaphoric on 'something'. The reasoning controls the implicit generality of the argument from one open sentence to another by tracking the free variable or pronoun, not by contemplating the relation between two sets of possible worlds. Some alternative theorists regard 'x' or 'it' as a sort of arbitrary name of an object, or even as a name of a sort of arbitrary object, but to control the implicit generality of the argument the reasoning must still operate at the finer-grained level. The same considerations apply to the Suppositional Procedure, when it verifies the open conditional 'If x is an F, x is a G.'[4]

The comparison between the Procedure and reasoning also helps explain a feature which may initially look troublesome. The Suppositional Procedure and Rule explicitly mention only *unembedded* occurrences of conditionals. What about embedded occurrences, where conditional sentences figure as constituents of more complex sentences? Surely a story needs to be told about cognition of them too. But consider the crucial role of the usual logical connectives, including a conditional one, in mathematical reasoning. At least to a first approximation, it is adequately codified by the introduction and elimination rules for each connective in a standard system of natural deduction. The introduction and elimination rules for a given connective explicitly mention only its unembedded occurrences, where it occurs as the main connective of the given sentence; that applies in particular to the natural deduction rules for the conditional. Nevertheless, when the rules for all the connectives are combined, the result is an adequate background logic for mathematical reasoning, which handles embedded occurrences of conditionals just as well as embedded ones. The Suppositional Rule serves the analogous purpose for 'if' in the wider setting of general language use, where non-logical considerations usually dominate. There is no special lacuna for embedded conditionals. Of course, they may occur under other connectives whose associated epistemology remains to be understood, but that problem is not for an account of conditionals to resolve.

---

[4] See section 6.5 for more on the relation between the semantic and cognitive levels, in light of the specific semantics to be proposed for 'if'.

The role of the Suppositional Procedure and Rule is epistemological rather than purely semantic. Nevertheless, they play a key role in semantic theorizing—if the Suppositional Conjecture is true. For semanticists continually check what their theories predict about the status of sample sentences against pre-theoretic native speaker judgements. Given the Suppositional Conjecture, those pre-theoretic judgements will normally be products of the Suppositional Procedure. If the Procedure is a less than fully reliable heuristic, the standard data for semantic theorizing will also be less than fully reliable. In that case, even when native speakers strongly agree on the status of a sample sentence in a specified context, they may all be wrong, having applied the heuristic to an unfavourable case. We must be careful not to dismiss good theories on the basis of bad data.

2.3 Suppositions and updating

Describing what is in effect the Suppositional Procedure, Frank Ramsey famously wrote in a footnote (1929: 143):[5]

> If two people are arguing 'If A, then C?' and are both in doubt as to A, they are adding A hypothetically to their stock of knowledge and arguing on that basis about C.

We can regard supposing A as a simulated or offline analogue of receiving the new information A, and developing the supposition as a simulated or offline analogue of updating one's knowledge and beliefs on the new information. For example, when I suppose that it will snow tomorrow, what I do is in some ways like what I do when I read a reliable forecast that it will snow tomorrow. Within the scope of the supposition, I accept 'Warm clothes will be needed tomorrow.' If the news comes as a surprise, I may have to reject some propositions I previously accepted, such as 'Light clothes will do for tomorrow.'

The analogy between developing a supposition and updating on new information helps make our ability to develop suppositions less mysterious. For it is not mysterious *that* we are able to update on new information, even though much remains to be explained about *how* we do it. Intelligent life is impossible without the ability to update on new information. The natural hypothesis is that we re-employ some of the very same abilities offline when we develop a supposition, although again much remains to be explained about how we do that.

Of course, when I read that it will snow tomorrow, I may not believe the forecast. I may form the belief that it was *forecast* to snow tomorrow, but not the belief that it *will* snow tomorrow. In that case, I am not updating on the information that it will snow tomorrow. I am updating only on the information that it was forecast to snow tomorrow. Thus, when I update on new information, I *believe* the new information. Obviously, by contrast, when I develop the supposition that it will snow tomorrow, my attitude to the proposition that it will snow tomorrow is not belief, but merely acceptance on that very supposition.

---

[5] Schematic letters adjusted to present notation.

Less obviously, there are additional, subtler asymmetries between updating on new information and developing a supposition.

Typically, though not always, when one learns something, one also learns that one has learned it. For example, when I get up in the morning, look out of the window, and see new snow, I come to know that it snowed overnight, but typically on reflection I also come to know that I know that it snowed overnight. Although natural models of knowledge include cases in which one knows without being in a position to know that one knows, they also provide a vast range of cases in which one does know that one knows. Even when one forms a false belief, one typically also comes to know that one has that belief, or at the very least to believe that one has it (without believing that it is false, of course). But when one supposes something, one typically does *not* simulate knowing or believing it to the point of supposing *that* one knows or believes it.[6]

For example, before getting up, if I lie in bed supposing just that it snowed overnight, I do not, from that supposition, conclude that I know that it snowed overnight. For otherwise I could apply the Suppositional Procedure to conclude, *outside* the supposition:

(1)   If it snowed overnight, I know that it snowed overnight.

But I also know (2a), and indeed (2b):

(2a)   I do not know that it snowed overnight.
(2b)   If it snowed overnight, I do not know it.

For I know that I have not yet looked out of the window to check. But from (1) and (2a) or (2b) I can presumably conclude (3):

(3)   It did not snow overnight.

Such thinking would be absurd. It would be equally absurd to reach the same conclusion by analogous thinking about one's belief or confidence rather than one's knowledge, from (4) and (5a or (5b) in place of (1) and (2):

(4)   If it snowed overnight, I am confident that it snowed overnight.
(5a)   I am not confident that it snowed overnight.
(5b)   If it snowed overnight, I am not confident of it.

Presumably, of course, if it snowed overnight, then sooner or later I *shall* know that it snowed overnight. But that is just a feature of the example. By contrast, there is no reason to accept (6) or even (7):

(6)   If a spider ran over the bed in the night, I shall know that a spider ran over the bed in the night.

(7)   If a spider ran over the bed in the night, I shall believe that a spider ran over the bed in the night.

---

[6] The point goes back to van Fraassen 1980; he credits it to Richmond Thomason. See also Edgington 1995: 270 and Willer 2010.

There is no CCTV in the bedroom. Thus, from the supposition that a spider ran over the bed during the night, I do not conclude that anyone will ever know or believe that a spider ran over the bed during the night.

More acceptable is this conditional:

(8)   If a spider ran over the bed in the night, I have supposed that a spider ran over the bed in the night.

After all, I know that I *have* supposed that a spider ran over the bed in the night, and the content of the supposition does not cast doubt on my knowledge. Nevertheless, when one supposes something, the fact that one has supposed it often does not belong in the development of that very supposition. We can see that by considering examples with variables, such as (9):

(9)   If *s* is a snowflake, I have supposed that *s* is a snowflake.

If (9) were compelling for structural reasons, so should be its universal generalization (10):

(10)   For every snowflake *s*, I have supposed that *s* is a snowflake.

But (10) is obviously false, and not at all compelling. For almost every snowflake *s*, I have never identified *s* at all, or had any thought specifically about it, let alone *supposed* something specifically about it. Only in exceptional circumstances does the proposition that someone has taken some psychological attitude—whether knowledge, belief, or mere supposition—to the proposition that *A* belong in the development of the supposition that *A* itself. Normally, it is no part of what is being supposed.

We can reach a closely related conclusion by noting that the Ramsey Test must not be interpreted as a proposal to test 'If *A*, *C*' by testing 'If I know that *A*, I know that *C*', or 'If I believe that *A*, I believe that *C*', or 'If I suppose that *A*, I suppose that *C*', or anything of that kind. For any such proposal merely generates an infinite regress: the next step it would enjoin is to test the more complex conditional by testing 'If I know that I know that *A*, I know that I know that *C*', or 'If I believe that I believe that *A*, I believe that I believe that *C*', or 'If I suppose that I suppose that *A*, I suppose that I suppose that *C*', and so on, potentially ad infinitum. Such psychologizations of the Ramsey Test are pointless. Instead, the key to the intended test, in effect the Suppositional Procedure, is the offline simulation of online updating, by recruiting to the imagination an appropriate selection of the same cognitive capacities.

Similar considerations apply to more impersonal constructions. For example, the Suppositional Procedure should not be understood as endorsing any schema like (11)–(14) (where in each case the epistemic operator takes narrow scope):

(11)   If *A*, it is known that *A*.
(12)   If *A*, it is thought that *A*.
(13)   If *A*, it is obvious that *A*.
(14)   If *A*, it is very probable on the evidence that *A*.

For examples like (15)–(18) are fine:

(15)   If the volcano is about to erupt, it is not known that it is.
(16)   If the volcano is about to erupt, it is not thought that it is.
(17)   If the volcano is about to erupt, it is not obvious that it is.
(18)   If the volcano is about to erupt, it is not very probable on the evidence that it is.

Claims such as (15)–(18) do not even feel paradoxical, as they would if our primary means of assessing conditionals rejected them.

The fallaciousness of principles such as (11)–(14) casts light on the appropriate sense of taking an attitude to something under a supposition. For instance, consider the attitude of *accepting as obvious*. The Suppositional Rule, applied to the attitude of taking as obvious, mandates accepting 'If A, C' as obvious unconditionally just in case one conditionally accepts C as obvious on the supposition A. In particular, for C = A, the Rule mandates unconditionally accepting 'If A, A' as obvious just in case one conditionally accepts A as obvious on the supposition A. Presumably, unconditionally accepting 'If A, A' as obvious is mandatory. So conditionally accepting A as obvious on the supposition A should also be mandatory. Thus, since schema (13) is hopeless, we must not equate accepting A as obvious under a supposition with accepting 'A is obvious' under that supposition. Similarly, we must not equate accepting A as *probable* under a supposition with accepting 'A is probable' under that supposition. Rather, we should understand accepting A as probable under a supposition as something like assigning A a high *conditional probability* on the supposition. Similarly, we should understand accepting A as obvious under a supposition as something like assigning A a conditional probability of 1 on the supposition.

For the attitude of knowledge, knowing C under the supposition A is something like knowledgeably deriving C from A, which is compatible with the falsity of both A and C. That characterization is just approximate: the derivation may be closer to an imaginative exercise than to a formal deduction, and other suppositions and background information may play a more active role in it than A does. In general, we may think of attitudes under suppositions as offline analogues of the corresponding online attitudes.

Of course, a speaker who says things like (15)–(18) may *also* say things like (19)–(22), in the very same circumstances, while still regarding an eruption as improbable:

(19)   If the volcano is about to erupt, I know that there will be immense devastation.

(20)   If the volcano is about to erupt, I think that there will be immense devastation.

(21)   If the volcano is about to erupt, it is obvious that there will be immense devastation.

(22)   If the volcano is about to erupt, it is very probable on the evidence that there will be immense devastation.

Third-person analogues of (19)–(20) are also possible, where the speaker is talking about someone known to regard an eruption as improbable:

(23) If the volcano is about to erupt, Mary knows that there will be immense devastation.

(24) If the volcano is about to erupt, Mary thinks that there will be immense devastation.

In cases such as (19)–(24), the natural hypothesis is that, on the intended reading, the attitude expression takes wide scope, with the 'if' clause as part of the content of the attitude ascribed. For instance, (19) is to be understood as (25):

(25) I know that if the volcano is about to erupt, there will be immense devastation.

Analogous accounts can be given of (20)–(24).

However, wide scoping does not cover all cases. For instance, I may think it very unlikely that Kim is a spy, and assert (26):

(26) If Kim is a spy, we shall never know that he is.

I may go on to calmly assert:

(27) If Kim is a spy, I am very surprised that he is one.

But (27) is hardly to be understood as (28):

(28) I am very surprised that if Kim is a spy, Kim is a spy.

I am not surprised at a logical truth. Nor does (27) predict my future surprise when Kim is unmasked as a spy, since I think that he never will be unmasked. Instead, asserting (27) may be an attempt to convey an attitude of offline great surprise—great surprise under the supposition that Kim is a spy (compare Dietz 2019). Roughly: in online surprise, one comes to accept something one previously regarded as highly improbable; in offline surprise, one comes to accept under a supposition something one still regards online as highly improbable. Sometimes, we have to stretch our vocabulary to describe the subtleties of our mental life. An alternative explanation of (19)–(24) is that they involve similar stretching.

These remarks may help convey the intended spirit of the Suppositional Rule and Procedure. But they are also susceptible of more formal treatment, which they receive in the next chapter.

3

Consequences of the Suppositional Rule

3.1 From conditional probabilities to conditional proof

The first task of this chapter is to elicit more precise and general consequences of the Suppositional Rule.

One main application of the Rule is to probabilistic attitudes, *credences*, often regarded as *degrees of belief* or *subjective probabilities*. The credences of a rational agent are supposed to satisfy standard axioms of the probability calculus, and so are measured by real numbers between 0 (the probability of a contradiction) and 1 (the probability of a tautology). Such an agent's credence in a disjunction of mutually inconsistent disjuncts is the sum of the agent's credences in the disjuncts, and logical equivalents share the same credence for the same agent at the time.

Credences have been understood in various radically different ways: as states operationally defined in terms of betting behaviour, as the fine-grained natural psychological reality underlying the coarse-grained folk term 'belief', and as outright beliefs about probabilities on one's evidence.[1] For present purposes, we need not decide between these views of credence. We can just work on the usual basis that, to a first approximation, one's credence in '*P*' can be estimated from one's response to the question 'How probable is it that *P*?'

A mathematical probability space is based on a set Ω, whose members are conceived as mutually exclusive, jointly exhaustive possible *outcomes*; subsets of Ω are *events*. Probabilities are ascribed to such events. Since outcomes function like possible worlds, events function like coarse-grained propositions, sets of possible worlds. Thus probabilities are in effect ascribed to propositions. However, the assumption that conditional sentences express propositions is controversial (Edgington 1986, 1995; Bennett 2003). For now, we can finesse that issue by ascribing probabilities to declarative sentences implicitly relative to a background context of utterance, rather than to propositions. One may, but need not, equate the probability of a sentence with the probability of the proposition it expresses in the given context. Thus logically equivalent sentences will have the same probability, whether or not they express the same proposition. We will return to the question whether conditional sentences express propositions in section 3.3.

The mathematical apparatus of probabilities may seem far too precise to be psychologically realistic for human agents. Proposals have been made for adjusting

---

[1] See Weisberg 2019 and Williamson 2019 for recent discussion.

Suppose and Tell: The Semantics and Heuristics of Conditionals. Timothy Williamson, Oxford University Press (2020).
© Timothy Williamson.
DOI: 10.1093/oso/9780198860662.001.0001

the framework to achieve greater realism, although they tend to involve a massive loss of mathematical power and tractability, without coming much closer to the psychological reality of actual human uncertainty. Anyway, we will stick to classical probabilities here. They provide a perspicuous, well-understood working model of uncertainty in simple cases. That is the best place to start in working through the consequences of the Suppositional Rule.

The Rule is easily applied to credences, for we have a natural, standard account of probabilities on a supposition, as noted in section 2.3.

Let Prob(X) be the probability of a sentence X, and Prob($X|Y$) the probability of X conditional on a sentence Y. Then Prob($X|Y$) is standardly equated with the ratio Prob(X and Y)/Prob(Y), in effect the proportion of Y-cases which are also X-cases (as measured by Prob). In conditioning on Y, we in effect eliminate all possibilities where Y fails and keep the proportions between the remaining possibilities constant, while recalibrating to make the new probabilities sum to 1. The probability of X on the supposition Y is naturally equated with the probability of X conditional on Y, Prob($X|Y$).

Of course, the ratio definition crashes when Prob(Y) = 0. One can extend conditional probabilities to at least some such cases, by treating conditional probability as primitive rather than defining it in terms of the ratio, while still requiring the ratio equation to hold whenever Prob(Y) > 0.[2] For simplicity, however, we will usually treat Prob($X|Y$) as undefined when Prob(Y) = 0.

The Suppositional Rule for credences (SRC) tells us to equate our unconditional credence in 'If A, C' with our credence in C conditional on A. We can therefore state its requirement thus:[3]

$$\text{SRC}\quad \text{Cred(if } A, C) = \text{Cred}(C|A)\quad (\text{when Cred}(A) > 0)$$

For example, how probable is it that if a fair die comes up more than 1, it comes up more than 3? SRC predicts the answer 3/5, which is indeed the natural answer.

SRC has a significant history. It was propounded by Ernest Adams, Brian Ellis, and Richard Jeffrey in the 1960s, and published by Robert Stalnaker in 1970.[4] It is sometimes called 'Stalnaker's Hypothesis', sometimes simply 'the Equation'. There is much experimental evidence that it fits speakers' assessments of probability well.[5] Theorists of conditionals tend to agree that it is a good fit. However, some cases are also known where the fit is less good. Moreover, as a general principle, SRC has some alarming theoretical consequences, explained in section 3.3.

Given the Suppositional Conjecture, SRC results from the proper application of our primary way of assessing conditionals. It does not follow that SRC holds for ideally rational beings, since they may have better ways of assessing conditionals than we have. Nor does it follow that SRC always holds for ordinary speakers of English,

---

[2] The approach of treating conditional probability as undefined but constrained by suitable axioms goes back to Popper 1936.

[3] Quotation marks will often be omitted for readability.

[4] See Bennett 2003: 58 and Stalnaker 1970.

[5] Evans and Over 2004 was a pioneering work in this respect; see Douven 2016: 66–71 for a more recent (positive) assessment.

since we may sometimes rely on secondary ways of assessing conditionals. Rather, SRC is the default for ordinary speakers.

One consequence of SRC is that 'If A, C' and 'If A, not(C)' are treated similarly to contradictories, for one's credences in them should sum to 1 (at least, when $\text{Cred}(A) > 0$):

$$\text{Cred}(\text{if } A, C) + \text{Cred}(\text{if } A, \text{not}(C)) = \text{Cred}(C|A) + \text{Cred}(\text{not}(C)|A)$$
$$= \text{Cred}(C|A) + 1 - \text{Cred}(C|A) = 1$$

As the probability of one conditional goes up, the probability of the other goes down, and vice versa. Thus one should neither confidently accept both (since then one's credences in them would sum to more than 1) nor confidently reject both (since then one's credences in them would sum to less than 1). Much of our practice seems to fit that prediction. The more confident I am that if she went out, she took an umbrella, the less confident I am that if she went out, she did not take an umbrella, and vice versa. However, as we shall see later, that default is defeasible.

SRC can be generalized. We can assess conditionals on background suppositions. For example, suppose that we have a fair 20-sided die. How probable is it that if it comes up more than 1, it comes up more than 3? The natural answer is now 17/19. Such examples can be multiplied without limit. We are at ease with assigning conditional probabilities to conditionals. The Suppositional Rule already covers such cases, since credences on suppositions are themselves attitudes. When we replace credences in SRC by credence conditional on a background supposition B, the result is SRCC:

SRCC $\text{Cred}(\text{if } A, C|B) = \text{Cred}(C|A \text{ and } B)$ (when $\text{Cred}(A \text{ and } B) > 0$)

One can recover SRC from SRCC by substituting a tautology for B. SRCC is also called 'Generalized Stalnaker's Hypothesis'.

David Lewis (1976) derived SRCC from SRC by arguing that the rationality of a pattern of credences should be preserved under the standard updating procedure of conditionalization. By contrast, SRCC is motivated here directly by the Suppositional Rule, generalizing the motivation for SRCC rather than treating it as derivative. Thus we can regard 'Cred' in SRC and SRCC as meaning exactly the same distribution of credences, not separated by any process of updating. SRCC does not do as well as SRC on current experimental evidence (Douven 2016: 74–5), but that is not very surprising: the extra complexity of SRCC, with three propositions rather than two to keep track of, may well cause more confusion. After all, we need *some* way of cognitively assessing conditionals on suppositions, since in complex decision-making we often have to make suppositions within suppositions: the idea that we do so by quite different means from those we employ in assessing them without suppositions is hardly plausible. Thus SRCC is the natural form of probabilistic assessment for conditionals on suppositions.

A further issue is that some theorists have used SRC to argue that conditionals lack truth-conditions and so do not express propositions. They hold that credence in a conditional is a genuine doxastic state even though it does not correspond to credence in a proposition (Edgington 1986, 1995). One consequence is that they deny that truth-functional operators such as negation, conjunction, and disjunction are well defined on conditionals. But on the ratio definition of conditional credence, the left-hand side of SRCC is defined in terms of credence in a conjunction: the numerator is $\mathrm{Cred}((\text{if } A, C) \text{ and } B)$. Of course, the claim that conjunctions such as 'The dog is large and, if hungry, aggressive' are ill-defined is somewhat implausible.

In any case, no-proposition theorists may accept SRCC by treating conditional credence as primitive, rather than defining it in terms of the ratio formula. On an austere version of the view, they may allow conditionals to occur as inputs to the unconditional credence function only as in SRC, and as inputs to the corresponding conditional credence function only as in SRCC. They can still require Cred to satisfy all standard principles of probability whenever the terms are well defined.

Treating conditional probability as primitive rather than defined by the ratio formula is arguably more realistic psychologically, since one key function of conditionals, which the Suppositional Procedure captures, is to enable us to access connections between antecedent and consequent *without* assessing the antecedent and consequent. For example, how probable is it that if you bet on the outcome of a fair coin toss, you will win? You can answer that question '1/2' without having any idea how probable it is that you *will* bet: psychologically, you do not go via the ratio definition.

A no-proposition theorist might deny that conditionals really have well-defined conditional probabilities, but such a view is highly revisionary of ordinary practice. More specifically, it is quite consistent with the status of SRCC as a consequence of the Suppositional Rule; it is merely an error theory about that aspect of the Suppositional Rule.

Given SRCC, opposite conditionals behave even more like contradictories, even if they do not express propositions, for one's credences in them conditional on any supposition should sum to 1, whenever the probabilities are well defined:

$$\mathrm{Cred}(\text{if } A, C|B) + \mathrm{Cred}(\text{if } A, \mathrm{not}(C)|B) = \mathrm{Cred}(C|A \text{ and } B) + \mathrm{Cred}(\mathrm{not}(C)|A \text{ and } B)$$
$$= \mathrm{Cred}(C|A \text{ and } B) + 1 - \mathrm{Cred}(C|A \text{ and } B) = 1$$

By similar reasoning, if D is a logical consequence of C, then one should have at least as much credence in 'If A, D' as in 'If A, C', on any supposition for which the probabilities are well defined:

$$\mathrm{Cred}(\text{if } A, C|B) = \mathrm{Cred}(C|A \text{ and } B) \leq \mathrm{Cred}(D|A \text{ and } B) = \mathrm{Cred}(\text{if } A, D|B)$$

Hence, if C and E are logically inconsistent, 'If A, C' and 'If A, E' behave like contraries, in the sense that one should not be confident of both, for 'Not(E)' is a logical consequence of C, so on any supposition for which the probabilities are well defined:

$$\mathrm{Cred}(\text{if } A, C|B) + \mathrm{Cred}(\text{if } A, E|B) \leq \mathrm{Cred}(\text{if } A, \mathrm{not}(E)|B) + \mathrm{Cred}(\text{if } A, E|B) = 1$$

So far, SRCC seems to be imposing a reasonable structure on credences.

The special case of SRCC for *maximal* credence merits separate attention:

SRCC1 $\mathrm{Cred}(\text{if } A, C|B) = 1 \Leftrightarrow \mathrm{Cred}(C|A \text{ and } B) = 1$ (when $\mathrm{Cred}(A \text{ and } B) > 0$)

By interpreting the epistemic modal 'must' in terms of credence 1, we can rewrite SRCC1 as an application of the Suppositional Rule to 'must':

$$\text{SRmust} \quad \text{Must}(\text{if } A, C|B) \Leftrightarrow \text{Must}(C|A \text{ and } B)$$

In other words, 'If A, C' must hold on the supposition B just in case C must hold on the supposition 'A and B'. We can restate this in terms of *conditional epistemic necessity*, understood as analogous to conditional probability: 'If A, C' is epistemically necessary conditional on B just in case C is epistemically necessary conditional on 'A and B'.

Similarly, SRCC also entails:

$$\text{SRCC} > 0 \quad \text{Cred}(\text{if } A, C|B) > 0 \Leftrightarrow \text{Cred}(C|A \text{ and } B) > 0 \quad (\text{when } \text{Cred}(A \text{ and } B) > 0)$$

By interpreting the epistemic modal 'may' in terms of credence greater than 0, we can rewrite SRCC > 0 as an application of the Suppositional Rule to 'may':

$$\text{SRmay} \quad \text{May}(\text{if } A, C|B) \Leftrightarrow \text{May}(C|A \text{ and } B)$$

In other words, 'If A, C' may hold on the supposition B just in case C may hold on the supposition 'A and B'. We can restate this in terms of *conditional epistemic possibility*, understood as analogous to conditional probability: 'If A, C' is epistemically possible conditional on B just in case C is epistemically possible conditional on 'A and B'.

Those interpretations of epistemic modals in terms of credence need refining. First, epistemic modality concerns what follows from or is compatible with what is *known*, rather than with what is *believed*. What *must* hold *does* hold, and what *does* hold *may* hold. 'Must' and 'may' are *epistemic* not *doxastic* modals. Second, in infinite probability spaces, some possibilities must have probability 0, for combinatorial reasons, so not even epistemic probability 1 entails truth, and truth does not even entail probability greater than 0 (Williamson 2007b). Thus SRmust and SRmay are better understood as analogous to SRCC than as strict entailments of it. However, epistemic modals can still be interpreted over the same space of possibilities over which epistemic probabilities are defined: what must hold is what holds in every epistemic possibility; what may hold is what holds in some epistemic possibility. By thus eliminating the dependence on SRCC, we can free SRmust and SRmay of the irksome constraint that 'A and B' must have positive probability, which was needed only to guarantee that the probabilities were well defined. Instead, we can understand the conditional epistemic modalities in SRmust and SRmay simply by a restriction on the quantifiers over possibilities, to those in which the given condition holds. With these glosses, both SRmust and SRmay are legitimate applications of the Suppositional Rule in their own right.

The conditional nature of the epistemic modalities in SRmust and SRmay is crucial. SRmust would be obviously hopeless if 'must($C|A$)' were understood as saying in effect that C is unconditionally or independently epistemically necessary if A is true. For when $C = A$ and B is a tautology, SRmust would then tell us, absurdly, that since 'If A, A' is always trivially epistemically necessary, any truth is unconditionally or independently epistemically necessary—a fallacy of just the sort noted at the end of section 2.3. Consequently, 'may($C|A$)' should not be understood as saying in effect that C is unconditionally or independently epistemically possible if A is true, for that would undermine the natural duality between 'may' and 'might', on which 'must($C|A$)' is equivalent to 'not(may(not(C)|A))' and 'may($C|A$)' to 'not(must(not(C)|A))'. Instead, given the quantificational truth-conditions, the conditional epistemic necessity of C on A is equivalent to the unconditional epistemic necessity of the material conditional $A \supset C$, and the conditional epistemic possibility of C on A is equivalent to the unconditional epistemic possibility of the conjunction 'A and C'.

Now consider a context in which unconditional epistemic necessity reduces to logical truth, and conditional epistemic necessity to logical consequence. Thus we can replace 'Must($Y|X$) by '$Y \vdash X$'. Hence SRmust becomes SR⊢:

$$\text{SR}\vdash \quad B \vdash \text{ if } A, C \Leftrightarrow A \text{ and } B \vdash C$$

Let BB be a finite set (possibly empty) of ordinary declarative sentences. Then BB is equivalent to a single sentence B, in effect its conjunction. Thus SR⊢ yields an equivalent form with multiple premises:

$$\text{SR}\vdash\Leftrightarrow \quad BB \vdash \text{ if } A, C \Leftrightarrow BB; A \vdash C$$

Now divide SR⊢⇔ into its right-to-left and left-to-right directions:

$$\text{SR}\vdash\Leftarrow \quad BB; A \vdash C \Rightarrow BB \vdash \text{ if } A, C$$
$$\text{SR}\vdash\Rightarrow \quad BB \vdash \text{ if } A, C \Rightarrow BB; A \vdash C$$

In other words, the Suppositional Rule tells us to treat 'if' as subject to these joint constraints.

Those last two rules may well look familiar. They are the standard introduction and elimination rules for a conditional operator in many formal proof systems of the kind known as *natural deduction*. They are also exactly the rules needed to support standard mathematical reasoning with 'if'.

SR⊢⇐ is also known as *conditional proof*, the deductive form of prospective reasoning for conditionals. When one has deduced C from the supposition A, given background assumptions BB, it lets one discharge the assumption A and conclude 'If A, C' on just the background assumptions. It is the introduction rule for 'if' because it lets one argue from a deductive relation involving only unconditional sentences to a deductive relation involving a conditional sentence. In brief, the introduction role takes one *to* conditionals.

Similarly, SR⊢⇒ is the elimination rule for 'if' because it takes one *from* conditionals: it lets one argue from a deductive relation involving a conditional sentence to a deductive relation involving only unconditional sentences. In fact, SR⊢⇒ is equivalent to the standard rule of *modus ponens*:

$$\text{MP} \quad A; \text{if } A, C \vdash C$$

To establish the equivalence rigorously, we need some *structural rules* for ⊢, rules not specific to any particular vocabulary in the sentences of the object language. Here are the three standard structural rules, where double letters stand for (possibly empty) sets of sentences, single letters for single sentences, and set-theoretic notation is avoided for the sake of readability:

Assumptions	$A \vdash A$
Monotonicity	$AA \vdash C \Rightarrow AA; BB \vdash C$
Cut	$[AA \vdash B \text{ and } B; CC \vdash D] \Rightarrow AA; CC \vdash D$

The rule of Assumptions says that deductive consequence is reflexive. The Monotonicity rule says that adding extra premises never breaks deductive connections, by contrast with inductive and default reasoning. The Cut Rule is a kind of generalized transitivity for deductive consequence; it is needed to chain together many short arguments into a long one, and is unreflectively assumed in most mathematical proofs.

We can now establish the equivalence of SR$\vdash \Rightarrow$ with MP. To derive the latter from the former, let BB in SR$\vdash \Rightarrow$ comprise just 'If A, C'; then we have the left-hand side by Assumptions and the right-hand side is MP. Conversely, just feeding the left-hand side of SR$\vdash \Rightarrow$ and MP into Cut yields the right-hand side of SR$\vdash \Rightarrow$ (the sentence cut is 'If A, C').

The derivability of *modus ponens* from the Suppositional Rule shows that the Rule supports *retrospective* as well as prospective assessments of conditionals, for *modus ponens* supports the predictions from a conditional which may be retrospectively verified or falsified, when one verifies the minor premise A, turning the supposition into known fact (see section 2.2).

Of course, the plausibility of conditional proof and *modus ponens* is by no means confined to mathematical contexts. Indeed, their plausibility in mathematical contexts is just a special case of their much more general plausibility. Even those who allege counterexamples against one or both of them usually acknowledge the plausibility of the rules, which they seek to explain away. Their plausibility is just what the Suppositional Conjecture would lead one to expect.

The strong analogy between SRCC and the natural deduction rules for the conditional confirms the naturalness of allowing auxiliary assumptions in SRC (Stalnaker's Hypothesis) and generalizing it to SRCC (Generalized Stalnaker's Hypothesis). For although one can consider an artificially restricted version of conditional proof, with auxiliary premises forbidden, it lacks much of the unrestricted version's power. It fails to support much normal reasoning both inside and outside mathematics, for we often have to make suppositions within suppositions, when working through a branching tree of apparently open possibilities.

The derivation of SR$\vdash \Leftrightarrow$ from SRmust is not really needed, for we can obtain SR$\vdash \Leftrightarrow$ directly from the Suppositional Rule itself, simply by applying it to the attitude of treating something as a logical consequence, with side premises BB. One advantage of this more direct route is that it frees us from the constraint that BB must be a *finite* set of premises. There is no such restriction on SR$\vdash \Leftrightarrow$. Thus the Suppositional Rule yields the very rules we standardly need for deductive reasoning with a conditional.

3.2 Deductive paradoxes for the Suppositional Rule

Unrestricted conditional proof and *modus ponens* together impose tight constraints on a conditional: they make it equivalent to the material conditional $\supset$, where '$A \supset C$'

is true just in case either A is false or C true, and is false otherwise. For $\supset$ obeys *modus ponens*: $A; A \supset C \vdash C$. Thus by conditional proof for 'if', $A \supset C \vdash$ if A, C. Conversely, $\supset$ also obeys conditional proof; applying it to *modus ponens* for 'if': if $A, C \vdash A \supset C$.

This is not good news for the Suppositional Rule. The present worry is not that the material reading of the natural language 'if' is highly controversial. Rather, the more urgent concern is that the material reading does not fit other cases of the Suppositional Rule, in particular SRC. For when Cred obeys the probability axioms, $\text{Cred}(A \supset C) = \text{Cred}(C|A)$ only in two very special cases: when $\text{Cred}(A \supset C) = 1$ and when $\text{Cred}(A) = 1$. Otherwise $\text{Cred}(A) > 0$ and $\text{Cred}(C|A) < \text{Cred}(A \supset C)$, for:

$$\text{Cred}(\text{not}(A \supset C)) = \text{Cred}(A \text{ and not}(C)) < \text{Cred}(A \text{ and not}(C))/\text{Cred}(A)$$
$$= \text{Cred}(\text{not}(C)|A)$$

So:

$$\text{Cred}(C|A) = 1 - \text{Cred}(\text{not}(C)|A) < 1 - \text{Cred}(\text{not}(A \supset C)) = \text{Cred}(A \supset C)$$

This is in effect an internal tension in the Suppositional Rule, between what it says about attitudes to logical consequences and what it says about intermediate credences.

We can see the tension from a different angle by considering the special case of SSRC for *minimal* credence:

SRCC0 $\text{Cred}(\text{if } A, C|B) = 0 \Leftrightarrow \text{Cred}(C|A \text{ and } B) = 0$ (when $\text{Cred}(A \text{ and } B) > 0$)

SRCC0 is the analogue of SRCC1 for probability 0 in place of probability 1. Just as SRCC1 turned out to be strongly analogous to the principle SR$\vdash$ about the logic of 'if', via an analogy between logical consequence ($\vdash$) and probability 1, so SRCC0 is strongly analogous to the principle SR$\vdash^-$ about the logic of 'if', via an analogy between logical inconsistency ($\vdash^-$) and probability 0:

$$\text{SR}\vdash^- \quad B \vdash^- \text{ if } A, C \Leftrightarrow A \text{ and } B \vdash^- C$$

Just as SR$\vdash$ is equivalent to SR$\vdash\Leftrightarrow$, so SR$\vdash^-$ is equivalent to SR$\vdash^-\Leftrightarrow$, where BB is a finite set of sentences:

$$\text{SR}\vdash^-\Leftrightarrow \quad BB \vdash^- \text{ if } A, C \Leftrightarrow BB; A \vdash^- C$$

As with SR$\vdash\Leftrightarrow$, we can lift the restriction to finite BB by deriving SR$\vdash^-\Leftrightarrow$ directly from the Suppositional Rule. Either way, we can also separate the right-to-left and left-to-right directions of SR$\vdash^-\Leftrightarrow$:

$$\text{SR}\vdash^-\Leftarrow \quad BB; A \vdash^- C \Rightarrow BB \vdash^- \text{ if } A, C$$
$$\text{SR}\vdash^-\Rightarrow \quad BB \vdash^- \text{ if } A, C \Rightarrow BB; A \vdash^- C$$

The introduction rule S$\vdash^-\Leftarrow$ says that when we can deductively rule out C on the basis of A and other assumptions, we can deductively rule out 'If A, C' on the basis of those other assumptions alone. The elimination rule SR$\vdash^-\Rightarrow$ says that when we can deductively rule out 'If A, C' on the basis of assumptions, we can deductively rule out C on the basis of A and those other assumptions.

Like $SR\vdash\Leftrightarrow$, we can obtain $SR\vdash^-\Leftrightarrow$ directly from the Suppositional Rule itself, simply by applying it to the attitude of treating something as deductively ruled out, with side premises BB.

Of course, if $\vdash^-$ were simply interpreted as the negation of $\vdash$, $SR\vdash^-\Leftarrow$ and $SR\vdash^-\Rightarrow$ would simply be equivalent to $SR\vdash\Rightarrow$ and $SR\vdash\Leftarrow$ respectively, so nothing would have been added. But that is analogous to interpreting $\vdash^-$ in terms of probability less than 1. That is not the intended interpretation of $\vdash^-$, which is analogous to interpreting it in terms of probability 0. For $BB \vdash^- C$ is to mean that BB is inconsistent with C, not that BB fails to entail C. But even on that intended interpretation of $\vdash^-$ in terms of inconsistency, the rules have some pull.

Nevertheless, the 'if' rules for $\vdash^-$ are in tension with those for $\vdash$. Here is one case. Let A and C be any declarative sentences. Since the 'if' rules for $\vdash$ make material conditionals entail the corresponding 'if' conditionals, as already seen, we have:

(1) $not(A) \supset C \vdash$ if $not(A), C$

But the three sentences '$not(A)$', '$not(A) \supset not(C)$', and C form an inconsistent triad by *modus ponens* for $\supset$, so:

(2) $not(A); not(A) \supset not(C) \vdash^- C$

But $SR\vdash^-\Leftarrow$ on (2) yields (3):

(3) $not(A) \supset not(C) \vdash^-$ if $not(A), C$

The semantics of $\supset$ yields (4a) and (4b):

(4a) $A \vdash not(A) \supset C$
(4b) $A \vdash not(A) \supset not(C)$

By Cut, (1) and (4a) yield (5):

(5) $A \vdash^-$ if $not(A), C$

Similarly, (3) and (4b) yield (6) by a Cut-like rule linking $\vdash$ and $\vdash^-$:

(6) $A \vdash^-$ if $not(A), C$

The required Cut-like structural rule is this:[6]

$$Cut\vdash^- \quad [AA \vdash B \text{ and } B; CC \vdash^- D] \Rightarrow AA; CC \vdash^- D$$

For if B entails AA, and B and CC are jointly inconsistent with D, then AA and CC are jointly inconsistent with D (where all the connections are deductive). This is a disastrous result, for (5) and (6) together make the same sentence ('If $not(A)$, C') both a deductive consequence of A and deductively inconsistent with A, which is to make A itself logically inconsistent. But nothing very special was assumed about A!

---

[6] Less naturally but more concisely, one can also derive (5) by $SR\vdash\Leftarrow$ from the classically vacuously valid principle that a contradiction entails everything: A, $not(A) \vdash$ C. Similarly, one can also derive (6) by $SR\vdash^-\Leftarrow$ from the corresponding classically vacuously valid principle for $\vdash^-$: A, $not(A) \vdash^-$ C. The longer derivations in the text show that there is no need to go via the vacuous case.

.

Of course, no-proposition theorists who deny that conditionals embed coherently under logical operators will not allow A or C to be a conditional sentence, since the argument involves embedding both of them under 'not' and 'if'. Still, the rules make any simple declarative sentence inconsistent.

We can eliminate the premise in (5) and (6) by substituting any logical truth T for A, so:

(7) $T \vdash$ if not(T), C

(8) $T \vdash^-$ if not(T), C

(9) $\vdash T$

Then Cut on (5) and (7) yields (10), and Cut$\vdash^-$ on (6) and (8) yields (11):

(10) $\vdash$ if not(T), C

(11) $\vdash^-$ if not(T), C

Thus the same sentence is both provable and refutable: the rules themselves are inconsistent. But it is worth remembering (5) and (6), since we can instantiate them to have an everyday premise, with its everyday negation as the antecedent of the conditional.

We can simplify the derivation of an inconsistency by using a Monotonicity rule for $\dashv$ instead of Cut$\vdash^-$:

$$\text{Monotonicity}\vdash^-\quad AA \vdash^- C \Rightarrow AA; BB \vdash^- C$$

Adding extra premises does not remove a deductive inconsistency. Again, let A be any declarative sentence. By Assumptions:

(12) $A \vdash A$

Since A is deductively inconsistent with not(A):

(13)   not(A) $\vdash^- A$

By Monotonicity, (12) yields (14):

(14) $A,$ not(A) $\vdash A$

Similarly, by Monotonicity$\vdash^-$, (13) yields (15):

(15) $A,$ not(A) $\vdash^- A$

By SR$\vdash\Leftarrow$, (14) yields (16):

(16) $A \vdash$ if not(A), A

By SR$\vdash^-\Leftarrow$, (15) yields (17):

(17) $A \vdash^-$ if not(A), A

This is another disastrous result, for (16) and (17) together make the same sentence ('If not(*A*), *A*') both a deductive consequence of *A* and deductively inconsistent with *A*, which is to make *A* itself logically inconsistent. Again, nothing very special was assumed about *A*. Of course, no-proposition theorists who deny that conditionals embed coherently under logical operators will again not allow *A* to be a conditional sentence, since the argument involves embedding it under 'not' and 'if'. Still, in this way too, the rules make any simple declarative sentence inconsistent.

The inconsistency essentially depends on the interaction of the rules for 'if' with those for $\vdash$ and $\vdash^-$. Neither SR$\vdash\Leftarrow$ and SR$\vdash\Rightarrow$ nor SR$\vdash^-\Leftarrow$ and SR$\vdash^-\Rightarrow$ produce any such formal inconsistency by themselves. That is clear for SR$\vdash\Leftarrow$ and SR$\vdash\Rightarrow$, since both are valid when 'if' is interpreted as a material conditional, as already observed. To see that SR$\vdash^-\Leftarrow$ and SR$\vdash^-\Rightarrow$ are equally harmless by themselves, we can interpret '*XX* $\vdash^-$ *Y*' as equivalent to '*XX* $\vdash$ not(*Y*)', which fits its intended interpretation, and 'If *X*, *Y*' as the conjunction '*X* and *Y*' which does not. Thus SR$\vdash^-\Leftarrow$ and S$\vdash^-\Rightarrow$ are interpreted as these two rules:

$$\text{R1} \quad BB; A \vdash \text{not}(C) \Rightarrow BB \vdash \text{not}(A \text{ and } C)$$
$$\text{R2} \quad BB \vdash \text{not}(A \text{ and } C) \Rightarrow BB; A \vdash \text{not}(C)$$

Both R1 and R2 are classically valid. That the interpretation of 'if' is so far from its intended one does not matter for the purpose of showing the formal consistency of SR$\vdash^-\Leftarrow$ and SR$\vdash^-\Rightarrow$, even when combined with the structural rules.

The pairing of conjunction with the material conditional as readings of 'if' is anyway less odd, structurally, than it may seem. Both readings agree that when *A* is true, 'If *A*, *C*' has the same truth-value as *C*. They disagree over the cases where *A* is false, but both treat them uniformly. On the material conditional reading, whenever *A* is false, 'If *A*, *C*' is true. On the conjunctive reading, whenever *A* is false, 'If *A*, *C*' is false.

The inconsistency proofs do *not* show that we have not really been implicitly using the Suppositional Rule. Nor do they show that the Rule does not really have the consequences we took it to have. We may have been using an inconsistent rule for 'if' all along. After all, both derivations involve applications of the rules which do not readily come to mind, because they go via mutually inconsistent suppositions (*A* and 'Not(*A*)'), even though the overall result is to derive an inconsistency from the most harmless of premises. The problem is severe, but from a practical perspective easy to overlook, like semantic paradoxes such as the Liar. There may be a glitch in our primary heuristic for assessing conditionals, just as the semantic paradoxes reveal a glitch in our way of assessing ascriptions of truth and falsity.

The tensions just identified in the Suppositional Rule do *not* depend on any assumption that 'if' conditionals express propositions, and are not resolved by denying any such assumption. Although the rules for $\vdash$ force 'if' to be equivalent to $\supset$, that is an *outcome* of applications of the Rule, not an imposition on it from outside. Of course, 'if' conditionals were treated as entering into deductive relations ($\vdash$ and $\vdash^-$), but those relations were not assumed to be characterized in terms of truth-preservation or the like. For all the argument required, those relations might be purely proof-theoretic. Indeed, following Ernest Adams (1965, 1975), no-proposition theorists have gone to great lengths to explain how there can nevertheless be a logic of conditionals. Nor did the derivations embed 'if' conditionals in more complex sentences of the object-language, as would be problematic on the no-proposition view. Denying that 'if' conditionals express propositions in no way reduces the urgency of the problem posed by the derivations for the Suppositional Rule.

One might guess that the glitch in the Rule is quite local, confined to the handling of inconsistent suppositions. But that is over-optimistic. If the problem arose only in cases of inconsistent suppositions, the probabilistic application SRCC of the rule would avoid it, since the conditional probabilities could be left undefined in those cases. That is not so: SRCC faces a closely related problem, discussed in the next section.

3.3 Probabilistic paradox for the Suppositional Rule

David Lewis initiated a large literature on combinatorial problems facing Stalnaker's Hypothesis (SRC), Generalized Stalnaker's Hypothesis (SRCC), and other variations on the same theme. The proofs are usually interpreted as showing, or attempting to show, that no assignment of propositions to conditional sentences satisfies the stipulated constraints. However, that is to underestimate what the proofs achieve. They also make trouble for the idea that conditionals do not express propositions but can still be assigned probabilities in accordance with the proposed equation, even though they are not probabilities of truth.[7]

Lewis's original argument (1976) can be applied to SRCC. We will rework it to allow for the option of assigning probabilities to conditionals according to SRCC without taking them to express propositions.

Consider a restricted language with two types of sentence. Type 1 sentences are assumed to express propositions; they include three mutually inconsistent sentences P, Q, and R, and their truth-functional combinations. Type 2 sentences are not assumed to express propositions; they are 'if' conditionals whose antecedent and consequent are type 1 sentences. Type 2 sentences are not subject to further embedding. Let Cred be a probability distribution over sentences of both types such that $\text{Cred}(P) > 0$, $\text{Cred}(Q) > 0$, $\text{Cred}(R) > 0$. Conditional probabilities are taken as primitive, but as obeying standard principles of conditional probability whenever they are defined. Type 2 sentences are permitted in the X position of $\text{Cred}(X|Y)$ but not in the Y position. Then Cred does not satisfy SRCC.

Here is a proof. Assume that SRCC holds. Then:

$$\text{Cred}(\text{if}(P \text{ or } Q), P|(P \text{ or } Q) \supset P) = \text{Cred}(P|(P \text{ or } Q) \text{ and } ((P \text{ or } Q) \supset P))$$
$$= \text{Cred}(P|P) = 1$$

Therefore, by SRCC again and the principle that $\text{Cred}(X) \leq \text{Cred}(Y)$ whenever $\text{Cred}(Y|X) = 1$:

$$\text{Cred}((P \text{ or } Q) \supset P) \leq \text{Cred}(\text{if}(P \text{ or } Q), P) = \text{Cred}(P|P \text{ or } Q)$$

---

[7] See for example Lewis 1976 and 1986, Hájek 1989, 2011, 2012, Hájek and Hall 1994, Bennett 2003: 60–77, Rumfitt 2013, Kaufmann 2015.

But we saw in section 3.2 that $\text{Cred}(A \supset C) \leq \text{Cred}(C|A)$ only when either $\text{Cred}(A) = 1$ or $\text{Cred}(A \supset C) = 1$. Thus either $\text{Cred}(P \text{ or } Q) = 1$ or $\text{Cred}((P \text{ or } Q) \supset P) = 1$. But, P, Q, and R are mutually incompatible by hypothesis, so 'P or Q' entails 'not(R)' and '(P or $Q) \supset P$' entails 'not(Q)', hence:

$$\text{Cred}(P \text{ or } Q) \leq \text{Cred}(\text{not}(R)) < 1 \text{ and } \text{Cred}((P \text{ or } Q) \supset P) \leq \text{Cred}(\text{not}(Q)) < 1.$$

This contradicts the disjunction just established. Thus SRCC does not hold.

Nothing in that argument requires 'if' conditionals to be treated as expressing propositions, yet it still shows that SRCC is unacceptably restrictive. We usually face more than two live, mutually exclusive possibilities.

Following Lewis, there is a large literature on ways of restricting SRCC to avoid the triviality results, and proofs of further triviality results from weaker assumptions. We need not go into details, for SRCC is the natural heuristic with which to calibrate credences in conditionals by conditional probabilities. It combines the required generality, by covering conditionals under suppositions, with the required simplicity, by avoiding the clutter of gerrymandered restrictions.

The triviality results for credences cannot quite be strengthened to proofs of outright inconsistency like those in section 3.2 for the deductive rules. The restriction to cases where the conditional probabilities are well defined is just enough to rescue SRCC from that even worse fate. In fact, we can show that SRCC is satisfiable when only two mutually exclusive possibilities are relevant.

Here is the proof. Let W be a set of just two worlds, and treat propositions as subsets of W. Let $[X]$ be the proposition expressed by the sentence X. We assign propositions to conditional sentences thus:

$$\text{When } [A] \cap [C] \neq \{\}, \text{ let } [\text{if } A, C] = [A \supset C].$$
$$\text{When } [A] \cap [C] = \{\}, \text{ let } [\text{if } A, C] = \{\}.$$

Propositions are assigned to unconditional sentences, and to truth-functional combinations of sentences (including conditional sentences) in a standard compositional way. Let Prob be any probability distribution over W; set $\text{Cred}(X) = \text{Prob}([X])$ for every sentence X. Then SRCC is satisfied, as we show thus (considering only cases where Cred is well defined):

If $[A] = W$, then $\text{Cred}(\text{if } A, C|B) = \text{Cred}(C|B) = \text{Cred}(C|A \text{ and } B)$.

If $[A] \neq W$ and $[A] \cap [C] \neq \{\}$, then $[A] \subseteq [C]$, so $\text{Cred}(\text{if } A, C|B) = 1 = \text{Cred}(C|A \text{ and } B)$.

If $[A] \cap [C] = \{\}$, then $\text{Cred}(\text{if } A, C|B) = 0 = \text{Cred}(C|A \text{ and } B)$.

This shows that the use of *three* mutually exclusive cases with nonzero probability (P, Q, R) in the triviality proof was the best possible: one cannot get the number down to two or one, let alone to zero.

Still, even though the trivialization implied by SRCC does not amount to outright inconsistency, it is bad enough to show that SRCC is unsatisfiable under reasonable conditions. Since denying that conditionals express propositions does not block the trivialization, the no-proposition view loses its main theoretical motivation.

The no-proposition view also fails to give good explanations of the linguistic data, as has often been noted. Many sentences with embedded 'if' seem to make perfectly good sense and to express propositions of a perfectly ordinary kind. Many universally quantified indicative sentences with conditionals are cases in point. Here is just one example, where 'if' is embedded under both conjunction and a universal quantifier:

(18) Everything was put in the suitcase if it belonged to Mary and on the table if it belonged to John.

That seems to make perfectly good sense, and to be equivalent to (19), whose truth-conditions are not in doubt:

(19) Everything that belonged to Mary was put in the suitcase and everything that belonged to John was put on the table.

Sometimes no-proposition theorists simply claim that we 'interpret' a sentence with an embedded conditional as another sentence with no embedded conditional, for example (18) as (19), even though on their view the reinterpretation is unwarranted by the compositional semantics of English. In the absence of any explanation as to *why* we reinterpret one sentence as another, such a strategy inherits the ad hoc character of the repairs it postulates. Moreover, the phenomenology of smoothly reading (18) as equivalent to (19) is quite unlike that of guessing at the intended interpretation of a string of words that makes no literal sense.

We can parse (18) thus, in overall form:

(20) Everything$_x$ ((if x belonged to Mary, x was put in the suitcase) and (if x belonged to John, x was put on the table))

To evaluate (18)/(20), as we surely can, we must be able to evaluate its matrix (the open formula following 'Everything$_x$'), and so the conditionals (21) and (22), for each value of the variable 'x' in the domain:

(21) If x belonged to Mary, x was put in the suitcase.
(22) If x belonged to John, x was put on the table.

For (18)/(20) to be true, as it may well be with a suitably restricted domain of contextually relevant things, (21) and (22) must also be true for each value of 'x' in the domain, not least for those on which the antecedent is false. The component conditionals must also be true for values of 'x' of which the speaker has no idea, and which play no role in structuring the speaker's credences or epistemic and doxastic state more generally.[8] Such values of variables are often contextually relevant, and falsify our rash universal generalizations, or verify our existential ones: I can truly say 'There are more things in heaven and earth than are dreamt of in my philosophy'.

---

[8] This presents a challenge for expressivist frameworks for compositional semantics on which sentences express constraints on the speaker's mental state, rather than propositions about the world, within which attempts have been made to make sense of indicative conditionals as entering more freely into complex embeddings without expressing propositions (Yalcin 2007, Moss 2018). Even over a finite domain, one can coherently give high credence to an existential generalization but credence 0 to every instance of it one can entertain. See also Williamson 2007a: 16–17.

To deny that conditionals such as (21) and (22) have truth-conditions obstructs the natural understanding of generalizations such as (18). As we already saw, it also fails to solve the paradoxes created by the Suppositional Rule. In what follows, indicative conditionals will therefore be assumed to be meaningfully embeddable in more complex sentences in all the ways standard in natural languages. For example, the quantified sentence (18) can be freely embedded, just like any other ordinary declarative sentence.

3.4 The Suppositional Rule for complex attitudes

Sections 3.2 and 3.3 considered paradoxical consequences of the Suppositional Rule when applied to credences and deductive attitudes respectively. It also has paradoxical consequences when applied to other attitudes. However, before we consider them, we must enlarge our understanding of what the Suppositional Rule implies.

We can use logical operators to define new attitudes from old ones. For example, from negation and a non-probabilistic attitude of *acceptance* we can define a corresponding non-probabilistic attitude of *rejection*, by stipulating that to reject something is just to accept its negation (under the same suppositions, if any). Some authors reserve the word 'reject' for a negative attitude unmediated by an added negation in the content, but that terminological issue should not obscure the legitimacy of the attitude just defined, whatever we call it. Such newly defined attitudes allow us to draw further consequences from the Suppositional Rule. For it involves taking *any* cognitive attitude to 'If A, C' just in case one takes that attitude to C on the supposition A.

We use the symbol $\|$ for an arbitrary cognitive attitude; $AA \| C$ means that the attitude $\|$ is taken to C on the set of suppositions AA. We can therefore rewrite the Suppositional Rule as a generalization of the natural deduction rules for the conditional, for any attitude $\|$, sentences A, C, and set of sentences BB:

$$\text{SR} \quad BB; A \| C \quad \text{just in case} \quad BB \| \text{if } A, C$$

For any attitude $\|$ (such as acceptance), we can now define a corresponding negative attitude $\|^{not}$ of 'rejection', in the way just explained:

$$\|^{not} \quad AA \|^{not} C \quad \text{just in case} \quad AA \| \text{not}(C)$$

In other words, to take the negative attitude to something is just to take the positive attitude to its negation (under the same suppositions).

The combination of SR and $\|^{not}$ has significant consequences. It makes negating the consequent of an indicative conditional equivalent to negating the whole conditional, in the sense that exactly the same attitudes to them are mandated on any given set of suppositions. For SR and $\|^{not}$ together yield this chain of mandated equivalences, for any attitude $\|$, sentences A, C, and set of sentences BB:

$$
\begin{array}{llll}
BB \| (\text{if } A, \text{not}(C)) & \text{just in case} & BB; A \| \text{not}(C) & (\text{by SR}) \\
& \text{just in case} & BB, A \|^{not} C & (\text{by } \|^{not}) \\
& \text{just in case} & BB \|^{not} (\text{if } A, C) & (\text{by SR}) \\
& \text{just in case} & BB \| \text{not}(\text{if } A, C) & (\text{by } \|^{not})
\end{array}
$$

In this sense, given the Suppositional Rule and the way of defining negative attitudes $||^{\mathrm{not}}$, the conditional commutes in the consequent with negation:

CCCN $BB \,||\,$ (if A, not(C))   just in case $BB \,||\,$ not(if A, C)

In particular, we can apply CCCN to the attitude of acceptance, written $||^{\mathrm{a}}$. Since anything is accepted on a set of suppositions to which it belongs, we have:

(23) BB; (if A, not(C))$||^{\mathrm{a}}$ (if A, not(C))
(24) BB; not(if A, C)$||^{\mathrm{a}}$ not(if A, C)

But then, by applying opposite directions of CCCN to (23) and (24) respectively, we have:

(25) BB; (if A, not (C))$||^{\mathrm{a}}$ not(if A, C)
(26) BB; not(if A, C)$||^{\mathrm{a}}$ (if A, not(C))

Thus, still given SR and $||^{\mathrm{not}}$, negating a conditional and negating its consequent are also equivalent in the additional sense that each is accepted on the supposition of the other and any given set of further suppositions. This equivalence, unlike CCCN, is specific to the attitude of acceptance, for not every cognitive attitude is to be taken to something on a set of suppositions to which it belongs; *rejection* is an obvious counterexample.

For most purposes, then, negating a conditional is tantamount to negating its consequent. This idea has attracted theorists of conditionals in natural language. For example, Dorothy Edgington tentatively proposes that 'A is to $\neg A$ as "If A, B" is to "If A, $\neg B$"', and she asserts that '"It's not the case that if A, B" has no clear established sense distinguishable from this' (1995: 283).[9] As noted in section 3.1, there is considerable evidence that we tend to conform to SRC (the Suppositional Rule for credences), which requires credences for 'If A, C' and 'If A, not(C)' to sum to 1, so in that respect we treat 'If A, not(C)' like the contradictory of 'If A, C'.

A striking feature of the derivations of both CCCN and (25)–(26) is that they are independent of the logic of 'not', of which they exploit no characteristic principle. That suggests generalizing them to other operators. To reach other non-trivial truth-functions, we must generalize from one-place operators like negation to many-place operators.

A salient candidate is *conjunction*. The analogue of CCCN is the conditional commuting in the consequent with conjunction, for any attitude $||$, sentences A, C, D, and set of sentences BB:

CCCC $BB \,||\,$ (if A, (C and D))   just in case $BB \,||\,$ ((if A, C) and (if A, D))

In other words, conjoining conditionals with the same antecedent is equivalent to conjoining their consequents. That sounds plausible. Just as (25) and (26) were derived from CCCN, so we can derive (27) and (28) from CCCC:

(27) BB, (if A, (C and D)) $||^{\mathrm{a}}$ ((if A, C) and (if A, D))
(28) BB, ((if A, C) and (if A, D)) $||^{\mathrm{a}}$ (if A, (C and D))

---

[9] See also Dummett 1982: 82 for the view that a subjunctive conditional is to be negated by negating its consequent.

These are natural outcomes of the Suppositional Rule. If you accept a conjunction on some suppositions, you can apply the elimination rule for conjunction to accept each of its conjuncts, on the same suppositions. Conversely, if you accept each of the conjuncts on some suppositions, you can apply the introduction rule for conjunction to accept the conjunction, again on the same suppositions. The Suppositional Rule then takes one the rest of the way to (27) and (28). Of course, the introduction rule for conjunction does not work for a probabilistic standard of acceptance, with a threshold for acceptance less than 1, since a conjunction may be less probable than each of its conjuncts, but it still works for *full acceptance*, which gives us what we want in (27) and (28). When scientists develop the consequences of a hypothesis, they seem quite willing to conjoin different consequences which they have separately extracted from it.

However, those considerations are specific to conjunction. For example, disjunction does not satisfy the elimination rule for conjunction: you can accept a disjunction without accepting all its disjuncts, even without accepting any of them. You saw the coin land, but you were too far away to see which way it came up. You accept 'It came up heads or tails', but you do not accept 'It came up heads', nor do you accept 'It came up tails' (nor do you accept their negations). But (27) and (28) were derived from CCCC, the analogue of CCCN. If CCCC is properly analogous to CCCN, its rationale should not depend on the logic of conjunction, any more than the rationale for CCCN depends on the logic of negation. Indeed, such a rationale for CCCC should smoothly generalize to a rationale for the corresponding principle about disjunction, the conditional commuting in the consequent with disjunction:

CCCD $\quad BB \parallel (\text{if } A, (C \text{ or } D)) \quad$ just in case $\quad BB \parallel ((\text{if } A, C) \text{ or } (\text{if } A, D))$

In other words, *dis*joining conditionals with the same antecedent is equivalent to *dis*joining their consequents. Just as (27) and (28) were derived from CCCC, so we can derive (29) and (30) from CCCD:

(29) $BB, (\text{if } A, (C \text{ or } D)) \parallel^a ((\text{if } A, C) \text{ or } (\text{if } A, D))$
(30) $BB, ((\text{if } A, C) \text{ or } (\text{if } A, D)) \parallel^a (\text{if } A, (C \text{ or } D))$

Perhaps surprisingly, these generalized expectations can be implemented. Just as deriving CCCN required defining negative attitudes, so deriving CCCC and CCCD requires defining conjunctive and disjunctive attitudes respectively. These new forms of complex attitude will be attitudes to *sequences* of sentences (or, through them, to sequences of propositions). For simplicity, we just consider attitudes to ordered pairs, but the generalization to longer sequences is straightforward. Here are the definitions:

$\parallel^{\text{and}} \quad AA \parallel^{\text{and}} C, D \quad$ just in case $\quad AA \parallel C \text{ and } D$
$\parallel^{\text{or}} \quad AA \parallel^{\text{or}} C, D \quad$ just in case $\quad AA \parallel C \text{ or } D$

In other words, to take the conjunctive attitude to some things is just to take the original attitude to their conjunction (under the same suppositions); to take the disjunctive attitude to the things is just to take the original attitude to their disjunction (again, under the same suppositions). For instance, to accept a plurality conjunctively is to accept its conjunction; to accept the plurality disjunctively is to accept its disjunction. These definitions have singular attitudes on the right-hand side, the definiens, so any initial mystery about the plural attitudes on the left-hand side, the definiendum is cleared up by the definition itself.

We need the Suppositional Rule to apply to plural attitudes as well as singular ones. That is entirely within the spirit of the rule. It mandates taking whatever attitude one takes to some sentences conditionally on the supposition A unconditionally to the corresponding conditionals with those sentences as consequents and A as the antecedent (with the same background suppositions). For instance, one is to treat C and D as *mutually incompatible* conditionally on the supposition A just in case one treats 'If A, C' and 'If A, D' as unconditionally mutually incompatible. Here is the Rule stated explicitly in plural form, where $A, C_1, \ldots, C_n$ are any sentences, BB any set of sentences, and $||$ any attitude:

SR+ $BB, A \,||\, C_1, \ldots, C_n$ just in case $BB \,||\, (\text{if } A, C_1), \ldots, (\text{if } A, C_n)$

Here is the derivation of CCCC from SR+ and $||^{\text{and}}$, analogous to the derivation of CCCN from SR and $||^{\text{not}}$, for any attitude $||$, sentences A, C, D, and set of sentences BB:

$BB \,||\, \text{if } A, (C \text{ and } D)$	just in case	$BB, A \,||\, C \text{ and } D$	(by SR+)
	just in case	$BB, A \,||^{\text{and}}\, C, D$	(by $||^{\text{and}}$)
	just in case	$BB \,||^{\text{and}}\, (\text{if } A, C), (\text{if } A, D)$	(by SR+)
	just in case	$BB \,||\, (\text{if } A, C) \text{ and } (\text{if } A, D)$	(by $||^{\text{and}}$)

By substituting 'or' for 'and' throughout, one obtains a parallel derivation of CCCD from SR+ and $||^{\text{or}}$. These derivations make no use of any characteristic principles of the logic of conjunction or disjunction. We then obtain the equivalences (27)–(28) and (29)–(30) from CCCC and CCCD respectively, as already observed.

The main point of these derivations of CCCC and CCCD is to check that both principles follow simply from the Suppositional Rule as applied to a suitable variety of attitudes, without appeal to any special principles of the logic of conjunction or disjunction. For, as we also noted, the principle that the conditional commutes with conjunction in the consequent is pre-theoretically very plausible, without explicit appeal to the Suppositional Rule, and all truth-functions, including disjunction, are definable in terms of conjunction and negation. Thus, by repeatedly applying the commutativity principles for conjunction and negation, we can derive the corresponding commutativity principle for *any* truth-function, at least in its defined form. In the case of disjunction, we could anyway argue for the equivalence of the De Morgan equivalents of 'if A, (C or D)' and '(if A, C) or (if A, D)', in other words, 'if A, not(not(C) and not(D))' and 'not(not(if A, C) and not(if A, D))'. From that, the equivalence of the explicitly disjunctive versions is a comparatively small step. One can also derive the new commutativity principle, along the same lines as CCCC and CCCD.

The commutativity principles enable us to derive some much-discussed theses in the logic of conditionals. One is *Conditional Excluded Middle*, this schema:

CEM   (if A, C) or (if $A, \text{not}(C)$)

For, given that 'if' commutes with 'not' in the consequent, CEM is equivalent to (31):

(31)   (if A, C) or not(if A, C)

But (31) is just a special case of the classical law of excluded middle ('B or not(B)'), and classical logic is not here in question. Equally, we can argue for CEM from the principle that 'if' commutes with 'or' in the consequent, for it makes CEM equivalent to (32):

(32)   if $A, (C$ or not$(C))$

Again given classical logic, (32) has a tautologous consequent, and so should be accepted.

The position is similar for *Conditional Non-Contradiction*, this schema:

CNC   not((if A, C) and (if $A, \text{not}(C)$))

For, given that 'if' commutes with 'not' in the consequent, CNC is equivalent to (33):

(33)   not((if A, C) and not(if A, C))

But (33) is just a special case of the classical law of non-contradiction ('not(B and not(B))'), and classical logic is not here in question. Equally, we can argue for CNC from the principle that 'if' commutes with 'and' in the consequent, for it makes CNC equivalent to (34):

(34)   not(if $A, (C$ and not$(C))$

Again given classical logic, the conditional in (34) has a contradictory consequent, which one might take to merit rejection on any supposition.

Together, CEM and CNC guarantee that 'not(if A, C)' is materially equivalent to 'if $A, \text{not}(C)$'. For CEM guarantees that 'not(if A, C)' materially implies 'if $A, \text{not}(C)$', and CNC guarantees the converse material implication. Thus they jointly yield a version of the principle that the conditional commutes in the consequent with negation.

How do the commutativity principles fare on various interpretations of 'if'? There is a trivial interpretation on which all such principles hold: it treats the antecedent as idle, and 'if A, C' as having the same semantics as C itself. Thus for any n-place sentential operator O and sentences $A, C_1, \ldots, C_n$, 'if $A, O(C_1, \ldots, C_n)$' has the same semantics as '$O(C_1, \ldots, C_n)$', which in turn has the same semantics as '$O(($if $A, C_1), \ldots, ($if $A, C_n))$', provided that O itself has a compositional semantics, in the sense that the semantics of the output sentence '$O(B_1, \ldots, B_n)$' is always determined just by the semantics of the input sentences $B_1, \ldots, B_n$. Negation, conjunction, and disjunction all have a compositional semantics in that sense. The trivial reading also validates both CEM and CNC.

That unintended interpretation of 'if' suffices to show that the commutativity principles *by themselves* generate no inconsistency. However, it is obviously inadequate, since it implies that 'if A, A' is acceptable only if A is acceptable, which is false: one can rationally accept 'If there is life in other galaxies, there is life in other galaxies' without accepting 'There is life in other galaxies.' In such cases, the acceptability of 'if A, A' is a trivial upshot of the Suppositional Rule, since A is acceptable on the supposition A, so this unintended interpretation does not even validate the full consequences of the Suppositional Rule, which therefore exceed the commutativity principles. Thus we restrict attention to interpretations of 'if' which vindicate the acceptability of 'if A, A'.

The material interpretation of 'if' validates the commutativity principles for both the binary conjunction and disjunction operators, as well as 'if A, A'. For when A is true, it makes the truth-value of any conditional with antecedent A the truth-value of the consequent, so both 'if A, (C and D)' and '(if A, C) and (if A, D)' have the same truth-value as 'C and D', while 'if A, (C or D)' and '(if A, C) or (if A, D)' have the same truth-value as 'C or D' ; when A is false, the material interpretation makes any conditional with antecedent A true, so 'if A, (C and D)', '(if A, C) and (if A, D)', 'if A, (C or D)', and '(if A, C) or (if A, D)' are all true; either way, commuting 'if' with conjunction or disjunction in the consequent is logically guaranteed to preserve truth-value. Thus the material interpretation validates the principles that the conditional commutes with binary conjunction and disjunction in the consequent. It also validates CEM. However, it does not fully validate the commutativity principle for negation. When A is true, the material interpretation also gives both 'if A, not(C)' and 'not(if A, C)' the truth-value of 'not(C)'. But when A is false, it makes 'if A, not(C)' true and 'not(if A, C)' false, so it fails hopelessly to validate the principle that the conditional commutes with negation in the consequent. It also invalidates CNC, because it makes both 'If A, C' and 'If A, not(C)' true when A is false. Thus the material interpretation fails to validate fully the applications of the Suppositional Rule at issue in this section.

A truth-functional interpretation of 'if' which validates the commutativity principle for negation, as well as 'if A, A', is the material *biconditional* reading, on which 'if A, C' is true or false according to whether the truth-values of A and C are the same or different. This reading makes both 'if A, not(C)' and 'not(if A, C)' true or false according to whether the truth-values of A and C are different or the same. Consequently, it also validates both CEM and CNC. However, the material biconditional principle invalidates the commutativity principles for both conjunction and disjunction. For let C and D have opposite truth-values, so 'if A, C' and 'if A, D' also have opposite truth-values. Hence 'C and D' and '(if A, C) and (if A, D)' are false, while 'C or D' and '(if A, C) or (if A, D)' are true, irrespective of the truth-value of A. Hence, when A is false, 'if A, (C and D)' is true while '(if A, C) and (if A, D)' is false, and 'if A, (C or D)' is false while '(if A, C) or (if A, D)' is true. Thus the material biconditional interpretation also fails to validate fully the applications of the Suppositional Rule at issue in this section.

The approach that arguably comes closest to validating the applications of the Suppositional Rule in this section is Stalnaker's (1968); Williams (2010) and Dorr and Hawthorne (2018) hold relevantly similar views. Such a semantics is modal; formulas are assigned truth-values at worlds. Each model is equipped with a selection function f, which maps each non-empty set X of worlds and world w to a world

$f(X, w)$ in X (informally, one might regard $f(X, w)$ as the closest world in X to w).[10] Let [A] be the set of worlds in the model at which A is true. Stalnaker ensures that [A] is non-empty for every sentence A by stipulating that each model includes an 'absurd world' λ where all sentences whatsoever are true. By contrast, the evaluation of truth-functional combinations at worlds other than λ is standard. His rule for the conditional is that the truth-value of 'if A, C' at w is the truth-value of C at $f([A], w)$. Informally, we can regard $f([A], w)$ as the world we imagine where A is true when we suppose A from the perspective of w. As a result, this framework has many consequences friendly to the applications of the Suppositional Rule at issue in this section.

Obviously, Stalnaker's semantics makes 'if A, A' true at every world. Moreover, it makes a conjunction true at a world (possible or impossible) just in case every conjunct is true at the world, which ensures that 'if A, (C and D)' and '(if A, C) and (if A, D)' coincide in truth-value at every world. Thus the conditional commutes in the consequent with conjunction. Similarly, a disjunction is true at a world (possible or impossible) just in case some disjunct is true at the world, which ensures that 'if A, (C or D)' and '(if A, C) or (if A, D)' coincide in truth-value at every world. Thus Stalnaker's conditional also commutes in the consequent with disjunction.

Stalnaker's semantics *almost* validates commutativity in the consequent with negation too. For whenever w is a world other than λ, and X contains at least one world other than λ, $f(X, w)$ is required to be a world other than λ too. Thus when the sentence A expresses a possibility, so [A] contains at least one possible world, $f([A], w)$ is required to be possible too. Informally: when we make a possible supposition from a possible perspective, we should not imagine it in an impossible way. Now 'not(C)' is true at a possible world just in case C is not true at that world. Thus, when A expresses a possibility and w is possible, the truth-values of C and 'not(C)' at the possible world $f([A], w)$ are opposite, so the truth-values of 'if A, C' and 'if A, not(C)' back at the world w are also opposite, so the truth-values of 'not(if A, C)' and 'if A, not(C)' at w are the same.

Unfortunately, this nice pattern breaks down when A expresses an impossibility. For then [A] is simply {λ}, so $f([A], w)$ must be λ, so both C and 'not(C)' are true at $f([A], w)$, so both 'if A, C' and 'if A, not(C)' are true at any possible world w, for any sentence C, while 'not(if A, C)' is not true. Thus CNC fails in this special case, though CEM holds unrestrictedly, since at least one of C and 'not(C)' is true at any possible world, possible or impossible. Nevertheless, the upshot is that Stalnaker's conditional does not fully commute in the consequent with negation.

Stalnaker's semantics can be tweaked to validate CNC as well as CEM. For one can drop λ, treat the truth-functors classically at all worlds, and make the brute stipulation that $f(\{\}, w) = w$ for every world w. Thus whenever A expresses an impossibility, 'if A, C' has the same truth-value as C. That is in effect to treat counterpossible conditionals as don't-cares. But it makes 'if A, A' impossible whenever A is impossible. Since it invalidates 'if A, A', this tweaked semantics also violates a blatant requirement of the Suppositional Rule.

---

[10] The text slightly simplifies Stalnaker's original presentation, for instance by ignoring his accessibility relation between worlds, but not in ways that affect the present issues.

Tweaked or untweaked, Stalnaker's approach faces obvious worries about how the selection function f is determined. There is the problem of equally good candidates. For example, let A be 'the coin is tossed.' If A is false at w, what determines whether $f([A], w)$ is a world where the coin comes up heads or one where it comes up tails? Worse, there is also the problem of sequences not containing their own limit. For example, let the rod R be exactly m long in w, and B be 'R is longer than m.' Consider any world x in $[B]$. R is exactly $m + \delta$ long in x; but in another world y in $[B]$, R is only $m + \delta/2$ long, so y seems to beat x as a candidate to be $f([B], w)$, since y is closer than x to w in the relevant respect. Thus every candidate is bettered by another candidate. Stalnaker handles these problems by suggesting that it can be indeterminate which selection function is operative, even though it is determinate that some selection function or other is operative. Thus CEM is still determinately true, because each selection function verifies it, even though it may be indeterminate which disjunct of CEM is true, because some selection functions verify one disjunct while other selection functions verify the other disjunct. The theory of *supervaluationism* supplies the apparatus to implement this approach.[11]

Theorists who like a modal treatment of indicative conditionals but dislike the arbitrariness of selection functions may prefer to generalize over a plurality of relevant worlds at which the antecedent is true, perhaps regarding the worlds as epistemically rather than objectively possible. Thus they regard the indicative conditional as an epistemically strict conditional, true just in case *all* relevant antecedent-worlds are also consequent-worlds. The strict reading does at least validate 'If A, A'. However, it invalidates CEM, since some but not all antecedent-worlds may be consequent-worlds. Moreover, CNC remains invalid, since counterpossibles with mutually contradictory consequents are both vacuously true.

Claudio Pizzi has shown how to validate CNC within the modal approach by strengthening strict implication to *consequential implication* (Pizzi 1977; Pizzi and Williamson 1997, 2005). The idea is that 'if A, C' is true just in case A strictly implies C and A and C have the same *modal status*, in the sense that the propositions they express are either both necessary, or both contingent, or both impossible. In other words, the truth-condition is this: every antecedent-world is a consequent-world, and if there are consequent-worlds, there are antecedent worlds, and if there are only consequent-worlds, there are only antecedent-worlds. To see that this truth-condition validates CNC, suppose that 'If A, C' and 'If A, not(C)' are both true. Then every A-world is both a C-world and a 'not(C)'-world, so there are no A-worlds. Since both conditionals must have consequents with the same modal status as their antecedent, there are no C-worlds and no 'not(C)'-worlds, which is a contradiction, since the actual world is either a C-world or a 'not(C)'-world. The consequential reading also validates 'If A, A', for A always both strictly implies itself and has the same modal status as itself.

However, consequential implication is even worse than strict implication for CEM. When A strictly implies neither of C and 'not(C)', it also consequentially implies neither of them. Moreover, consequential implication creates additional problems for CEM. When one of A and C expresses an impossibility or necessity, while the other expresses a contingency, A differs in modal status from both C and 'not(C)', so all these cases are counterexamples to CEM.

Moreover, despite the feeling that CEM is metaphysically ungrounded, the natural language data suggest that much of our ordinary thought and talk respects CEM. Imagine yourself contemplating two hypotheses about a fair coin, not knowing whether it will be tossed:

(35) If it is tossed, it will come up heads.
(36) If it is tossed, it will come up tails.

Consider (35) and (36) as uttered in your context. If an oracle tells you that (35) is false, it seems natural for you to conclude that (36) is true. But the strict and consequential readings make that inference fallacious. For it ignores the possibility that the coin is never actually tossed, in which case both heads and tails are symmetric open possibilities. Thus the relevant counterfactual worlds where the coin is tossed should include some where it comes up heads and some where it comes up tails. Consequently, on those readings, both (35) and (36) are false. By contrast, CEM licenses the inference, via disjunctive syllogism: eliminating the first disjunct leaves you with the second (here we can treat 'not heads' as tantamount to 'tails', given that the coin was tossed).

Similarly, if the oracle instead tells you that (35) is *true*, it seems natural for you to conclude that (36) is false. That is further confirmation that much of our ordinary thought and talk respects CNC too.

The strict and consequential readings also invalidate the principle CCCD that the conditional commutes in the consequent with disjunction, for they disallow the deduction from (37) to (38):

(37) If it is tossed, (it will come up heads or it will come up tails).
(38) (If it is tossed, it will come up heads) or (if it is tossed, it will come up tails).

The strict reading does at least validate the principle CCCC that the conditional commutes in the consequent with conjunction, for every A-world is a 'C and D'-world just in case every A-world is a C-world and every A-world is a D-world. By contrast, the consequential reading invalidates even that principle. For if C expresses a contingency and D a necessity, 'If (C and D), (C and D)' is true on the consequential reading, while 'If (C and D), D' is false, since its antecedent and consequent differ in modal status.

Notably, *none* of the interpretations of 'if' just surveyed validates both the unrestricted commutativity principles and the unrestricted schema 'If A, A'. There is a more general reason for that, as we shall soon see.

3.5 The inconsistency of the Suppositional Rule for complex attitudes

Anything is to be accepted on the supposition of itself. Therefore, by the Suppositional Rule, 'If A, A' is always to be accepted, for any meaningful sentence A. As a special case of that schema, we have for any meaningful sentence A:

[I]   If $(A$ and not$(A))$, $(A$ and not$(A))$

For meaningful sentences are closed under conjunction and negation. Therefore, by applying commutativity in the consequent with conjunction (CCCC), we can derive [II] from [I]:

[II]   (If $(A$ and not$(A))$, $A)$) and (if $(A$ and not$(A))$, not$(A))$

By applying commutativity in the consequent with negation (CCCN) to the second main conjunct of (II), we can derive [III] from [II]:

[III]   (If $(A$ and not$(A))$, $A)$ and not(if $(A$ and not$(A))$, $A)$

But [III] is an explicit contradiction; it conjoins 'if $(A$ and not$(A))$, A' with its own negation.

For example, less formally: if it is raining and not raining, it is raining and not raining; so if it is raining and not raining, it is raining; but also, if it is raining and not raining, it is not raining, so it is not the case that if it is raining and not raining, it is raining; that is an outright contradiction.

The resources from conditional logic needed for the derivation can be pared down to three schematic principles: 'If A, A' to get [I], the inference rule from 'If A, $(B$ and $C)$' to '(If A, $B)$ and (if A, $C)$' to get from [I] to [II], and the inference rule from 'If A, not(C)' to 'Not(if A, $C)$' (in effect, CNC) to get from [II] to [III]. Since the Suppositional Rule delivers all the required resources, this is another direct proof of its inconsistency.

One initial reaction to the paradox [I]–[III] may be that it does not matter, because we have independent evidence that conditionals with self-contradictory antecedents are pathological. After all, probabilities conditional on a contradiction are normally undefined. Moreover, indicative conditionals normally presuppose that their antecedents are epistemically possible.

However, that initial reaction is too quick. Of course, to utter a conditional with a self-contradictory antecedent out of the blue is likely to cause puzzlement: what point can uttering it have? But some pointless utterances are meaningful, and even true; Queen Anne *is* dead. Moreover, uttering an indicative conditional with a self-contradictory antecedent is not always pointless. There was a theoretical point to propounding the paradox [I]–[III]. To take a different example, people who know the logician Graham Priest are not taken aback when he asserts something like (39):

(39)   If the Liar sentence is true and not true, dialetheism is vindicated.

Priest, a sane, rational native speaker of English, regards many contradictions, including the antecedent of (39), as epistemically possible (indeed true); in asserting (39), he may easily make a relevant contribution to a philosophical conversation. Even classical logicians and mathematicians may assert an indicative conditional in English with an inconsistent antecedent, in the course of proving the negation of the antecedent by *reductio ad absurdum*. In some cases, both they and their audience already regard the antecedent as epistemically impossible, when an old or new proof of an already known result is being explained; they are merely treating the antecedent as open for purposes of the proof.

In less technical conversation, imagine an utterance of (40), with the indefinite articles used to express a universal generalization:

(40)   If Mary hates a man and a woman doesn't hate John, the man is nastier than the woman.

In context, Mary may be clearly in the domain for 'a woman', and John clearly in the domain for 'a man'. Thus (41) is an instance of (40):

(41)   If Mary hates John and Mary doesn't hate John, John is nastier than Mary.

Since (40) is intended as a universal generalization, it is true only if every instance of it is true. Hence when (40) is true, (41) is also true. But when a sentence crashes, it is not true. Thus when (40) is true, (41) does not crash. Since (40) is true in some situations, (41) need not crash.

To take a simpler case, (42) understood as a universal generalization is clearly true:

(42)   If one tower is tall and one tower is not tall, the former is taller than the latter.

Here is an instance of (42):

(43)   If the Eiffel Tower is tall and the Eiffel Tower is not tall, the Eiffel Tower is taller than the Eiffel Tower.

The truth of (42) requires the truth of (43). Similarly, in natural mathematical English, one can truly say about the natural numbers:

(44)   If m belongs to an initial segment S and n does not belong to S, m is less than n.

But (44) entails (45) by universal instantiation:

(45)   If 7 belongs to S and 7 does not belong to S, 7 is less than 7.

Thus the paradox [I]–[III] cannot be so easily dismissed. Normal intelligent use of English commits us to treating some indicative conditionals with self-contradictory antecedents as not only intelligible but true.

Although most proponents of CEM restrict CNC in such cases, they sometimes argue for CEM in ways which also commit them to CNC quite generally. Moreover, since they normally take for granted that conditionals commute in the consequent with conjunction, they are vulnerable to the paradox [I]–[III]. In particular, there is an argument for CEM just like one given by Robert Williams (2010); he gives it for subjunctive rather than indicative conditionals, but the difference plays no role in his argument. He uses considerations from James Higginbotham (1986) and Kai von Fintel and Sabine Iatridou (2002).

Here is the argument. Consider pairs such as (46) and (47):

(46)   No student passed if they goofed off.
(47)   Every student failed if they goofed off.

In this context, failing is equivalent to not passing. Then (46) and (47) seem to be equivalent. Moreover, they can be regimented as (46a) and (47a) respectively:

(46a)   [no x: student x] (if x goofed off, x passed)
(47a)   [every x: student x] (if x goofed off, not(x passed))

Generalizing this pattern, we have the equivalence of (48) and (49):

(48)　[no x: Fx] (if Gx, Hx)
(49)　[every x: Fx] (if Gx, not(Hx))

But '[no x: Fx] A' is quite generally equivalent to '[every x: Fx] not(A)'. Thus, in particular, (48) is quite generally equivalent to (50):

(50)　[every x: Fx] not(if Gx, Hx)

Thus (49) and (50) are quite generally equivalent. But that requires 'if Gx, not(Hx)' and 'not(if Gx, Hx)' to be quite generally equivalent. For given any value o of the variable 'x', we can always interpret F to be true of o and nothing else, in which case 'if Gx, not(Hx)' and 'not(if Gx, Hx)' for that value of 'x' are equivalent to (49) and (50) respectively. Although the conclusion usually drawn in the literature is only CEM, the argument delivers something stronger: that the conditional commutes in the consequent with negation (CCCN), which requires CNC as well as CEM.

Very similar arguments conclude that the conditional commutes in the consequent with conjunction (CCCC). For pairs such as (51) and (52) seem to be equivalent:

(51)　Every student sang and danced if they passed.
(52)　Every student sang if they passed and every student danced if they passed.

They can be regimented as (51a) and (52a) respectively:

(51a)　[every x: student x] (if x passed, (x sang and x danced))
(52a)　[every x: student x] (if x passed, x sang) and [every x: student x] (if x passed, x danced)

Generalizing this pattern, we have the equivalence of (53) and (54):

(53)　[every x: Fx] (if Gx, (Hx and Ix))
(54)　[every x: Fx] ((if Gx, Hx) and (if Gx, Ix))

By reasoning parallel to that above, the general equivalence of (53) and (54) requires the equivalence of 'if Gx, (Hx and Ix)' and '(if Gx, Hx) and (if Gx, Ix)' for any value of the variable 'x'. Thus the conditional commutes in the consequent with conjunction.

Once we recognize that the commutativity principles generate contradictions, we cannot simply pick and choose which conclusions they support, restricted to avoid inconsistency—for example, by taking them to support unrestricted CEM but only restricted CNC. The source is tainted; that fact must qualify our view of all its products. Of course, the taint of paradox does not make it evidentially worthless, any more than perceptual error motivates abandoning perception as a source of evidence. Rather, we need to take a more critical attitude towards our sources of evidence, which is a normal part of a scientific approach. That includes understanding their principles of operation, and the implications for the scope and limits of their reliability. For apparent equivalences like those of (46) to (47) and (51) to (52), the relevant source is the Suppositional Rule.

Of course, arguments like those just given are not the only support for the commutativity principles. CEM and CNC also gain support from probability judgements, as discussed in sections 3.1 and 3.4. For example, we tend to treat the probabilities of 'if *A*, *C*' and 'if *A*, not(*C*)' as summing to 1. But the relevant probability judgements too are explained by the Suppositional Rule, and more specifically by our tendency to equate the probability of 'if *A*, *C*' with the conditional probability of *C* on *A*, which was also seen to generate untenable results, by the triviality proofs. Indeed, the paradoxical consequences of the Suppositional Rule for credences in section 3.3 and for complex attitudes in this section are in a way complementary, for the latter concern only conditionals with logically inconsistent antecedents, while the former concern only conditionals with logically *consistent* antecedents, since probabilities conditional on logically inconsistent suppositions are normally undefined. The paradoxical consequences of the Suppositional Rule for deductive attitudes in section 3.2 concern both kinds of conditional. Thus merely making an exception of conditionals with logically inconsistent antecedents fails to resolve the underlying problem.

We can draw a further moral from the way in which the Suppositional Rule supports implicitly inconsistent commutativity principles. We should be suspicious of the tempting idea that epistemic modals obey such principles: that 'It must be that if *A*, *C*' is somehow equivalent to 'If *A*, it must be that *C*', and that 'It may be that if *A*, *C*' is somehow equivalent to 'If *A*, it may be that *C*.' Even if such principles result from legitimate applications of the Suppositional Rule, they may well be in cases of just the sort where the Rule is most unreliable.

3.6 Plain counterlogicals

A *counterlogical* is a conditional with a logically inconsistent antecedent, like those in the [I]–[III] paradox (section 3.5). Logical considerations pressure semantic theories of conditionals to make all counterlogicals true. For the schema 'if *A*, *A*' seems trivially correct, whether *A* is consistent or not; the Suppositional Rule endorses that pre-theoretic impression. Moreover, this principle of single-premise deductive closure in the consequent seems plausible:

$$\text{SPCC} \quad C \vdash D \Rightarrow \text{if } A, C \vdash \text{if } A, D$$

The Suppositional Rule arguably endorses SPCC too, since iterated deductive reasoning is as reliable a way of developing a supposition as any. But, at least in classical logic, an inconsistent premise entails any conclusion. Thus, whenever *A* is inconsistent, $A \vdash C$, for any *C*, so by SPCC (if *A*, *A*) $\vdash$ (if *A*, *C*). But $\vdash$ (if *A*, *A*) as above, so $\vdash$ (if *A*, *C*) by the Cut Rule from section 3.1. In other words, all counterlogicals are logical truths.

As observed in section 3.1, the Suppositional Rule yields standard natural deduction rules for the conditional, of which SPCC is an easy consequence. For they yield *modus ponens*, and so *A*; (if *A*, *C*) $\vdash$ *C*; therefore, when $C \vdash D$, Cut gives *A*; (if *A*, *C*) $\vdash$ *D*, so by conditional proof (if *A*, *C*) $\vdash$ (if *A*, *D*). These results are robust with respect to the background standard of logical consequence and logical truth represented by $\vdash$. The rules just invoked hold on both very narrow conceptions of logicality and much broader conceptions. The rest of this section will employ a rather loose standard of logicality, but similar arguments apply to much tighter standards.

Despite such strong reasons for regarding the Suppositional Rule as forcing the logical truth of counterlogicals, other aspects of the Rule seem to pull in the opposite direction. Consider examples like these:

(55)   If nothing is self-identical, everything is self-identical.
(56)   If parthood is non-transitive, parthood is transitive.
(57)   If the Russell set both is and is not a member of itself, dialetheism explodes.
(58)   If intuitionistic logic is the right logic, every mathematical hypothesis is either true or false.

By orthodox logic, broadly conceived, (55)–(58) are all counterlogicals, so by the preceding considerations they should be logical truths. But if they occurred in a student paper, they might well attract the grader's red pen. Indeed, it is far from obvious that the Suppositional Procedure would endorse them. However orthodox the grader's own logical outlook, (55)–(58) are about what holds *if that outlook fails* in a specified way. To assess them properly, using the Suppositional Procedure, one must take the antecedent seriously, and assess the consequent from an appropriately unorthodox perspective. In the case of (55)–(58), doing so requires one to reject the consequent, in that hypothetical spirit. Thus each conditional will be assessed as false, even if its consequent is *in fact* a logical consequence of its antecedent—by the correct (orthodox) logic, but not by the unorthodox sort of logic its antecedent demands. A formal semantic framework appropriate to the evaluation of such conditionals may well need logically impossible worlds, or something like that: in which case, the more the merrier. Or so the usual arguments go.

In effect, such arguments describe the application of the Suppositional Procedure to indicative counterlogicals. They also rely in effect on its correctness. In doing so, they are of course in good company. Indeed, despite their counterlogicality, examples like (55)–(58) are in a way peculiarly well-suited to applying the Procedure. For the deviation stipulated by the antecedent from orthodoxy is of such an abstract theoretical character that one is forced hypothetically to adopt some sort of unorthodox logic in order to suppose it seriously; one can then assess the consequent in the light of that logic. In some cases, the unorthodox logic is explicitly specified, as in (58). On the intuitionistic approach to logic, the universal generalization 'Every mathematical hypothesis is either true or false' is false, so the Suppositional Procedure supports assessing (58) itself as false. In other cases, the unorthodox approach to logic lies very close to hand, as with (57). Dialetheism, the view that there are true contradictions, typically rejects the claim that it explodes (in the sense of entailing everything) as just plain false, and not also true, so the Suppositional Procedure supports assessing (57) itself as just plain false, and not also true. In still other cases, the unorthodox logic is characterized only negatively and unspecifically, as with (55) and (56). Nevertheless, given that in those cases the unorthodox logic is not meant to endorse contradictions, the antecedents of (55) and (56) hint at unorthodox logics on which their respective consequents are false, so the Suppositional Procedure supports rejecting (55) and (56) themselves as false.

However, that the Suppositional Rule supports the rejection of some counterlogicals does *not* mean that the Rule does not, after all, support the derivation of the paradox [I]–[III]. In particular, the relevant commutativity principles CCCN and CCCC for negation and conjunction were derived in section 3.4 from the Suppositional Rule on very general structural grounds, without appeal to principles specific to the classical logic of negation and conjunction as contrasted with other operators of the same grammatical type. Thus rejecting the classical logic of negation and conjunction would not block the derivation of the paradox [I]–[III].

Of course, since the Suppositional Rule is implicitly inconsistent, it indirectly supports incompatible attitudes to the same conditional or conditional-involving argument, though typically one direction of support will be more salient psychologically than the others.

Those who reject some counterlogicals like (55)–(58) as false are also in no position to reject the paradox [I]–[III] on the grounds that counterlogicals are meaningless. For what is meaningless is neither true nor false. Indeed, our ability to apply the Suppositional Procedure non-trivially to assess examples like (55)–(58) provides good evidence of their meaningfulness. Even when one is quite certain that a theory is wrong, one may still understand the theory well enough to develop it as a hypothesis appropriately, by drawing out its consequences in the way its proponents intend. Thus one may be able to apply the Suppositional Procedure non-trivially to conditionals with the theory as antecedent.

Philosophers who envisage judgement on a supposition exclusively in probabilistic terms may underestimate the applicability of the Suppositional Procedure to counterlogicals and similar examples. It can make sense even when probabilities conditional on the antecedent are undefined. For instance, Dorothy Edgington claims 'there is no thought which begins "if I don't exist now": this is a non-starter' (1995: 265). But, although I am quite sure that I do exist now, I can still intelligibly and reasonably use the Suppositional Procedure to reach conclusions such as this:

(59)   If I don't exist now, thinking occurs without a thinker.

Even if I lack well-defined credences conditional on A, I can still make well-defined assessments of other kinds on the supposition A. Not all our attitudes are probabilistic.

Of course, our ability to apply the Suppositional Procedure to a conditional does not guarantee that its verdicts on that conditional are *correct*. We have already seen ample reason to treat the Procedure as fallible. In particular, the paradox [I]–[III] shows that it can generate inconsistent results when applied to counterlogicals. Philosophers who uncritically appeal to examples like (55)–(58) to show that there are false counterlogicals presumably fail to realize that their verdicts on them are likely to result from applying a heuristic which is provably inconsistent, not least in its application to counterlogicals. Without a more sophisticated methodology, one cannot expect to reach adequately supported verdicts on such cases.

There is further reason to doubt the reliability of the Suppositional Procedure for conditionals with philosophically wayward antecedents. For example, most semantic (and non-semantic) theories of conditionals validate conditionals of the form 'If A, A': even if A is true only at an impossible world w, A is still true at w. As already noted, the Suppositional Procedure seems to agree: surely one is warranted in asserting *A* on the supposition of *A* itself, so the Procedure supports one in asserting 'If *A*, *A*' on no supposition. But now consider the hyper-sceptical theory that nothing is assertible (on a supposition or otherwise). Try applying the Suppositional Procedure to this:

(60)   If nothing is assertible, nothing is assertible.

Suppose that nothing is assertible. According to that hypothesis, that nothing is assertible is itself not assertible. Hence, by the Suppositional Procedure, (60) is itself not assertible. Indeed, asserting (60) can feel like a refusal to take its antecedent seriously, in something like the way in which asserting one of (55)–(58) can feel like a refusal to take *its* antecedent seriously. But must we assume that the Suppositional Procedure is correct in such a recherché case? That natural semantic theories of conditionals validate (60) suggests a negative answer.

At this stage of the argument, the appropriate attitude to counterlogicals is to suspend judgement on their truth-value. Theory-driven reasoning suggests that they are all true, but it is not yet strong enough to be conclusive. At first sight, some of them look true, while others look false. However, counterlogicals are one main case where our primary pre-theoretic way of assessing conditionals has turned out to be inconsistent. Given those problems, it would be naïve to take appearances uncritically at face value in a special case so marginal to normal use of the language, for example by offering them as clear counterexamples to a proposed semantics of conditionals.

What is far from plausible is that the ordinary meaning of the word 'if' has been crafted to give special treatment to counterlogicals: that would be the tail wagging the dog, a dysfunctional complication favouring a tiny minority of theoretical uses over the robust simplicity needed for most practical applications. A far more plausible assumption is that the semantic behaviour of counterlogicals is just a byproduct of the general patterns governing the semantic behaviour of other conditionals, not something for which special provision has been made. Thus it is good methodological practice to concentrate on conditionals with less bizarre antecedents in determining our best semantic theory of conditionals, and then follow its verdicts on the semantics of counterlogicals.

3.7 Inconsistent linguistic practices

This chapter has explained various ways in which the Suppositional Rule is inconsistent. Such inconsistency does not imply inconsistency in the Suppositional *Conjecture*. It is consistent to hold that people implicitly rely on inconsistent rules in speaking and understanding their native language.[12] In that sense, linguistic competence may be inconsistent.

Such a view is not at all unprecedented. For an extended example, we may consider the ordinary use of the words 'true' and 'false', and their analogues in other languages. It is governed by schematic principles like these:

---

[12] See Eklund 2002 for discussion.

TRUE   '*P*' is true if and only if *P*.

FALSE   '*P*' is false if and only if not(*P*).

Here declarative sentences are to be substituted for '*P*'. For example, the statement 'There was a sea battle at Salamis' is true if and only if there was a sea battle at Salamis; it is false if and only if there was no sea battle at Salamis. To avoid problems with context-dependent statements, we must interpret the statement '*P*' as if made in the same context as the corresponding instances of TRUE and FALSE, for *your* statement in the words 'I am hungry' is true if and only if *you* are hungry, rather than if and only if *I* am hungry.

Plato and Aristotle were already aware of the importance of principles like TRUE and FALSE. Nevertheless, as the ancient Greeks also knew, they have paradoxical consequences. By standard reasoning, they have inconsistent instances when tricky Liar-like sentences are substituted for '*P*'. For example, consider:

The underlined statement is not true

When we substitute the meaningful sentence 'The underlined statement is not true' for '*P*' in TRUE, this is the result:

(61)   'The underlined statement is not true' is true if and only if the underlined statement is not true.

But the underlined statement *is* 'The underlined statement is not true', so (61) is equivalent to (62):

(62)   The underlined statement is true if and only if the underlined statement is not true.

But (62) is of the form '*Q* if and only if not(*Q*)', which is inconsistent in classical logic. One can derive a similar paradox from FALSE, by putting 'false' in place of 'not true' in the underlined statement.

Centuries of research on the semantic paradoxes, in both the medieval period and the past hundred years, have shown how desperately hard it is to devise a satisfactory treatment of them, one that avoids inconsistency and less formal absurdities without imposing over-draconian restrictions on ordinary uses of 'true' and 'false'. In light of that track record of recurrent disappointment, it is not remotely plausible that any solution to the semantic paradoxes, an escape clause, is somehow written into our ordinary understanding of 'true' and 'false'. To overcome the paradoxes, we must build new theories, not uncover old ones already hidden in our heads.

Some theorists blame the semantic paradoxes on classical logic, not on TRUE and FALSE, but for present purposes the upshot is the same: our ordinary practice has relied on reasoning with jointly inconsistent rules, whether they are rules for reasoning with 'true' and 'false', or rules for reasoning with 'if' and 'not', or some other rules for reasoning. We need not know *which* of our rules are inconsistent to know that some of them are, perhaps jointly.

Despite the semantic paradoxes, in practice our ordinary use of 'true' and 'false' is robustly stable. Such examples demonstrate that a practice can combine several striking features. First, a stable practice can rely on inconsistent rules. Second, participants can be immersed in the practice with no inkling that its rules are inconsistent. Third, through theoretical inquiry participants can gain good evidence that their stable practice relies on inconsistent rules (the appendix, 3.8, discusses another example with similar features). Our stable practice of using 'if' is turning out to have all three features. First, it relies on the Suppositional Rule, which is inconsistent. Second, we ordinarily use 'if' with no inkling that in doing so we rely on an inconsistent rule. Third, through theoretical inquiry we can gain good evidence for the Suppositional Conjecture, good evidence that our practice of using conditionals relies on the Suppositional Rule and so is inconsistent.

According to some philosophers, principles like TRUE and FALSE, despite their inconsistency, are built into competence with the words 'true' and 'false'. That would be a precedent for regarding the Suppositional Rule, despite its inconsistency, as built into competence with 'if'. However, the possibility of inconsistent competence puts more pressure on the very idea of competence.

Consider a native speaker of English who uses the word 'if' in the normal way, respecting the Suppositional Rule, and relying primarily on the Suppositional Procedure in her prospective assessments of conditionals. If anyone is competent with 'if', she is. She becomes an expert on the logic, psychology, and linguistics of conditionals, and discovers by experimentation and theoretical reasoning both that she is relying on the Suppositional Rule and that it is inconsistent. She consciously modifies her practice to achieve consistency. She no longer relies on the Suppositional Rule itself. What she relies on instead is some consistent, complex modification of it, which she devised through a long process of trial and error. At first, she still has a *disposition* to use the original Suppositional Rule, which she consciously inhibits. Gradually, however, she becomes habituated to the revised rule, which she uses increasingly automatically. She loses even the disposition to use the original Rule. But she still remembers vividly what it was like to rely on the original Rule. She can use the Suppositional Procedure in imagination whenever she wants to, although she does not automatically accept the results. She knows that most other speakers of English rely on the original Rule. Does such enlightenment constitute loss of competence with 'if'?

Our imagined expert on conditionals still *knows how* to use the original Rule for 'if', but that is a very weak condition for competence. Someone who uses 'red' as if it meant *green* and 'green' as if it meant *red* will normally be counted as incompetent with those words, even though he also knows how to use 'red' as if it meant *red* and 'green' as if it meant *green*, just as you know how to use 'red' as if it meant *green* and 'green' as if it meant *red*—you know how to pretend to be someone who is making just that mistake. The distinction between 'competence' and 'performance' is too blunt an instrument to clarify the position of our expert on conditionals.

Irrespective of the competence/performance distinction, the inconsistency of the Suppositional Rule limits its capacity to constrain, by a principle of charity, the semantics of conditionals. *No* semantics can fully validate an inconsistent rule. Nevertheless, we can still hope for the semantics to validate the Rule as far as possible. The trouble is that the inconsistencies in sections 3.2–6 do not reveal exactly where the Rule goes wrong, *which* of the constraints it mandates the semantics violates.

However, as we shall see later in the book, that does not make it impossible to find out.

Inconsistencies in the Suppositional Rule also limit its capacity to guide our assessments of conditionals. In problematic cases, we find ourselves conflicted, pulled in opposite directions, and we resort to more or less ad hoc repair strategies. That also happens with our assessments of applications of 'true' and 'false' in semantic paradoxes, but the problematic cases for conditionals are more widespread and often arise in more natural ways. That is further reason to be cautious of reading too much into experimental studies of the use of conditionals, despite their obvious relevance in principle. When we have theoretical reasons to expect a variety of alternative processes to be at work in assessments of conditionals, aggregated data from heterogeneous samples may well present a confusing picture.

3.8 Appendix: tolerance principles as fallible heuristics

This appendix explains how another family of philosophical paradoxes may result from our reliance on useful but fallible heuristics.

When using a vague term, one can easily have the impression that some differences are too small to make any difference to its application. For example, a difference of one grain seems too small to make the difference between a heap and no heap. Thus we are tempted to assume that when one grain is removed from a heap of sand, leaving the rest undisturbed, what remains is still a heap. Such rules are known as *tolerance principles*: the vague term can tolerate small differences.[13] Notoriously, tolerance principles generate sorites paradoxes, because many small differences can add up to a difference along the same dimension too large for the vague term to tolerate. Remove enough grains, one by one, and no heap is left.

On some views of vagueness, competence with a vague term requires a disposition to find instances of a tolerance principle for it compelling. If so, I am not competent with any vague term, since I have no such disposition. Nevertheless, in everyday life, applying a tolerance principle is often reasonable. That suggests an understanding of tolerance principles as *heuristics* for applying vague terms, reliable but not perfectly reliable. They are cognitively efficient because they obviate the need for reassessing the application of the vague term from scratch to a case which differs only trivially from one already assessed. After recognizing a heap and removing one grain, you need not take another look at what is left to determine whether it still counts as a heap. You just apply the tolerance principle. When one unreflectively applies a tolerance principle in everyday life, the lurking threat of paradox need never occur to one.

A complicating factor with vague terms is their tendency to shift in reference as the conversational context develops and new applications of them are charitably accommodated. However, such shifts are not inevitable. For example, when Mary tells John 'Bring me a pot of red paint!', what matters is the reference of 'red' in *Mary*'s context;

---

[13] Crispin Wright (1976) introduced the term 'tolerance principle', although he does not endorse all the claims that have been made for the status of such principles.

for John to expand the reference of 'red' in supposedly obeying the instruction would be inappropriate.

In what follows, the reference of the relevant vague terms will be taken to remain constant over the discourse. This is the most interesting case because the most resistant to tolerance principles, one-off instances of which are often easy to accommodate once contextual shifts are allowed. As noted in section 1.3, referential constancy is typically the appropriate assumption for discourse which integrates information originally linguistically encoded outside the immediate conversational context, at earlier times or by other people. Chapter 5 will discuss the issue of referential constancy in more depth and detail. In any case, as theorists of vagueness have observed, one cannot accumulate ascriptions of the vague term all the way along the sorites series. Sooner or later one comes to a shade one cannot count as 'red'. Thus, even with contextual variation in reference, tolerance principles turn out to be fallible: they can postpone the moment of truth, but not indefinitely. To understand more clearly what is going on, we do best to focus on the simpler case without contextual variation.

How reliable are tolerance principles? On predominant views of vagueness, when tolerance principles are formulated as universal generalizations, sorites paradoxes show them to be just false: they generate contradictions from clear cases of the application of the vague term and clear cases of its non-application. However, in everyday practice what we use is not the universal generalization but instances of a rule of inference. For a simple example, let us pretend that the application of 'heap' depends only on the number of grains. The false universal generalization is this (where the variable 'n' ranges over natural numbers):

HEAP$_U$ For every n, if n grains make a heap, $n - 1$ grains make a heap.

The rule of inference is this:

HEAP from: n grains make a heap
to: $n - 1$ grains make a heap

A natural way to understand the reliability of an inference rule like HEAP is as the probability of the conclusion, conditional on the premise. For we want to know how good the rule is at preserving truth from premise to conclusion: in what proportion of cases where the premise holds does the conclusion also hold? By taking values of the variable 'n' as proxies for cases, we can reduce that question to the question: for what proportion of numbers n in the domain such that n grains make a heap do $n - 1$ grains also make a heap? Given the simplifying assumption that all those numbers are equally likely to arise, the answer to that question is just the conditional probability of the conclusion on the premise. For this purpose, instances of the rule with a false premise are irrelevant.

Some non-classical theories of vagueness undermine the distinction between truth and falsity so deeply that trying to estimate the reliability of a rule like HEAP has no obvious sense. To make progress, we will consider views on which there is a least number k of grains to make a heap, so n grains make a heap whenever $k \leq n$ and n grains do not make a heap whenever $n < k$. For example, on an epistemicist view of vagueness, there is a true exact specification of the value of 'k', but the vagueness of 'heap' prevents us from knowing what it is: many different exact specifications are consistent with what we know. Even more elusively, on a supervaluationist view of vagueness, usually treated as the leading rival to epistemicism, many different exact specifications of the value of 'k' are consistent with our vague meaning of 'heap'; nevertheless, on each such specification, there is such a number k. On the former approach, the problem with 'k' is epistemic; on the latter, it is semantic, but both permit ordinary mathematical reasoning with 'k', even though it is vague what its value is.

Keeping the example realistic and mathematically straightforward, we assume that the domain of contextually relevant natural numbers is finite—say, all those less than a million—and treat all of them as equally probable values of the schematic letter 'n' in HEAP. In reality, there may be many more candidate-heaps with twenty grains than with twenty thousand, but for simplicity we can ignore such variations: taking them into account would not change the big picture. We also assume that the domain contains all relevant candidates to be k, the only value of 'n' for which HEAP has a true premise and false conclusion. By hypothesis, HEAP has a true premise for $1{,}000{,}000 - k$ values of 'n'. Consequently, the reliability of HEAP is $(999{,}999 - k)/(1{,}000{,}00 - k)$. For example, if $k = 5{,}000$, the reliability is $994{,}999/995{,}000$. If $k = 50$, the reliability is $999{,}949/999{,}950$. The latter value for 'k' is more plausible: fifty grains can be nicely heaped up. In short, HEAP is *almost* perfectly reliable. Only in the exceptional circumstances of a sorites series is HEAP liable to get one into trouble. As a heuristic for applying 'heap' in realistic circumstances, HEAP is excellent.

Does vagueness in 'k' undermine such calculations of reliability? Even if we only know an upper bound on k, say $k \le 10{,}000$, we can calculate a lower bound on the reliability: it is at least $989{,}999/990{,}000$, already very high. Any reasonable view of vagueness should make room for some such approximate knowledge articulated in vague terms, on pain of making vague language and thought useless in ways they obviously are not. For example, on an epistemicist account of vagueness, we can know that $k \le 10{,}000$ because it is true for every value of 'k' compatible with what we know. Similarly, on a supervaluationist account of vagueness, we can know that $k \le 10{,}000$ because it is true for every assignment of a value to 'k' consistent with its vague meaning, even though none of them is uniquely correct. Estimating the tolerance principle's reliability in the way described does not commit one to any one view of vagueness.

To vary the example, we can consider a case where the space of possibilities may be treated as continuous rather than discrete. Here is a tolerance principle for the vague term 'about 6 o'clock' (where the independent variables 't' and 't^*' range over times in a twelve-hour cycle):

> ABOUT6   from: t is about 6 o'clock
> t^* is within a minute of t
> to: t^* is about 6 o'clock

Suppose that the probability distribution over possible values of 't' is uniform, so its value is equally likely to fall in any two periods of the same length. In the same spirit as before, suppose for the sake of argument that in the given context the times which qualify as 'about 6 o'clock' are just those between 5:55 and 6:05. Thus, given the first premise of ABOUT6, the probability of t being between 5:56 and 6:04 is 8/10;

conditional on any such value and the second premise of ABOUT6, the probability of the conclusion is 1. Also given the first premise, the probability of t being between 6:04 and 6:05 is 1/10; conditional on any such value and the second premise, the probability of t^* being about 6 o'clock falls linearly from 1 to ½ as t goes from 6:04 to 6:05 (a time within a minute of 6:05 is just as likely to be before 6:05 as after it). Similar reasoning applies to times between 5:55 and 5:56. A standard probability calculation then shows that the reliability of ABOUT6 is $(8/10 \times 1) + 2 \times (1/10 \times (1 + ½)/2) = 19/20$. So ABOUT6 is very reliable too, though less spectacularly reliable than HEAP. On average, the conclusion is false one time out of twenty when the premises are true.

We can easily ratchet up the tolerance principle's reliability by narrowing the margin of tolerance in the second premise of ABOUT6, as a proportion of the period of times about 6 o'clock. For example:

ABOUT$_s$6   from: t is about 6 o'clock
t^* is within a second of t
to t^* is about 6 o'clock

Formally, ABOUT$_s$6 is just as capable as ABOUT6 of generating a sorites paradox, though the required sorites series of times will be longer. By a calculation just like the previous one, the reliability of ABOUT$_s$6 is $(598/600 \times 1) + 2 \times (1/600 \times (1 + ½)/2) = 1199/1200$. On average, the conclusion of ABOUT$_s$6 is false less than one time out of a thousand when the premises are true.

Although the numbers were obviously chosen with some arbitrariness, they are not unrepresentative. The calculations show clearly that tolerance principles can make highly reliable heuristics, even though the corresponding universal generalizations are just false. Moreover, by reducing the margin of tolerance in the second premise, we can make the reliability as high as we like, short of 1. No wonder we tend to rely on such principles.

In many tolerance principles for supposedly observational terms, the second premise employs a standard of indiscriminability. Here is a standard example (where the independent variables 'x' and 'y' range over determinate shades of colour, and naked-eye indiscriminability in colour is meant):

RED   from: x is red
y is indiscriminable from x
to: y is red

A reasonable conjecture is that the reliability of RED is not radically different from that of ABOUT6. It could be measured experimentally, given suitable operational tests for 'red' and 'indiscriminable'. For example, one could consider a large set of determinate shades of colour, distributed more or less evenly over the colour sphere, and treated as equiprobable. The question is then: if you repeatedly pick x and y independently from the set, over a long series of trials, what proportion of cases verifying both premises also verify the conclusion?[14]

---

[14] Thanks to Will Davies and Igor Douven for interesting discussion of ways of experimentally measuring the reliability of RED.

When sorites paradoxes are presented as arguments, rules of inference such as HEAP, ABOUT6, and RED are often replaced by multiple instances of a corresponding conditional schema, treated like axioms:

HEAPif If n grains make a heap, $n - 1$ grains make a heap.
ABOUT6if If t is about 6 o'clock and t^* is within a minute of t, t^* is about 6 o'clock.
REDif If x is red and y is indiscriminable from x, y is red.

The argument is then driven forward by repeated applications of *modus ponens*, starting from a case where the vague term clearly applies. Each conditional can in turn be derived from the corresponding rule of inference, by an application of conditional proof. Of course, estimating the probability of such indicative conditionals in natural language depends on the semantics of 'if', which is at stake in this book.

Our immediate concern, however, is just to make the sorites paradoxes maximally challenging, for which purpose we need whatever reading of 'if' minimizes the strength, and so maximizes the probability, of the conditionals, while still validating *modus ponens*, without prejudice to the meaning of 'if' in English. That minimal reading of 'if' is the material one, so the conditionals become:

HEAP⊃ (n grains make a heap) $\supset$ ($n - 1$ grains make a heap)
ABOUT6⊃ (t is about 6 o'clock and t^* is within a minute of t) $\supset$ (t^* is about 6 o'clock)
RED⊃ (x is red and y is indiscriminable from x) $\supset$ (y is red)

As noted in section 3.2, $\text{Prob}(C|A) \leq \text{Prob}(A \supset C)$ whenever the probabilities are defined, and indeed $\text{Prob}(C|A) < \text{Prob}(A \supset C)$ when in addition $\text{Prob}(A) < 1$ and $\text{Prob}(A \supset C) < 1$, as would hold for at least some instances of the conditionals in a sorites paradox. Thus the material conditional versions of tolerance principles are even more reliable than the corresponding inference rule versions. Of course, the extra probability comes only from cases where the material conditional is true because its antecedent is false, and so does not concern the extra risk involved in moving from a true antecedent to the consequent. But the upshot is still that tolerance principles in material conditional form make at least as good heuristics as they do in inference rule form.

In brief, sorites paradoxes are another example where philosophical paradoxes may have arisen because users of reliable but not perfectly reliable heuristics misinterpret and overestimate their semantic status.

4

Heuristics within Heuristics

4.1 Heuristics for applying the Suppositional Rule

The Suppositional Rule is a high-level heuristic. It specifies what to do in extremely general and abstract terms. It tells us, when we take an attitude to C conditionally on the supposition A, to take the same attitude unconditionally to the indicative conditional 'If A, C'. But it does not specify *how* to determine our attitude to C on A in the first place. The natural expectation is that we shall need a hierarchy of lower-level heuristics to enable us to do so, many of them dependent on the specific content of A and C. Some of those lower-level patterns of implementation are too content-specific to be of distinctive interest for the theory of conditionals. But others involve more general strategies.

Some strategies for implementing the Suppositional Rule conflict with more direct applications of the Rule. In such cases, we may appear not to be applying the Rule. On closer investigation, however, these strategies turn out to involve less direct applications of the Rule. Thus the attempts to demonstrate its inapplicability have the opposite effect. This chapter discusses some examples.[1]

The focus here will be on probabilistic attitudes. In chapter 3, the denial that indicative conditionals express propositions turned out not to help with the problem it was meant to solve. Thus we will assume that a declarative sentence X does express a proposition [X]. We take credence in X to correspond to an assignment of a probability Prob([X]) to the coarse-grained proposition. We can therefore write the Suppositional Rule for these cases in the form of the equation SRP, which will be more convenient to work with later in the chapter:

SRP $\text{Prob}([\text{if } A, C]) = \text{Prob}([C]|[A])$

As observed in section 3.3, SRP makes trouble, especially when the two sides of the equation are allowed to share background suppositions, but for all that we often rely on it.

Daniel Rothschild describes an example where our probabilistic judgements seem to violate SRP. There is a defect for cars of a given make. For cars without the defect, if they crash (at the relevant speed), the airbag goes off. For cars with the defect, if they crash (at the relevant speed), the airbag does not go off. The defect also makes cars more likely to crash. According to Rothschild, 'It seems like the probability that

---

[1] Some readers may wish to skip this chapter and move straight to the next stage of the argument in chapter 5.

Suppose and Tell: The Semantics and Heuristics of Conditionals. Timothy Williamson, Oxford University Press (2020).
© Timothy Williamson.
DOI: 10.1093/oso/9780198860662.001.0001

if *this* car crashes its airbag will go off just is the probability that it lacks the defect (at least on one salient reading of the conditional)' (2013: 63).

The idea is this. Let A = 'This car crashes', C = 'This car's airbag goes off', and D = 'This car has the defect.' As Rothschild reads it, $\mathrm{Prob}([if\ A,\ C]) = \mathrm{Prob}([not(D)])$. Moreover, $\mathrm{Prob}([D]|[A]) > \mathrm{Prob}([D])$, because the information that the car crashes raises the probability that it has the defect, so $\mathrm{Prob}([not(D)]|[A]) < \mathrm{Prob}([not(D)])$. But $\mathrm{Prob}([not(D)]|[A]) = \mathrm{Prob}([C]|[A])$, because the cases where the car crashes and lacks the defect are exactly those where it crashes and the airbag goes off. Putting the pieces together:

$$\mathrm{Prob}([C]|[A]) = \mathrm{Prob}([not(D)]|[A]) < \mathrm{Prob}([not(D)]) = \mathrm{Prob}([if\ A,\ C])$$

Thus $\mathrm{Prob}([C]|[A]) < \mathrm{Prob}([if\ A,\ C])$, contrary to SRP.

Why should the probability that if this car crashes its airbag will go off, $\mathrm{Prob}([if\ A,\ C])$, coincide with the probability that it lacks the defect, $\mathrm{Prob}([not(D)])$? The reason is presumably that, given the background conditions of the example, BB, 'if A, C' is equivalent to 'not(D)'. BB establish two deductive connections. First, on the assumptions that the car has the defect and that it crashes, the airbag does not go off:

(1) $BB, D, A \vdash not(C)$

Second, on the assumptions that the car lacks the defect and that it crashes, the airbag does go off:

(2) $BB, not(D), A \vdash C$

By conditional proof, itself a corollary of the Suppositional Rule (section 3.1), we can derive (3) from (1) and (4) from (2):

(3) $BB, D \vdash if\ A, not(C)$

That is, on the assumption that the car has the defect: if it crashes, the airbag does not go off.

(4) $BB, not(D) \vdash if\ A, C$

That is, on the assumption that the car lacks the defect, if it crashes, the airbag does go off. Both (3) and (4) are entirely natural descriptions of the case. But in chapter 3 we also observed that in several ways the Suppositional Rule supports CNC, the principle of conditional non-contradiction, on which 'if A, not(C)' is inconsistent with 'if A, C', and so entails 'not(if A, C)'. Thus the Suppositional Rule allows us to derive (5) from (3):

(5) $BB, D \vdash not(if\ A, C)$

But (4) and (5) jointly make 'not(D)' equivalent to 'if A, C', given the background conditions BB: they stand or fall together. If the car lacks the defect, they both stand. If the car has the defect, they both fall. Since the background conditions are priced into the probability distribution Prob by hypothesis, Rothschild's claim that $\mathrm{Prob}([if\ A, C]) = \mathrm{Prob}([not(D)]$ follows. Thus the Suppositional Rule itself explains Rothschild's identification of probabilities. This is no alternative to the Rule, but instead just a slightly indirect way of applying it.

What is the point of the indirection? Conditionals are often tricky and elusive to think with. By using the more concrete claim 'This car lacks the defect' as a proxy for the conditional 'If this car crashes, its airbag will go off' we can simplify and facilitate our thinking about the case. It is no surprise that we often take advantage of such cognitive shortcuts, especially with a very general and abstract heuristic such as SRP.

Contrary to Rothschild's claim that the example involves an ambiguity in the conditional, it can be understood as a mere difference in cognitive strategy, with no need to postulate two readings. Of course, the choice of strategy makes a difference to the ensuing probability judgements, but that is yet another manifestation of the inconsistency already shown to be implicit in the Suppositional Rule.

4.2 The Divide-and-Rule strategy

As observed in section 4.1, working out the value of the conditional probability on the right-hand side of SRP is often no easy task. To accomplish it, we must often use lower-level heuristics.

One frequent difficulty is that the antecedent proposition [A] covers a heterogeneous variety of possibilities, some favourable to the consequent proposition [C], others unfavourable. To make the problem tractable, a promising heuristic is to partition the possibilities into a finite number of mutually exclusive and jointly exhaustive cells, each more or less homogeneous in relevant respects. The conditional probability within such a cell tends to be easier to estimate. One can first apply the Equation *locally*, to each cell separately, and then combine the results, each weighted by the probability of the respective cell, using the law of total probability. Call this the *Divide-and-Rule* strategy. There is evidence that we sometimes find the strategy natural in practice (Kaufmann 2004).

Of course, the Divide-and-Rule strategy is only one step down from the Equation itself; it merely reduces hard cases of the Equation to easier ones. Still lower-level heuristics will be needed for dealing with individual cells. This section deals only with the Divide-and-Rule strategy itself.

We first make the strategy more precise. The underlying probability space gives us a set Ω of specific possibilities (mini-worlds). A finite partition of Ω is, for some natural number n, a family $\{\Pi_i\}_{i \leq n}$ of $n + 1$ subsets of Ω, whose union is Ω and whose pairwise intersections are empty. For these purposes, the probability $\text{Prob}(\Pi_i)$ must be well-defined and non-zero for each cell Π_i. Thus, by elementary probability theory,

$$\sum_{i \leq n} \text{Prob}(\Pi_i) = 1$$

By the law of total probability:

(6) $\text{Prob}([\text{if } A, C]) = \sum_{i \leq n} \text{Prob}(\Pi_i)\text{Prob}([\text{if } A, C] | \Pi_i)$

For each $i \leq n$, we now assume SRP for the probability distribution conditionalized on the cell Π_i:

(7i) $\text{Prob}([\text{if } A, C] | \Pi_i) = \text{Prob}([C] | [A] \cap \Pi_i)$

For the conditional probability on the right-hand side of (7_i) to be well-defined, the antecedent $[A]$ must have an intersection of positive probability with Π_i, for any $i \leq n$. Putting (6) and $(7_0), \ldots, (7_n)$ together, we obtain an unconditional probability for the indicative conditional:

$$(8) \quad \text{Prob}([\text{if } A, C]) = \sum_{i \leq n} \text{Prob}(\Pi_i) \text{Prob}([C] | [A] \cap \Pi_i)$$

Of course, as we saw in section 3.3, SRP can hold only for very limited ranges of cases. However, in favourable circumstances, for given propositions $[A]$ and $[C]$, one can find a partition for which all the instances $(7_0), \ldots, (7_n)$, and (6) and (8) too, are jointly satisfiable by a suitable choice of proposition to be $[\text{if } A, C]$. More specifically, such favourable circumstances are these: for each $i \leq n$, $[A] \cap \Pi_i$ has positive probability, and so is non-empty, and is *uniform* with respect to $[C]$, in the sense that it is either a subset of $[C]$ or disjoint from $[C]$ (it cannot be both, since $[A] \cap \Pi_i$ is non-empty).

Here is the proof. Suppose that the specified favourable circumstances obtain. Now interpret the conditional in 'if A, C' as a form of strict implication for a modality restricted to the cells of the partition. In other words, for any world w and $i \leq n$:

if $w \in \Pi_i$, then $w \in [\text{if } A, C]$ if and only if $[A] \cap \Pi_i \subseteq [C]$.

Thus if $[A] \cap \Pi_i \subseteq [C]$ then $\Pi_i \subseteq [\text{if } A, C]$, so $\text{Prob}([\text{if } A, C] | \Pi_i) = 1 = \text{Prob}([C] | [A] \cap \Pi_i)$. Otherwise, $[A] \cap \Pi_i$ is disjoint from $[C]$, so $[\text{if } A, C]$ is disjoint from Π_i, so $\text{Prob}([\text{if } A, C] | \Pi_i) = 0 = \text{Prob}([C] | [A] \cap \Pi_i)$. Either way, (7_i) holds, for each $i \leq n$. Since (6) holds automatically, (8) also holds, as required.

So far, things look bright for the Divide-and-Rule strategy. But it has severe limitations. Although it applies the Equation locally and combines the results, the upshot is not in general equivalent to applying the Equation globally. To obtain the global conditional probability of $[C]$ on $[A]$ in terms of the partition, we can use the law of total probability starting with the probability distribution conditionalized on $[A]$:

$$(9) \quad \text{Prob}([C] | [A]) = \sum_{i \leq n} \text{Prob}(\Pi_i | [A]) \text{Prob}([C] | [A] \cap \Pi_i)$$

Compare the right-hand side of (9) term-by-term with the right-hand side of (8). The difference is in the probabilistic weights attached to the cells. Those in (9) are conditionalized on the antecedent $[A]$; those in (8) are not. Thus when the cells are not probabilistically independent of the antecedent, the right-hand sides of (8) and (9) may easily differ in value, so that (8) and (9) imply that $\text{Prob}([\text{if } A, C]) \neq \text{Prob}([C] | [A])$. An example will be given below. Thus buying the local instances of the Equation requires sacrificing its global version.

Another problem with the Divide-and-Rule strategy is that its results are not robust: they are very sensitive to the choice of partition. In particular, the interpretation of 'if' as a strict conditional used to establish the joint satisfiability of $(7_0), \ldots, (7_n)$ varies with the partition. This partition-sensitivity holds even for partitions which all satisfy the conditions for the strategy to work optimally: the intersection of the antecedent with each cell has positive probability and is uniform with respect to the consequent. To meet those strict conditions in full, we use a slightly artificial example (based on Lycan 2001 and Moss 2018).

Mr N and Ms S are two thinkers with exactly the same prior probability distribution over propositions and exactly the same evidence in all relevant respects. They know of a certain skyscraper. Occasionally there is a net around its roof, to protect builders. For both Mr N and Ms S, the epistemic position is this. It is 1 per cent probable that there is a net today (N). It is also 1 per cent probable that an unspecified individual X has suicidal tendencies (S). It is certain that anyone who jumps off without a net will die (as a result), and that no one who jumps off with a net will die (as a result). Psychiatric evidence also makes it certain that only those with suicidal tendencies jump off without a net, and that only those without suicidal tendencies jump off with a net. This combination has a non-zero probability: there is a net today (N), X has no suicidal tendencies ($\neg S$), and jumps off (J; thrill-seekers occasionally do). This other combination also has a non-zero probability: there is no net today ($\neg N$), X has suicidal tendencies (S), and jumps off (J). How probable is (10) ('if J, D')?

(10)   If X jumps off the roof today, X will die (as a result).

Mr N and Ms S use different partitions to work out the answer.

Mr N reasons thus. (i) Suppose that there is a net. Jumping off with a net is never fatal, so Prob($[D]||[J \wedge N]$) = 0. (ii) Suppose that there is no net. Jumping off with a net is always fatal. So Prob($[D]||[J \wedge \neg N]$) = 1. (iii) Therefore, by (8) applied to the $[N]/[\neg N]$ partition:

Prob($[\text{if } J, D]$) = Prob($[N]$)Prob($[D]||[J \wedge N]$) + Prob($[\neg N]$)Prob($[D]||[J \wedge \neg N]$) = Prob($[\neg N]$) = 99/100

Meanwhile, Ms S reasons thus. (i) Suppose that X has suicidal tendencies. By the psychiatric evidence, X will jump off only if without a net, so X will die. Hence Prob($[D]||[J \wedge S]$) = 1. (ii) Suppose that X has no suicidal tendencies. By the psychiatric evidence, X will jump off only with a net, so X will not die. Hence Prob($[D]||[J \wedge \neg S]$) = 0. (iii) Therefore, by (8) applied to the $S/\neg S$ partition:

Prob($[\text{if } J, D]$) = Prob($[S]$)Prob($[D]||[J \wedge S]$) + Prob($[\neg S]$)Prob($[D]||[J \wedge \neg S]$) = Prob($[S]$) = 1/100

By seemingly plausible reasoning from the same prior probability distribution and the same evidence, Mr N and Ms S have reached radically different conclusions about the probability that if X jumps off the roof today, X will die (as a result). Mr N finds (10) almost certainly true; Ms S finds (10) almost certainly false. Even on more realistic, less extreme assumptions about the conditional probabilities in their arguments, the difference between their conclusions would still be very large. Which of them, if either, is right? How probable is (10) really, on their shared evidence?

Each partition meets the conditions for optimal performance of the Divide-and-Rule strategy. The intersections of the cells with the antecedent, $[J \wedge N]$, $[J \wedge \neg N]$, $[J \wedge S]$, and $[J \wedge \neg S]$, all have positive probability, and all are uniform with respect to the consequent $[D]$, as shown in parts (i) and (ii) of the two lines of reasoning.

Since Mr N and Ms S have the same initial probability distribution over propositions and the same evidence, if they have reasoned correctly they should not disagree over the probability of any proposition. In that case, they must express different propositions by the conditional (10), so the notation '[if J, D]' is ambiguous. The difference between them is just that the $[N]/[\neg N]$ partition is salient to Mr N, while the $[S]/[\neg S]$ partition is salient to Ms S. This suggests a form of contextualism on which one parameter of a context is the salient partition; which proposition an indicative conditional expresses varies with that parameter. To handle contexts in which no partition is salient, we can assign the degenerate partition into a single cell, the whole of Ω. Thus Mr N and Ms S are speaking in different contexts. Mr N correctly assigns a probability of 99/100 to the proposition which (10) expresses in his context. Ms S correctly assigns a probability of 1/100 to the proposition which (10) expresses in her context. Neither is making a mistake, though they may be in danger of speaking past one another. Stefan Kaufmann (2004) has worked out the details of a systematic theory along such lines.

What such contextualism cannot preserve over all contexts is the global form of SRP for (10), Prob([if J, D]) = Prob([D]$||$[J]). For the contextualist account concerns the meaning of the indicative conditional, not the meaning of the probability operator 'Prob'. Given the global form of SRP, Mr N's conclusion implies that Prob([D]$||$[J]) = 99/100, while Ms S's conclusion implies that Prob([D]$||$[J]) = 1/100. Since Prob([D]$||$[J]) does not involve the indicative conditional, contextualism about the latter cannot reconcile opposing calculations of the value of the former. Instead, one concludes that Prob([if J, D]) $\neq$ Prob([D]$||$[J]) for at least one of the two postulated local readings of the indicative conditional.

To understand better what is going on, we can calculate the conditional probability independently by the law of total probability, as in (9), once for Mr N's partition and once for Ms S's. The base probability distribution for the calculation is Prob conditionalized on [J].

As Mr N noted, reasoning, Prob([D]$||$[$N \wedge J$]) = 0 and Prob([D]$||$[$\neg N \wedge J$]) = 1. Thus, by the law of total probability:

Prob([D]$||$[J]) = Prob([N]$||$[J])Prob([D]$||$[$N \wedge J$]) + Prob([$\neg N$]$||$[J])Prob([D]$||$[$\neg N \wedge J$])
= Prob([$\neg N$]$||$[J])

As Ms S noted, Prob([D]$||$[$S \wedge J$]) = 1 and Prob([D]$||$[$\neg S \wedge J$]) = 0. Thus, by the law of total probability:

Prob([D]$||$[J]) = Prob([S]$||$[J])Prob([D]$||$[$S \wedge J$]) + Prob([$\neg S$]$||$[J])Prob([D]$||$[$\neg S \wedge J$]) = Prob([S]$||$[J])

Together, these two calculations imply that Prob([$\neg N$]$||$[J]) = Prob([D]$||$[J]) = Prob([S]$||$[J]). But that further consequence just follows from Mr N and Ms S's shared assumptions, on which $\neg N \wedge J$, $D \wedge J$, and $S \wedge J$ are all equivalent: those who jump off without a net die, those who jump off and die have suicidal tendencies, and those with suicidal tendencies jump off only without a net.

As expected, the key difference between the new calculations and the old ones is in the weights. Mr N's original calculation used Prob([N]) and Prob([$\neg N$]); the corresponding new one uses Prob([N]$||$[J]) and Prob([$\neg N$]$||$[J]). Ms S's original calculation used Prob([S]) and Prob([$\neg S$]); the corresponding new one uses Prob([S]$||$[J]) and Prob([$\neg S$]$||$[J]).

In one extreme scenario, the shared evidence shows that almost all of those who jump off do so without a net. Thus Prob([$\neg N$]$||$[J]) is very close to 1, as are Prob([D]$||$[J]) and Prob([S]$||$[J]). Nevertheless, Ms S assigns a probability of 1/100 to the conditional that if X jumps off, X will die.

In a contrasting extreme scenario, the shared evidence shows that almost all of those who jump off do so with a net—thrill-seekers love the net, while those with suicidal tendencies prefer other methods. Thus Prob($[\neg N]|[J]$) is very close to 0, as are Prob($[D]|[J]$) and Prob($[S]|[J]$). Nevertheless, Mr N assigns a probability of 99/100 to the conditional that if X jumps off, X will die.

On the face of it, the difference between those opposite scenarios is quite relevant to the probability of (10). That X jumps off is evidence as to whether there is a net; Mr N and Ms S's reasoning ignores that consideration. The claim is that it is irrelevant on their current understandings of (10). That claim is odd. On reflection, how plausible is the proposed contextualism?

Salience is a fleeting feature. Imagine someone thinking about sentence (10), who keeps switching attention between the $[N]/[\neg N]$ and $[S]/[\neg S]$ partitions, and so between the judgements 'That is 99 per cent probable' and 'That is 1 per cent probable'. On a contextualist account, in principle that is no more puzzling a predicament to be in than the predicament of someone who keeps switching attention between a sphere and a cube, and so between the judgements 'That is a sphere' and 'That is a cube.' But the former predicament may well feel more puzzling than the latter.

A closer comparison with the (alleged) context-dependence of the indicative conditional is the context-dependence of epistemic modals such as 'may', 'might', and "must", on a straightforward account of their epistemic readings (discussed in section 1.3). Normal speakers are clear in effect that epistemic modals can easily make a sentence express a truth in some contexts and a falsehood in others. For example, a child in a treasure hunt truly says 'The treasure may be in the house and it may be in the garden', because both possibilities are compatible with what they know, but does not expect the same sentence to express a truth when uttered after the treasure is found, or whispered by the watching adults in the know (unless they are speaking projectively from the children's perspective).

The contextualist may reply that of course the postulated context-sensitivity of 'if' feels more puzzling than the context-sensitivity of perceptual demonstratives or epistemic modals, because the variation in context is subtler and more abstract; it is better hidden. The context-sensitivity of epistemic modals is in turn subtler and more abstract than that of perceptual demonstratives, but the relation of epistemic modals to some individual or group's present knowledge base is so central to their use and meaning that their context-sensitivity is effectively overt. By contrast, sensitivity to a partition is not central to the use or meaning of 'if'; uttering 'if' rarely draws attention to a tacit partition. A conditional with no partition-sensitivity could still function quite like our indicative conditional. Thus the alleged context-sensitivity of 'if' is covert.

There is a danger for contextualists in admitting that the context-sensitivity of 'if' is hidden. Their account is motivated primarily by *charity* in interpretation. Reasonably enough, they seek to avoid imputing inexplicable error to speakers and hearers. When two people are talking about objects in plain view, one saying 'That is a sphere', the other 'That is a cube', a natural hypothesis is that they are referring to different objects. The speakers themselves are likely to adjust accordingly: 'Oh, you mean this one.' But contextual variation so deeply hidden from speakers and hearers that they do not even implicitly adjust to it is itself a new source of error, especially when information in verbal form is transmitted or communicated from one context to another. The danger is that such covert contextualist accounts turn out, all things considered, *less* charitable than the invariantist accounts they were intended to supersede, and so undermine their own motivation.

How might that happen in the present case? When the [N]/[¬N] partition is salient to Mr N, he calculates a probability of 99 per cent for (10). Unaware of the partition-sensitivity of his calculation, he does not bother to recalculate when the context changes and [S]/[¬S] becomes the salient partition; he simply retains the probability of 99 per cent for the sentence (10), not realizing that it now expresses a different proposition. He may also inform others that (10) is 99 per cent probable, and they may take his word for it, without bothering to do the calculation themselves, even though the salient partition for them is [S]/[¬S]. Those are all errors, on the proposed contextualist account.

Equally, analogous errors may occur in the reverse direction. When the [S]/[¬S] partition is salient to Ms S, she calculates a probability of 1 per cent for (10). Unaware of the partition-sensitivity of her calculation, she does not bother to recalculate when the context changes and [N]/[¬N] becomes the salient partition; she simply retains the probability of 1 per cent for the sentence (10), not realizing that it now expresses a different proposition. She may also inform others that (10) is just 1 per cent probable, and they may take her word for it, without bothering to do the calculation themselves, even though the salient partition for them is [N]/[¬N]. Those too are all errors, on the contextualist account.

Of course, when the difference between the two perspectives is so extreme, it is more likely to be noticed. But it can be less extreme, and consequently harder to notice, even when the error is significant enough to matter for practical purposes.

An important feature of such cases is that the mathematics of probability is far from transparent to us. That is what gives so much scope to memory and testimony in probabilistic beliefs: even when we possess all the relevant evidence, we often make no attempt to calculate or recalculate the probabilities for ourselves. Even when we know how, it is typically costly in time and energy, and the risk of performance errors is high. We often prefer to get the answer by remembering it, or by being told it by someone else.

For most people, even most of those with a basic mathematical education, calculating is in general a slow and unreliable business. When the calculations involve the subtleties of conditional probabilities, we are often unsure or mistaken about *what* we ought to be calculating. Indeed, humans are notoriously liable to commit gross fallacies in probabilistic reasoning: for instance, the conjunction fallacy and the base rate fallacy. Even subjects who have been taught probability theory remain liable to commit those fallacies when caught off guard. The fallacies may be byproducts of mainly reliable heuristics, but the results are still errors.

Conscious probabilistic reasoning of any complexity does not come naturally to us. In everyday thought and talk, 'probable' may be closer in meaning to 'plausible' than to anything conforming to mathematical theories of probability, which are far more recent than words like 'probable'. The mathematical understanding of probability goes back only to the seventeenth century. Leibniz, one of the greatest minds of his age, made an elementary error in his written probabilistic reasoning: he thought that when two dice are cast, 11 and 12 are equally likely sums to result, because each can come about in only one way (5 + 6, 6 + 6). Statisticians are frustrated by high court judges' misunderstandings of Bayesian reasoning about DNA evidence.

Perhaps some forms of Bayesian probabilistic reasoning are hardwired into some of our sub-personal modules for perception and motor control, but that does not make such reasoning automatic at the level of conscious thought. In the semantics literature, examples are standardly presented in words. In particular, probabilistic examples are typically presented in an explicit verbal form which leaves the reader with little option but to resort to some sort of conscious reasoning.

Here is Stefan Kaufmann's presentation of the first example he uses to justify postulating readings of the indicative conditional whose probability is correctly calculated by the local method over a suitable partition:

> You are about to choose a ball from a bag. It could be one of two bags, X or Y. Bag X contains ten red balls, nine of them with a black spot, and two white balls. Bag Y contains ten red balls, one of them with a black spot, and fifty white balls. By virtue of additional evidence—say, the bag in front of you looks big—you are 75% sure that it is bag Y.

Kaufmann then invites readers to consider whether the strength of their belief in (11) is best characterized as 'high', 'fifty-fifty', or 'low' (2004: 584–5):

(11)   If I pick a red ball, it will have a black spot.

The scenario, and the challenge to estimate the probability of (11), are reminiscent of a high school mathematics problem, the kind that most of the class gets wrong.

When a student gets a mathematics problem wrong, by conventional standards, the usual reaction is not to seek a new context-sensitive semantics for the question on which the student's answer was, in its context, right after all. Having walked from their class in probability theory to their class in semantics, should they expect a much more sympathetic hearing?

The semantic explanation of a calculation with a non-canonical answer may occasionally have a point, but more often the correct explanation is cognitive rather than semantic; the student really did mean something false. That is so even when students in general have a systematic tendency to make a certain mistake, by conventional standards. The explanation may be that they are using a common heuristic in circumstances where it is unreliable.

Many semanticists (though not all) prefer semantic explanations. That is not surprising. Every discipline tends to favour its proprietary ways of thinking. That can include neglect of obvious alternatives distinctive of other disciplines. Although semanticists are usually alert to possible rival explanations from pragmatics, the danger here is neglect or dislike of potential explanations from less linguistic disciplines, such as psychology and cognitive science.

As already noted, the semantic approach can be motivated by a methodological principle of charity, but such an appeal is liable to be counterproductive when the semantic hypothesis postulates kinds of contextual variation to which ordinary users of the language are insensitive. Sentences are passed from one context to another by memory and testimony, without being reassessed in the new context, so errors result by the standards of the context-sensitive semantics. Neglect of this danger is subtly built into the standard methodology of semantics. For when readers are introduced to a sample sentence, in a described context, they are expected to assess it for themselves, rather than relying on memory or testimony. Of course, such assessments constitute crucial data for semantics, but over-concentration on them has the unintended side effect of occluding the widespread and scarcely avoidable practice of accepting verbally encoded information in derivative ways too.

The problem is exacerbated by the complexity of many contextualist semantic proposals, including those using the salient partition as a parameter of the context. The greater the computational complexity of the non-derivative semantic assessment of a sentence, the less able and willing language users will probably be to make such an assessment. They will prefer all the more strongly to rely on their own memory or the testimony of others, thereby increasing unreliability.

The methodological problem is not addressed by an idealization to perfectly rational speakers and hearers. For how do *we* know what *they* would say or do? In arriving at a semantic theory, we rely on *our* reactions to sample sentences, or the reactions of other living human language users. We are all far from perfectly rational. The question is how much our real-life reactions depend on fallible heuristics, computational errors, and the like. The answer depends on which semantic theory is correct. In the long run, we have to decide holistically between theoretical packages each of which draws its own line between semantics and heuristics, and so between perfect and bounded rationality.

As the phrase 'bounded rationality' signals, the relevant contrast is not between pure rationality and pure irrationality. The indicative conditional does not present its semantics transparently to native speakers of a natural language. Their ordinary understanding of 'if' does not make it transparent to them whether indicative conditionals have truth-conditions and, if so, what they are. After centuries of debate, the issue is still controversial. Moreover, if speakers and hearers are mistaken, the relevant mistakes are not random or idiosyncratic: they are systematic, and built into standard ways of using the language. A reasonable hypothesis is that they are explained by patterns of thought which, on the whole, are efficient and reliable for normal humans, with our computational limitations and our scant practical need for a theoretical understanding of our own language.

For the present case, Kaufmann (2004: 599–602) defends the contextualist approach by arguing that it is sometimes detrimental to base one's actions on the global version of the Adams-Stalnaker equation (SRP), and advantageous to use the local version, as in (8) above. His argument concerns a complex and artificial diachronic series of conditional bets, with asymmetries in the information available to the Dutch bookie and the punter. But its core is independent of the semantics of the indicative conditional. It is intended to show that sometimes the 'local' probabilistic relation between two propositions $[A]$ and $[C]$, given by the right-hand side of (8), is more relevant to action than the 'global' conditional probability $Prob([C]||[A])$, given by the right-hand side of (9). But even if so, it shows nothing about $Prob([if\ A,\ C])$ unless one already assumes that it is determined by some sort of conditional probabilities of $[C]$ on $[A]$. If, instead, the plausibility of the 'global' equation SRP is the artefact of a useful but less than fully reliable heuristic, the Suppositional Rule, then the 'local' variant of SRP can be understood as the result of applying a less than fully reliable sub-heuristic for feasibly implementing the over-arching heuristic—in effect, by ignoring the conditionalization of the cell weights on the antecedent. Intriguingly, in a footnote to a sympathetic discussion of an approach like Kaufmann's, Daniel Rothschild (2010, n.31) admits 'I'm slightly tempted to think this is an instance of cognitive error of some sort, akin to base-rate neglect.'[2]

4.3 Conditionals within conditionals

Since conditionals tend to trigger applications of the Suppositional Rule, conditionals embedded within conditionals will tend to trigger applications of the Rule within applications of the Rule.

Conditionals can occur embedded within the consequents of other conditionals, as here:

(12)   If the sample was salt, then if it was put in water, it dissolved.

The insertion of 'then' in (12) is just for the sake of readability. Conditionals can also occur embedded within the antecedents of other conditionals, as here:

(13)   If the sample dissolved if it was put in water, then it was soluble.

The insertion of 'then' in (13) is also just for readability. More complex embeddings are possible too, of course: conditionals within conditionals within conditionals, conditionals within other operators within conditionals, and so on.

On the Suppositional Conjecture, the primary way of assessing sentences like (12) and (13) involves making suppositions within suppositions. However, suppositions within suppositions are more easily understood as corresponding to conditionals embedded in the consequents of conditionals, not their antecedents, in the manner of (12) rather than (13). The two loci of embedding give rise to different cognitive phenomena and explanatory challenges. We take them in turn.

4.4 Embedding in the consequent

In mathematical proofs, it is standard to make one supposition within the scope of another, and so on with no particular limit. One first supposes A, and then supposes B, without cancelling, ending, or suspending the former supposition. As a result, one is supposing each of A and B, and so in effect their conjunction 'A and B' too. The online analogue is first learning A, and then learning B, without retracting A. As a result, one has learnt each of A and B, and so in effect their conjunction 'A and B' too. The point of making the two suppositions separately, rather than lumping them together and just making a single conjunctive supposition, is to enable the discharge of one supposition without the other. For example, on the supposition of A and of B,

---

[2] For later developments of Kaufman's approach see Kaufman 2009, 2015. For criticism of his assumptions see Douven 2008.

you may reach a contradiction; you can then apply the rule of *reductio ad absurdum* to discharge the assumption B, and assert 'not(B)' on the supposition of A alone.

Nevertheless, what you can derive from the two suppositions A and B together is just what you can derive from the single supposition 'A and B', since the standard rules for introducing and eliminating conjunction make them equivalent, as in systems of natural deduction. This also means that what you can derive, having supposed A, and then supposed B too, is just what you can derive, having supposed B, and then supposed A too. The online analogue is the idea that first learning just A, and then learning just B, should leave you with the same knowledge base as first learning just B, and then learning just A: what matters is *what* you have learnt, not *when* you learnt it. In the offline case, at any point in a proof there is one stock of assumptions, to which more can then be added. Such order-invariance may not be self-evident, but it is compelling in the mathematical context, and very plausible more generally.

Order-invariance validates 'Import-Export' principles to the effect that sentences of these forms are all mutually equivalent (at least when A, B, C are not relevantly context-sensitive):

If A, (if B, C)
If B, (if A, C)
If (A and B), C
If (B and A), C

The truth-tables for the material conditional and conjunction validate these Import-Export equivalences, by equating the truth-conditions for all four of these formulas:

$A \supset (B \supset C)$
$B \supset (A \supset C)$
$(A \text{ and } B) \supset C$
$(B \text{ and } A) \supset C$

Each formula is false when both A and B are true and C false, and true otherwise.

By contrast, many semantic theories of indicative conditionals invalidate the Import-Export equivalences. For example, on the view that the indicative conditional expresses some sort of strict conditional, the relevant conditionals can be formalized thus (where $\square$ expresses the appropriate sort of necessity, whatever that is):

$\square(A \supset \square(B \supset C))$
$\square(B \supset \square(A \supset C))$
$\square((A \text{ and } B) \supset C)$
$\square((B \text{ and } A) \supset C)$

Treating these forms as equivalent collapses the modality $\square$. For consider the special case where B is a tautology T, and C is A. Thus the third formula becomes $\square((A \text{ and } T) \supset A)$, which is a trivial logical truth, while the first formula becomes $\square(A \supset \square(T \supset A))$, which reduces to $\square(A \supset \square A)$ (both equivalences can be derived by standard reasoning in any normal modal logic). Hence such Import-Export principles imply that necessarily, all truths are necessary. Thus the strict conditional theory saves Import-Export only in the limiting case where it collapses into the material conditional theory.[3]

A more general effect of Import-Export is to validate analogues of the so-called 'paradoxes' of material implication, given the natural assumption that 'If D, E' holds whenever E is a truth-functional consequence of D—an assumption supported by the Suppositional Rule, given that using elementary deductive logic is legitimate in developing the supposition D. For then 'If A and B, A' holds, because A is a truth-functional consequence of 'A and B', so by Import-Export 'If A, (if B, A)' also holds, for any B. Similarly, 'If A and not(A), C' holds, because C is a truth-functional consequence of 'A and not(A)', so by Import-Export 'If not(A), (if A, C)' also holds, for any C.

The Suppositional Rule sharpens the 'paradoxes'. For the high probability of A does not guarantee the high conditional probability of A on B, nor does the high probability of 'Not(A)' guarantee the high conditional probability of C on A. For example, imagine that you hold a ticket in a large lottery; the draw has already been made, but the result has not yet been announced. Here is an instance of *modus ponens*, from the premises (14) and (15) to the conclusion (16):

(14)   If your ticket lost, then if your ticket won, it lost.
(15)   Your ticket lost.
(16)   If your ticket won, it lost.

Since (14) is an instance of the schema 'If A, (if B, A)' (with 'then' added only for readability), by hypothesis it has a quasi-logical status, and so is not in doubt. By the circumstances of the example, you are also more than 99 per cent confident of (15). But when (16) is assessed by the Suppositional Rule, and more specifically by its probabilistic case SRP, it performs disastrously, since the conditional probability of its consequent on its antecedent is zero. From this perspective, it looks like a case where you can rationally be at least 99 per cent confident in accepting the conjunction of the premises of an instance of *modus ponens*, but 100 per cent confident in rejecting its conclusion. If so, (14)–(16) is a counterexample to *modus ponens*, for the conclusion of a deductively valid argument is always at least as probable as the conjunction of its premises.

We can also construct similar problems with an embedded conditional whose antecedent is consistent with its consequent. For example, Dorothy Edgington (1995: 268) has a case where 'I believe that the match will be cancelled; for all the players have flu' but 'I don't believe that if the players make a very speedy recovery the match will be cancelled.' Consider the conditional 'If the match will be cancelled, then if the players make a very speedy recovery the match will be cancelled' in place of (14) and proceed as before. For simplicity, we will focus on (14)–(16).

Here is another instance of *modus ponens* very similar to (14)–(16), from the premises (17) and (18) to the same conclusion as before:

(17)   If your ticket did not win, then if your ticket won, it lost.

---

[3] For a recent critical discussion of Import-Export see Mandelkern 2018.

(18)   Your ticket did not win.
(19)   If your ticket won, it lost.

This works exactly like the previous case, except that (17) has its quasi-logical status because it is an instance of the schema 'If not(A), (if A, C)', rather than 'If A, (if B, A)' (again, with 'then' added only for readability). As before, we could also consider instances of the schema where the consequent of the embedded conditional (19) is barely consistent with the antecedent, but the conditional probability is still exception-ally low; we stick with (17)–(19) for simplicity. Applying the Suppositional Rule to (19) has the effect of treating (17)–(19) as another counterexample to *modus ponens*.

Of course, on the material reading of the indicative conditional, both (14)–(16) and (17)–(19) are valid arguments and straightforward instances of *modus ponens*. Indeed, the premises (14) and (17) are redundant: (16) follows validly from (15) alone and (19) (= (16)) from (18) alone. On that reading, (16) is at least as probable as (15) and (19) at least as probable as (18).

If one goes by the Suppositional Rule, and more specifically by SRP, one will interpret those results as decisive against the material reading. For by SRP the probability of (16)/(19) is the conditional probability of its consequent on its antecedent, which is zero. Presumably, then, one will also assess (14) and (17) by the same rule. But the Suppositional Rule endorses both (14) and (17), and more generally the schemas 'If A, (if B, A)' and 'If not(A), (if A, C)'. That is implicit in what has already been said about the Import-Export principles, but it is worth being explicit.

For 'If A, (if B, A)', the Suppositional Procedure works thus. First, one supposes A. Next, one supposes B, within the scope of the former supposition. On both suppo-sitions together, one of course accepts A. One then applies the Procedure to discharge the second supposition (B), by accepting 'If B, A' on the first supposition (A) alone. Finally, one applies the Procedure again to discharge the first supposition, by accepting 'If A, (if B, A)' on no suppositions.

Similarly, for 'If not(A), (if A, C)', one first supposes not(A). Next, one supposes A, within the scope of the former supposition). On both suppositions together, one accepts C, because anything follows from inconsistent premises (the assumed back-ground logic of negation, conjunction, and disjunction is classical). One then applies the Suppositional Procedure to discharge the second supposition (A), by accepting 'If A, C' on the first supposition (not(A)) alone. Finally, one applies the Procedure again to discharge the first supposition, by accepting 'If not(A), (if A, C)' on no suppositions.

In both cases, one applies the Suppositional Procedure first within a supposition, and then outside it. That is just as one would expect, given the form of the sentences under assessment. As in mathematical proofs, such embeddings are an ordinary part of complex reasoning.

Thus the natural basis for assigning minimal probability to (16)/(19) is a Rule which also assigns maximal probability to (14) and (17). Since (15) and (18) are almost certain by hypothesis, it follows that the arguments (14)–(16) and (17)–(19) must be invalid. Thus the Rule classifies some instances of *modus ponens* as grossly invalid. But we saw in section 3.1 that another application of the Rule, to deductive attitudes, supports *modus ponens*, in the form of SR⊢⇒! Consequently, we are in effect dealing with yet more manifestations of the inconsistency of the Suppositional Rule.

Could it be denied that (14)–(16) and (17)–(19) are genuine instances of *modus ponens*, on the grounds that the indicative conditional construction is context-sensitive? On such a view, the sentence 'If your ticket won, it lost' somehow expresses a different proposition when embedded as the consequent of a conditional in (14) and (17) from the one it expresses when standing alone as (16)/(19).

Of course, the fact that the sentence 'If your ticket won, it lost' is assessed once on a supposition (in the assessment of (14) or (17)) and once on no supposition (in the assessment of (16)/(19)) is not by itself any reason to claim that it expresses different propositions on the two occasions: that difference in the conditions of assessment is generic to *all* instances of *modus ponens*, so recycling it as a difference in content would deprive *modus ponens* of all genuine instances. Imagine someone simply entertaining the hypothesis 'If my ticket wins, it loses', before starting to assess it. Do we really have to find out his epistemic circumstances before we can learn what proposition he is entertaining?

We have already seen that contextualism about bare indicative conditionals has a significant cost, because it threatens to undermine our standard practice of freely passing them about in the same verbal form between radically different epistemic contexts, by memory and testimony. However, that point need not be pressed here, for in any case no form of contextualism will save the Suppositional Rule. The proofs of its inconsistency in chapter 3 can all be run with respect to a single context, and several of them do not involve embedded occurrences of conditionals. Invoking contextualism to save the Rule in the present case would be unmotivated, since we know independently that something is wrong with the Rule.

A more plausible view of cases such as (14)–(16) and (17)–(19) is that they trade on the characteristic limitations of the Suppositional Rule. For example, it may lure us into drastically underestimating the probability of the conclusion (16)/(19). These instances of *modus ponens* are valid, just as they should be, but our fallible heuristic for assessing conditionals makes them look otherwise.

Similar considerations apply to Vann McGee's notorious proposed counterexample to *modus ponens*, which also turns on an indicative conditional whose consequent is itself an indicative conditional. Here is the case (McGee 1985: 462; compare Adams 1975: 33):

> Opinion polls taken just before the 1980 election showed the Republican Ronald Reagan decisively ahead of the Democrat Jimmy Carter, with the other Republican in the race, John Anderson, a distant third. Those apprised of the poll results believed, with good reason:
>
> If a Republican wins the election, then if it's not Reagan who wins it will be Anderson.
>
> A Republican will win the race.
>
> Yet they did not have reason to believe:
>
> If it's not Reagan who wins, it will be Anderson.

How should we understand McGee's example?

As before, the natural basis for rejecting the conclusion is the Suppositional Rule: from the epistemic perspective McGee describes, the probability of Anderson winning conditional on Reagan not winning is very low. From the same perspective, the Suppositional Rule also endorses the major premise. In more detail: First, one supposes that a Republican will win the election. Next, one supposes that it's not Reagan, within the scope of the former supposition. On both suppositions together, one accepts that it will be Anderson, since that is what the two suppositions and the background information that Reagan and Anderson are the only Republican candidates jointly entail. One then applies the Rule to discharge the second supposition, by accepting 'If it's not Reagan who wins, it will be Anderson' on the first supposition ('A Republican wins') alone. Finally, one applies the Rule again to discharge the first supposition, by accepting the major premise on no suppositions.

Of course, if having good reason to believe a proposition consists in its having probability at least c on one's evidence, where c is a fixed threshold intermediate between 0 and 1, then one can lack good reason to believe a conjunction while still having good reason to believe each conjunct, since the conjunction of two propositions each of probability at least c may have probability less than c. In the same way, one can lack good reason to believe the conclusion of an instance of *modus ponens* while still having good reason to believe each premise. But that standard observation misses the heart of McGee's example. For, from the given perspective, the first premise is more than 99 per cent probable while the second premise is, say, 70 per cent probable, which requires a probability of more than 69 per cent for their conjunction. Since the conclusion of a deductively valid argument is at least as probable as the conjunction of the premises (on any probability distribution), the argument at issue is deductively valid only if its conclusion is more than 69 per cent probable from the given perspective. But if one applies the Suppositional Rule to 'If it's not Reagan who wins, it will be Anderson', one assigns it the probability of Anderson winning conditional on Reagan not winning, which from the given perspective is very low, say 5 per cent. Thus, relying on the Suppositional Rule, one can conclude that the argument is far from deductively valid: a genuine counter-example to *modus ponens*.

For the same reasons as before, the idea that the problem turns on context-sensitivity in the indicative conditional is poorly motivated. In addition to the other reasons, we can imagine someone, Mary, entertaining the hypothesis 'If it's not Reagan who wins, it will be Anderson.' We need not know her epistemic circumstances to grasp her hypothesis. Thus, given the background information, McGee's argument is equivalent to one from the premises (20) and (21) to the conclusion (22):

(20)   If a Republican wins the election, Mary's conjecture holds.
(21)   A Republican will win the election.
(22)   Mary's conjecture holds.

Syntactically, this is a simple instance of *modus ponens*, with no embedding of conditionals. If McGee's original example is a counterexample to *modus ponens*, so is this variant. But if *modus ponens* fails in such simple cases, it is in very bad shape.

Once again, a more plausible view of the case is that it trades on the characteristic limitations of the Suppositional Rule. It may lure us into drastically underestimating the probability of the conclusion 'If it's not Reagan who wins, it will be Anderson.' The instances of *modus ponens* are valid, just as they should be, but our fallible heuristic for assessing conditionals makes them look otherwise. For example, on the material reading, the conclusion is equivalent to the disjunction 'Either Reagan will win or Anderson will win', which is in turn equivalent to the second premise given the background information that Reagan and Anderson are the only Republican candidates. Other semantic theories of indicative conditionals will validate the argument in other ways.

One may still wonder how the Suppositional Rule can generate apparent gross counterexamples to *modus ponens*, when it assesses conditionals by the corresponding conditional probabilities. For consider the argument from 'If A, C' and A to C. By elementary calculations of probability:

$$\text{Prob}([C]) \geq \text{Prob}([C \text{ and } A]) = \text{Prob}([C]|[A])\text{Prob}([A])$$

Thus $\text{Prob}([C])$ is close to 1 whenever both $\text{Prob}([C]|[A])$ and $\text{Prob}([A])$ are close enough to 1. Hence, one might suppose, the conclusion C will be acceptable to any required degree whenever the premises 'If A, C' and A are acceptable enough. However, when C is itself a conditional, as in (14)–(16), (17)–(19), and McGee's example, applying the Suppositional Procedure step by step to 'If A, C' involves *not* assessing C by its probability, but by the conditional probability of its consequent on its antecedent, which is not the same thing. Thus the apparent failures of *modus ponens* with embedded conditionals have the same source as the problems which arise for simple conditionals. Fallible heuristics are liable to manifest their fallibility in many different forms.

4.5 Embedding in the antecedent

When the Suppositional Procedure is applied to a conditional whose antecedent contains another conditional, the inner conditional figures as a constituent of the sentence which expresses the initial supposition. For some purposes, this verbal form suffices, as when the inner conditional supplies the major premise for an instance of *modus ponens* used to develop the initial supposition. For mathematical proofs, that is enough. But it is not always enough. As already noted, drawing out the non-logical implications of the initial supposition often involves a rich imaginative exercise. One must imagine how things will be if the supposition holds. But the non-transparency of the truth-conditions of conditionals to us obstructs the imagination.

Here is an example. Compare these two sentences:

(23)   Either the sample dissolved or it was not put in water.
(24)   The sample dissolved if it was put in water.

On the material reading, (24) has the same truth-conditions as (23). Nevertheless, they present quite different challenges to the imagination.

On being told to imagine (23), it is pre-theoretically natural to feel faced with a choice: which disjunct? To imagine just that the disjunction holds, without going further, feels too abstract and merely verbal. We want to go further, and get more concrete, by imagining one or other disjunct, in other words, either to imagine that the sample dissolved, or to imagine that it was not put into water. When (23) is our initial supposition, doing justice to its disjunctive form may require us to imagine each disjunct in turn, and draw out their implications separately; that corresponds to the standard rule for developing disjunctions in the tableaux proof procedure for first-order logic. Such branching involves computational complexity, but it is a fair price to pay for eliminating the disjunctive aspect on each branch.

By contrast, on being told to imagine (24), it is not pre-theoretically natural to feel faced with any such choice. More generally, there is no pre-theoretically natural way of eliminating the conditional aspect of supposing (24). Of course, one can further suppose that the sample *was* put in water, and then conclude from (24) by *modus ponens* that it dissolved, but that does not engage the possibility that the sample was *not* put in water.

At this point, the creative imagination may take over. To imagine (24) more concretely, in non-hypothetical terms, one may imagine a categorical ground for the conditional, for instance an intrinsic property of the sample (such as being salt) which disposes it to dissolve in water. One may then ignore other possibilities, and treat the categorical ground as the truth-condition for the conditional. That will tempt one to accept various conditionals whose antecedent is (24), such as (13):

(13)   If the sample dissolved if it was put in water, then it was soluble.

Consider this sentence:

(25)   If the sample was not put in water, it was soluble.

On the material reading, (13) entails (25), since the antecedent of (25) ('the sample was not put in water') entails the antecedent of (13) ('the sample dissolved if it was put in water'). Since one may well have no reason to accept (25), this can look like an argument against the material reading.

Why not simply take the categorical ground reading of sentences like (25) as licensed by the semantics of English? Consider this extension of the example. The sample belonged to a large collection. A few of the samples have been put in water; most have not. As it happens, my trusted and trustworthy assistant has compiled a list of all those samples in the collection that were put in water and did *not* dissolve. He checks the list: the sample in question is not on it. He is therefore in a position to assert (24), or in other words (26):

(26)   If the sample was put in water, it dissolved.

His testimony can appropriately be invoked in a wide variety of contexts, including ones where solubility is in question. But it provides no good evidence that the sample was soluble. It may simply never have been put in water. Even for those who think that such uses of the conditional encode a modal connection, the modality is epistemic rather than physical; it yields no ascription of solubility. Despite appearances, the assertion of (13) was not really warranted. It depended on the appearance of a non-epistemic categorical ground reading of the conditional (24). The appearance is an illusion, a mere artefact of our imaginative processes in assessing indicative conditionals.

Our use of the Suppositional Procedure can make us overestimate conditionals' consequences. As a result, we may assent to false conditionals with a conditional embedded in the antecedent. That is just the sort of thing that happens when one relies on a fallible heuristic.

Sometimes, a conditional in the antecedent of a conditional leaves the imagination at a loss. We simply do not know how to apply the Suppositional Procedure. Some authors take this to mean that the compositional semantics of the language does not apply to such sentences (Gibbard 1981: 234–8, Dummett 1973: 351–4 and 1992: 171–2, Edgington 1995: 283–4). Gibbard gives the example:

(27) If Kripke was there if Strawson was there, then Anscombe was there.

One can easily feel unable to make anything of (27): one mentally glazes over. Of course, with a little effort one can invent something: Anscombe intends to disrupt any attempt by Kripke to make an alliance with Strawson. But such confabulation is quite different from the effort involved in grappling with a complex formula such as (28) (which is truth-conditionally equivalent to (27) on its material reading):

(28) Either it is not the case that either Strawson was not there or Kripke was there, or Anscombe was there.

But to treat (27) as having no literal meaning is an overreaction. The difference in our reactions to (27) and (28) is adequately explained by the difference between our heuristics for assessing conditionals and our heuristics for assessing negations and disjunctions. The former typically involves an imaginative exercise; the latter is more likely to involve something closer to a calculation from our assessments of the simple constituent sentences. Whether the truth-conditions of (27) and (28) really are the same is another question. Heuristics are not truth-conditions; they are more like verification-conditions, though not with the semantic status mistakenly awarded the latter by verificationist theories of meaning.

The imaginative nature of many applications of the Suppositional Procedure does not prevent it from playing a central role in a complex practice of using conditionals which holistically and non-transparently determines a fully compositional truth-conditional semantics for 'if'. Nor does the poor performance of the Procedure in handling conditionals embedded in the antecedent of a conditional make the outer conditional undefined. An analogy: our ordinary way of applying the term 'curly' is on the basis of unaided sight, but that does not render undefined its application to things too small to be seen unaided. By looking through a microscope, or at a computer simulation, we can learn that some of those things were curly all along, while others were non-curly all along. Analogously for 'if': when our ordinary heuristics are aided by theoretical devices from logic, linguistics, and cognitive science, we may be able to work out how 'if' applied all along in cases intractable to our ordinary heuristics alone.

We should therefore not follow Michael Dummett's ad hoc suggestion that some conditionals of the form 'If (B if A), then C' should be understood as saying 'If you accept B if A, you must also accept C', as with (29):

(29) If John should be punished if he took the money, then Mary should be punished if she took the money.

Of course, someone who asserts (29) may want to communicate that if you accept that John should be punished if he took the money, then you must also accept that Mary should be punished if she took the money, but that point is secondary to (29), which says something about the case itself, not about what anyone does or should accept about the case. In a natural scenario, the speaker has derived (29) from a generalization like (30):

(30) Either whoever took the money should be punished, or whoever took the money should not be punished.

But (30) is conditional-free. That it entails (29) indicates that the occurrences of 'if' in (29) are not ad hoc communicative devices, but are tightly constrained by the logical interaction of 'if' with quantifiers.

Dummett points out that the attempt to state truth-conditions for indicative conditionals homophonically involves the very sort of embedding at issue. The homophonic clause for 'if' in a compositional semantic theory of truth-conditions takes a form like this:

(31) 'If A, C' is true if, and only if, if A is true, C is true.

The right-to-left direction of the biconditional has an indicative conditional in the antecedent, while the left-to-right direction has it in the consequent.

Dummett is right that if conditionals with conditional antecedents are unintelligible, then (31) will not make them intelligible, because it will be unintelligible itself. However, that does *not* result simply from the homophonic use of 'if' on the right-hand side of the biconditional. Indeed, the same underlying problem arises even for a *non*-homophonic truth-conditional clause for 'if' using a synonym $\rightarrow$ for 'if' in a non-English meta-language, with a corresponding translation $\leftrightarrow$ of 'if and only if', which we can write thus:

(32) $\text{Tr}(\text{'if } A, C\text{'}) \leftrightarrow (\text{Tr}(\text{'}A\text{'}) \rightarrow \text{Tr}(\text{'}B\text{'}))$

The problem also depends on the use of the biconditional 'if and only if' or something with the same meaning as the main connective. That is unnecessary for the theory to deliver homophonic truth-conditions. One could equally well use a more theoretically motivated connective in the meta-language for the semantic clauses, say 'just in case', standing for some appropriate kind of equivalence:

(33) 'If A, C' is true just in case if A is true, C is true.

To be meaningful, (33) must still make some appropriate semantic demand on indicative conditional sentences, depending on exactly how 'just in case' works. But the Suppositional Rule need not be central to the use of 'just in case'. Thus we need not have the same problem in coming to grips with (33) as we had with (31) and (32).

Of course, once we apply a homophonic compositional semantic theory with a clause like (33) to conditional sentences with conditional antecedents in the object-language, we get truth-conditions for those sentences in just such pre-theoretically baffling terms, with theorems of this form:

(34)   'If (if A, B), C' is true just in case if (if A is true, B is true), C is true.

But homophonic semantic theories were never designed for purposes of language-learning, nor for purposes of facilitating the verification or falsification of object-language sentences. Rather, their purpose was more theoretical: to show how the truth-conditions of complex sentences are compositionally determined by the semantic contributions of their simple constituents. For that purpose, (34) will do well enough, at least as a first approximation—a fully developed semantic theory will surely have to track all sorts of parameters of evaluation omitted from (34) for simplicity.

If we want truth-conditions in more tractable form, we must go non-trivially non-homophonic. The material interpretation permits us to use (35), again as a first approximation:

(35)   'If A, C' is true just in case either A is not true or C is true.

We can even restore 'if and only if':

(36)   'If A, C' is true if and only if either A is not true or C is true.

Other semantic theories of 'if' will use other non-trivially non-homophonic right-hand sides.

But arguing for one of these theories over the others in the first place is not an exercise in pure semantics. Rather, much of the work must be done in meta-semantics. It is a matter of showing that the favoured interpretation is integral to the semantic and cognitive account which, all things considered, makes the best sense of our total practice of using conditionals. That is the aim of this book.

5
Conditional Testimony

5.1 Pooling testimony: variations on a case

In section 2.2, we noted a general secondary heuristic for assessing conditionals: accept them on the basis of others' testimony, under the same conditions for trust which you apply to testimony in non-conditional form. For purposes of communication, conditionals are nothing special. In particular, there is no general need to reword them to preserve their content. In that respect, they contrast with ordinary indexicals: when Ana says something using the word 'I', I substitute the name 'Ana' for 'I' in recording the information—it is about her, not about me. When she says something using the word 'if', no such substitution is required.

The secondary testimonial heuristic can pull in the opposite direction from the Suppositional Rule. Understanding this tendency acts as a salutary check on vain attempts to tailor the semantics of conditionals so as to validate the Rule. We can explore the phenomenon by analysing several progressively more challenging versions of an example. It is a variant of cases developed by Allan Gibbard (1981) and Jonathan Bennett (2003), though the morals they drew from it are very different from mine. Examples of this kind are also standardly invoked by *contextualists* who take the proposition expressed by a conditional to vary with epistemic features of the context in which it is used. One influential case in point is Robert Stalnaker (1984: 108–14).

Version I

There has been an accident at a dodgy nuclear power plant. Several warning lights are connected to a single detector beside the nuclear core. When the detector is working and detects overheating in the core, each light is red. When the detector is working and does not detect overheating in the core, each light is green. When the detector is not working, each light is red or green at random, independently of the others. A competent engineer, East, sees only the east light, which is red, and says:

(1)   If the detector is working, the core is overheating.

Another competent engineer, West, not in contact with East, sees only the west light, which is green, and says:

(2)   If the detector is working, the core is not overheating.

There is no presupposition failure; each engineer assigns the common antecedent a probability much greater than 0, though less than 1. Although both engineers have

Suppose and Tell: The Semantics and Heuristics of Conditionals. Timothy Williamson, Oxford University Press (2020).
© Timothy Williamson.
DOI: 10.1093/oso/9780198860662.001.0001

incomplete information, neither is in error. Indeed, by ordinary standards, East knows (1) and West knows (2), so (1) and (2) are both true.

But how can (1) and (2) both be true? They are opposite indicative conditionals, with the same antecedent and mutually contradictory consequents. CNC, the principle of Conditional Non-Contradiction, makes the conjunction of a pair of opposite conditionals such as (1) and (2) itself a contradiction. As seen in section 3.4, the Suppositional Rule supports CNC in various ways, concerning both probabilistic and ungraded attitudes. The most direct connection involves the plural attitude of *treating as incompatible*. Given that we treat C and D as incompatible on the supposition A, the Rule mandates treating 'If A, C' and 'If A, D' as unconditionally incompatible. Since we treat 'The core is overheating' and 'The core is not overheating' as incompatible on the supposition 'The detector is working' (as well as without it), the Rule directs us to treat (1) and (2) as incompatible. Indeed, when one assesses (1) and (2) directly, it is hard not to feel a tension between them.

The stock way to solve the problem is contextualist. Although (1) is true in East's context of utterance while (2) is true in West's context of utterance, they are not both true together in any context. Thus the violation of CNC is merely apparent.

The only relevant difference between the two contexts is epistemic: East and West have access to different information. East's evidence rules out one set of possibilities; West's evidence rules out another set of possibilities. Such examples have been used to motivate the surprising idea that the indicative conditional is epistemic in meaning. The point is not the mere truism that the same indicative conditional may vary in evidential status from one speaker to another at the same time; that applies to virtually all declarative sentences. Rather, it is the more contentious claim that the same indicative conditional may vary *in truth-value* from one speaker to another, depending just on differences in their epistemic position—where this contextual variation in the content of the whole conditional does not originate in any contextual variation in the content of either the antecedent or the consequent.

One limitation of contextualist solutions is worth stressing. *They cannot fully validate the Suppositional Rule.* For the inconsistency and trivialization proofs for the Rule are structural and quite general. They apply even when by hypothesis all sentences are assessed with respect to the same context. Although contextualist approaches allow one to reconcile the truth of (1) and (2) in their respective contexts of utterance with the unrestricted validity of CNC, that must be at the expense of invalidating other consequences of the Rule. No reason has yet been suggested for prioritizing the rescue of CNC over the rescue of those other consequences. That already suggests that a contextualist solution may be under-motivated. For a non-contextualist approach might rescue applications of the Rule more urgent than CNC, in some sense to be explained.

Version II

As before, but now a central controller is trying to establish what the situation is. The two engineers are trustworthy, and the controller trusts them. They text in their reports to her. East's report is simply (1), and West's report is simply (2). The controller is a bureaucrat; she does not know how the two engineers came to their judgements. She simply accepts both reports. She even repeats them to herself. She then reasonably and correctly concludes (3) from (1) and (2):

(3)   The detector is not working.

What (1) and (2) express in the controller's mouth depends on her context, not directly on the contexts of the engineers. Anyway, they intended their reports for her context, not just for their own. East expects the controller to accept (1), and West expects her to accept (2), whatever other reports she may receive. Thus (1) and (2) are sometimes acceptable together in a single context, even though they are opposite conditionals, contrary to both the Suppositional Rule and contextualist accounts designed to accommodate CNC.[1]

Contextualists have a comeback. For the controller can reach the conclusion (3) *without* accepting (1) and (2) as uttered in her own context. She can reason thus. Since East is trustworthy, his report (1) is true as uttered in his own context. Therefore, by elementary conditional reasoning (supported by the Suppositional Rule), (4) is false as uttered in East's context:

(4)   The detector is working and the core is not overheating.

But (4) is just a conjunction; it lacks any shiftiness associated with indicative conditionals. Hence (4) is also false as uttered in the controller's context. But the falsity of (4) amounts to the truth of the *material* conditional analogue (5) of (1):

(5)   The detector is working $\supset$ the core is overheating.

Thus (5) is true as uttered in the controller's context. By exactly parallel reasoning about West's report, the material conditional analogue (6) of (2) is also true as uttered in the controller's context:

(6)   The detector is working $\supset$ the core is not overheating.

But (5) and (6) themselves jointly entail the controller's conclusion (3), as required. She has no need to endorse the engineers' indicative conditionals (1) and (2) on her own account. Moreover, once she accepts (3), she has ruled out the common antecedent of (1) and (2), so they both suffer presupposition failure and should not be endorsed in her context.

Such a contextualist resolution of the problem is not altogether satisfying, for several reasons. It shifts the main cognitive burden to the material conditionals (5) and (6), while insisting that they are not semantically equivalent to the indicative conditionals (1) and (2). Thus the engineers' use of indicative conditionals is made to look merely decorative. The contextualist story about the controller's implicit reasoning also seems unnecessarily and implausibly elaborate. A simpler and more natural thought process is for her just to accept (1) and (2) themselves, suppose that the

---

[1] This is a serious problem for Frank Jackson's view (1979, 1987) that indicative conditionals have the material truth-conditions but implicate that the consequent is highly probable on the antecedent. Such an implicature must fail for at least one of a pair of opposite conditionals such as (1) and (2). 'If A, C' and 'If A, not(C)' cannot both be robust in Jackson's sense with respect to acceptance of A, unless the speaker is ready to accept both C and 'not(C)' on accepting A.

detector *is* working, reach a contradiction by two applications of *modus ponens* (which the Suppositional Rule endorses), and then use *reductio ad absurdum* to discharge the assumption and conclude that the detector is *not* working, (3). The contextualist reconstruction has to postulate similar reasoning, just with material rather than indicative conditionals, but in addition a preliminary process of sanitizing the original indicative conditionals to make the results mutually consistent.

On the simpler account, the controller accepts (1) and (2) together, at least for the brief passage of time it takes her to reach the conclusion (3). That is enough to raise the challenge: how can her reasoning for (3) be sound if it rests on mutually inconsistent premises, (1) and (2)? Moreover, the claim that the controller has no use for (1) and (2) once she has reached the conclusion (3) is not accurate. She may use both in explaining to her assistants how she has reached the conclusion, saying something like this: 'As we know from East, if the detector is working, the core is overheating. But, as we know from West, if the detector is working, the core is not overheating. So the only possibility left is that the detector is not working.' That sounds like an intelligent explanation, rather than the words of someone too stupid to realize that she is contradicting herself.

Another range of cases where we accept opposite conditionals together is in mathematical proofs by *reductio ad absurdum* (for more discussion see Rumfitt 2013). For example, in the course of a standard proof that there is no largest prime number, mathematicians accept both 'If p is the largest prime, $p! + 1$ is composite' and 'If p is the largest prime, $p! + 1$ is prime', using 'if' with its normal meaning. Unless both opposite conditionals stand, the proof fails.

Similar examples occur in philosophical argument, without special symbols. In a famous passage, Bertrand Russell argues: 'Naïve realism leads to physics, and physics, if true, shows that naïve realism is false. Therefore naïve realism, if true, is false; therefore it is false' (Russell 1950: 15). Though the argument may rest on philosophically questionable premises, it is not linguistically infelicitous. Russell stably accepts 'naïve realism, if true, is false', but of course he also stably accepts the tautology 'naïve realism, if true, is true'. Thus he stably accepts two conditionals with the same antecedent and mutually inconsistent consequents.

But *how* can we override the Suppositional Rule in such cases, if the heuristic is built into cognitive mechanisms associated with 'if'? Compare it with perceptual recognitional capacities. They are typically automatic, yet we sometimes override them. A liquid looking exactly like water may automatically trigger our recognitional capacity for water, but we can override it when an authoritative source tells us that the liquid is not water but something quite different. A lookalike may automatically trigger my recognitional capacity for a friend, although I override it because I know that she is a thousand miles away. The disposition to judge 'That's water' or 'That's her' is activated, but then inhibited by the greater weight the agent assigns to another epistemic source. Similarly, (1) and (2) together may trigger the controller's recognitional capacity for conflicting conditionals, but she overrides it on the authority of the engineers. Even if her disposition to judge 'They can't both be true' is activated, it is then inhibited by the greater weight she assigns to her technical experts.

For now, the dialectical position is this: attempts to motivate contextualist accounts of 'if' by appeal to such examples are unconvincing, because they rely on

CNC, whose plausibility can be explained as an artefact of a heuristic, in cases where it is unreliable. Moreover, in such cases, the heuristic can in practice be overridden by weightier epistemic sources, such as authoritative testimony and mathematical proof. By itself, this is not an argument that 'if' is *not* semantically context-sensitive, just that cases of a sort standardly invoked to show that it *is* semantically context-sensitive do no such thing.

Version III

There is another problem for the envisaged contextualist account of how conditional testimony is pooled in version II, by recourse to a material conditional weaker than 'if'. It does not cover all the relevant cases. For consider a variant of the example where West's report is not (1) but (7):

(7) If the detector is working, either the reactor was not built to plan or the core is overheating.

Similarly, West's report is not (2) but (8):

(8) If the detector is working, either the reactor was not built to plan or the core is not overheating.

Clearly, (7) and (8) are mutually consistent, since they both follow from the obviously consistent claim (9), by the uncontentious principle of deduction in the consequent (which the Suppositional Rule endorses):

(9) If the detector is working, the reactor was not built to plan.

The consequent of (9) is logically equivalent to the conjunction of the consequents of (7) and (8) by elementary truth-functional logic (P is equivalent to '(P or Q) and (P or not(Q)'). Thus, conversely, the controller having accepted (7) and (8) can validly deduce (9). Pre-theoretically, she seems fully entitled to assert (9).

How can contextualists reconstruct the controller's reasoning? She assumes that (7) is true as uttered in East's context and that (8) is true as uttered in West's context. As in Version II, she can derive the truth of the corresponding material conditionals as uttered in her own context:

(10) The detector is working $\supset$ ((the reactor was not built to plan) or (the core is overheating)).

(11) The detector is working $\supset$ ((the reactor was not built to plan) or (the core is not overheating)).

From (10) and (11), she can unproblematically deduce (12):

(12) The detector is working $\supset$ the reactor was not built to plan.

But that does not yet explain how the controller is entitled to assert (9), the indicative conditional corresponding to the material conditional (12). On a contextualist account, they are not semantically equivalent.

Contextualists are not yet out of moves. For there is the following contextualist line of thought. When the controller fully accepts the material conditionals (10) and (11), she treats the conjunction of the antecedent with the negation of the consequent as epistemically excluded, and so in effect (10) and (11) as not just true but *epistemically necessary*. In other words, she implicitly accepts (13) and (14) in her own context, where 'must' is a contextually shifty epistemic modal operator:[2]

(13)   Must(the detector is working $\supset$ ((the reactor was not built to plan) or (the core is overheating))).

(14)   Must(the detector is working $\supset$ ((the reactor was not built to plan) or (the core is not overheating))).

But epistemic necessity is closed under truth-functional consequence; from (13) and (14), (15) follows by the standard normal logic of a necessity operator:

(15)   Must(the detector is working $\supset$ the reactor was not built to plan).

Thus, given that (13) and (14) are true as uttered in the controller's context, so is (15). The point of this argument is that, for many contextualists, although the truth of the material conditional does not suffice for the truth of the corresponding indicative conditional, the epistemic necessity of the material conditional *does* normally suffice for the truth of the indicative conditional. This applies to Stalnaker's account of indicative conditionals. It also applies to Kratzer's analysis of 'if' as not an operator in its own right but just a restrictor introducing a restriction on other operators (Kratzer 1981, 1986, 2012). For, Kratzer holds, in bare conditionals, with no overt operator to be restricted, 'if' typically introduces a restriction on a covert epistemic 'must'. Some other accounts yield a similar connection between 'if' and 'must' by other means (Gillies 2010). Thus, since (15) is true in the controller's context, so is the indicative conditional (9), just as required.

How do such considerations apply to version II? In exactly the same way, a contextualist can argue that the controller will treat the material conditionals (5) and (6) as epistemically necessary, and so (16) and (17) as true in her context:

(16)   Must(the detector is working $\supset$ the core is overheating).
(17)   Must(the detector is working $\supset$ the core is not overheating).

From (16) and (17), (18) follows, again by the closure of epistemic necessity under truth-functional consequence:

(18)   Must(the detector is not working).

Thus, given that (16) and (17) are true as uttered in the controller's context (in version II), so is (18).

However, there is a hitch. For if the epistemic necessity of a material conditional suffices quite generally for the truth of the corresponding indicative conditional, then the truth of (16) and (17) in the controller's context suffices for the truth of both (1) and (2) in her context, which is exactly what the contextualist strategy was designed

---

[2]   In English, epistemic 'must' implies that the evidence for an assertion is somehow indirect. For simplicity, this aspect of 'must' is ignored; for a recent discussion of it see Mandelkern 2019. As often in the literature, 'must' will be treated as a plainer epistemic necessity modal.

to avoid. To tighten the screw, just as (18) follows from (16) and (17) by the closure of epistemic necessity under truth-functional consequence, so does (19):

(19) Must(the detector is working ⊃ (the core is overheating and the core is not overheating)).

But if the epistemic necessity of a material conditional suffices for the truth of the corresponding indicative conditional, then the truth of (19) in the controller's context suffices for the truth of (20) in her context:

(20) If the detector is working, the core is overheating and the core is not overheating.

Those who dislike the conjunction of (1) and (2) may dislike (20) even more. If the controller applies the Suppositional Procedure to (20), and she rejects a contradiction even on the supposition that the detector is working, she will reject (20).

In response, contextualists are likely to qualify the claim that the epistemic necessity of a material conditional suffices for the truth of the corresponding indicative conditional. They may argue that it fails for the special case where the antecedent is epistemically impossible. Since the truth of (18) in the controller's context amounts to the epistemic impossibility of (16), (17), and (19) for her, that solves the immediate problem.

This contextualist account of how conditional testimony is pooled in versions II and III improves on the previous contextualist account, given under version II, at least in its greater explanatory generality. Nevertheless, it does not fundamentally improve the envisaged contextualist's dialectical position, as described under version II. Attempts to motivate contextualist accounts of 'if' by appeal to the relevant examples are still unconvincing, because they rely on CNC, whose plausibility can be explained as an artefact of a heuristic, in cases where it is unreliable; moreover, in such cases, the heuristic can in practice be overridden by weightier epistemic sources, such as authoritative testimony and mathematical proof. By itself, this is not an argument that 'if' is *not* semantically context-sensitive, just that cases of the sort standardly invoked to show that it *is* semantically context-sensitive do no such thing.

However, the revised contextualist account faces a further challenge. The role it assigns to epistemic necessity ('must') fits the case of perfect trust in the informants. What happens when trust is less than perfect?

Version IV

Things are as in version III, except that the controller feels a tiny sliver of doubt about the engineers' testimony. Reasonably enough, her trust in them is very slightly less than perfect. Her confidence in each of the indicative conditional reports (7) and (8) is a fraction less than 100 per cent. As a result, the corresponding material conditionals (10) and (11) are *not* epistemically necessary for the controller; (13) and (14), which ascribe epistemic necessity to them, are false in her context. The same goes for the indicative conditional (9) ('If the detector is working, the reactor was not built to plan'), which is equivalent to the conjunction of (7) and (8). The controller's confidence in (9) is slightly less than 99 per cent. As a result, the corresponding material conditional (12) is not epistemically necessary for her; (15), which ascribes epistemic necessity to it, is false in her context.

A contextualist might retort that since the controller does not fully trust the engineers' reports, on which (9) depends, she has no business asserting (9). But that obvious point does not go very far. For although the controller is not *certain* of the reports' truth, she still accepts them both tentatively, with a high degree of confidence just short of 100 per cent. She should still be able to make deductions from them and accept the conclusions with at least as much confidence as she has in the conjunction of the reports, (7) and (8). In particular, since (9) is a deductive consequence of (7) and (8), she should still be able to accept (9) tentatively, with a high degree of confidence just short of 1. For to postulate a radical discontinuity between perfect and near-perfect trust is to make communication by testimony a far less stable practice than it actually is. Plausibly, it is subject to *graceful degradation*: under near-perfect trust, the hearer should still be able to obtain an approximation to the benefits of perfect trust. There should be a probabilistic analogue of the story about epistemic necessity.

At first sight, such a probabilistic story is easy to tell. The controller has very high credence just short of 1 in each of (7) and (8), almost equally high credence in their conjunction, and at least as high credence as that in (9). But assume that her credence in each conditional goes with her conditional credence in the consequent on the antecedent, in accordance with the Suppositional Rule. Assume also that she puts very little credence in the antecedent. The trouble is this. Her credence in the conditional all depends on the tiny bit of her probability space where the antecedent holds, which may overlap with the tiny bit she assigns to the possibility that the engineers are not trustworthy. Thus she may end up with very little credence in the conditionals (7), (8), and (9), even though she is *almost* certain that the engineers spoke truly. This may happen even though she is almost certain of the material conditionals (10) and (11) corresponding to (7) and (8), which on a contextualist account is the only relevant non-epistemic aspect of the engineers' testimony transferrable to the controller's context.

For example, the controller may have independent inside information strongly indicating that the reactor *was* built to plan. Once she has taken in the engineers' reports, she may distribute her credences thus, where D = 'The detector is working, O = "The core is overheating", and R = 'The reactor was built to plan':

Credence(not(D)) = 98 per cent
Credence(D and R and O) = 0.9 per cent
Credence(D and R and not(O)) = 0.9 per cent
Credence(D and not(R) and O) = 0.1 per cent
Credence(D and not(R) and not(O)) = 0.1 per cent

Consequently:

Credence($D \supset$ (not(R) or O)) = Credence($D \supset$ (not(R) or not(O))) = 99.1 per cent
Credence($D \supset$ not(R)) = 98.2 per cent
Credence(not(R) or $O|D$) = Credence(not(R) or not(O)$|D$) = 55 per cent
Credence(not(R)$|D$) = 10 per cent

Thus the controller has credence 99.1 per cent in each of the material conditionals corresponding to the engineer's reports (7) and (8), and credence 98.2 per cent in the material conditional corresponding to the conclusion (9) which follows from (7) and (8). With respect to the material conditionals, her trust in the engineers is near-perfect, just as the contextualist strategy requires. Nevertheless, her credence in the conditional (9), which equals her conditional credence that the reactor was not built to plan on the supposition that the detector is working, is only 10 per cent. Her doubt in each engineer's report is 45 per cent, and the two doubts add up because they are mutually exclusive.

In brief, the position is this. Contextualist accounts of how conditional testimony is pooled depend on the idea that epistemic necessity (by contrast with epistemic possibility) is preserved or extended as available information accumulates. Thus each engineer's report transmits its epistemic necessity to the controller, despite the differences in their epistemic positions. That works under conditions of perfect trust (and trustworthiness), but fails under conditions of imperfect trust. One epistemic counterpossibility blocks epistemic necessity. The natural strategy for contextualists is to fall back on probabilistic attitudes to conditionals. But if the probability of a conditional is anything like the conditional probability of the consequent on the antecedent, as the Suppositional Rule requires, that fallback strategy fails too, because those conditional probabilities can hugely magnify small changes in probability space. High conditional epistemic probability is often not preserved or extended when imperfectly trusted testimony is pooled. The result is further tension between our primary heuristic for indicative conditionals, the Suppositional Rule, and our secondary heuristic, their use in testimony under normal conditions. The primary heuristic has us focus on conditional probabilities, the secondary heuristic sometimes has us sweep them aside.

A contextualist might object that the controller should not accept (9), even tentatively, because for her the negation of its consequent is more likely to be true, given its antecedent. She should also refrain from accepting (7) and (8), since they jointly entail (9), and are epistemically on a par. But the objection does little more than reiterate the Suppositional Rule, without explaining why it should be given priority over testimony. The controller is, quite reasonably, almost certain that the engineers spoke truly. Of course, she can accept (21) and (22) without difficulty:

(21)   East almost certainly spoke truly in uttering (7).
(22)   West almost certainly spoke truly in uttering (8).

But that is not the issue. Her overriding concern is with the physical state of the reactor, more specifically that of the detector and above all the core. Her interest in the engineers' speech acts is instrumental to that. The point is to extract from (21) and (22) their bearing on her main concern. To achieve that, she must do something like tentatively accepting (7) and (8), and then slightly more tentatively accepting (9) too. That is how she updates her credences in the relevant propositions on the engineers' testimony.

Of course, the controller can assign high probabilities (short of 1) to the *material* conditionals corresponding to (1)–(3) or (7)–(9) in version IV. But that is not the point, for contextualists do not in general accept the material reading. The challenge to them is to explain how the controller can assign high probabilities (short of 1) to the non-material conditionals they take (1)–(3) and (7)–(9) to express. For otherwise they wrongly classify her attitude to those conditionals as anomalous. More generally, when conditional testimony is accepted with imperfect trust from one epistemic context to another, what prevents a catastrophic loss of epistemic probability?

The dialectical position for contextualist accounts has now shifted. Examples of pooled testimony like those above have often been used to motivate such accounts. The discussion of versions I–III showed that such examples lend themselves to an anti-contextualist interpretation. That undermined a standard way of motivating contextualism. The discussion of version IV has gone further, by showing that contextualist accounts of 'if' face a more general problem. They have systematic difficulty in explaining the probabilistic acceptance of conditional testimony across epistemically different contexts under conditions of imperfect trust. By contrast, non-contextualist accounts of 'if' face no corresponding difficulty. This is direct evidence against contextualist accounts.

Version V

Things are as in version IV, except that the controller is fully confident that the reactor *was* built to plan, even on the supposition that the detector is working. She can then eliminate the disjunct 'the reactor was not built to plan' from the consequents of the engineers' report, which reduces (7) to (1) and (8) to (2). She still accepts all these conditionals. Putting (1) and (2) together, as in version III, she can then accept (20), with the same antecedent and a self-contradictory consequent ('If the detector is working, the core is overheating and the core is not overheating'). The spirit in which she accepts (20) is like that in which she accepts (23):

(23)   If the detector is working, I'm the Pope.

Neither (20) nor (23) accords with the Suppositional Rule. On the supposition that the detector is working, she accepts neither a contradiction nor that she is the Pope. On learning just that the detector was working, she would not update her credences according to (20) or (23). Of course, (20) lacks the rhetorical flourish of (23), while (23) lacks the testimonial support of (20), but both illustrate the possibility of rational confidence in an indicative conditional with a not obviously absurd antecedent but an obviously absurd consequent, despite the lurking awareness in effect that it violates the Suppositional Rule. Despite its not obviously absurd consequent, the controller accepts (9) in a similar spirit, though more tentatively, for there too lurks the awareness that it violates the Rule.

5.2 Testimony: generalizing the case

Versions I–III of the nuclear reactor case are similar in structure to earlier examples in the literature, of which the best known and most colourful is Allan Gibbard's Sly Pete case (Gibbard 1981: 231):

Sly Pete and Mr Stone are playing poker on a Mississippi riverboat. It is now up to Pete to call or fold. My henchman Zack sees Stone's hand, which is quite good, and signals its content to Pete. My henchman Jack sees both hands, and sees that Pete's hand is rather low, so that Stone's is the winning hand. At this point, the room is cleared. A few minutes later, Zack slips me a note which says 'If Pete called, he won,' and Jack slips me a note which says 'If Pete called, he lost.'

Here Zack and Jack correspond to East and West in the nuclear reactor, and the narrator, their boss, to the controller. Gibbard's case lacks the perfect epistemic symmetry between the two observers, which Bennett introduced in a case more similar to the present one (Bennett 2003: 85–6).[3]

Although Gibbard's and Bennett's examples are well known, they have generally been used for a very different purpose: to argue that either indicative conditionals do not express propositions at all, or which propositions they express is ultra-sensitive to their linguistic and epistemic context. The background assumption is CNC, the principle of conditional non-contradiction, as explained under version I.

A sign of what has gone wrong is Stalnaker's claim that 'to play their methodo-logical role, conditionals must be too closely tied to the agents who utter them for those conditionals to express propositions which could be separated from the contexts in which they are accepted' (1984: 111). On the contrary, the need to pool information from different sources, including information in conditional form, indicates that to play their full methodological role, conditionals must be capable of being freely passed about through memory and testimony amongst contexts in which they are questioned, contexts in which they are fully accepted, and contexts in which they are slightly less than fully accepted. That implies that conditionals must *not* in general be closely tied to the agents who utter them. Their basic role is to encode information about any aspect of the world, not to express the speaker's psychological state. The engineers' job is to inform the controller about the reactor, not to express themselves.[4]

Theorists of conditionals have misread the significance of Sly Pete and similar cases because they have implicitly relied too much on the Suppositional Rule, prioritizing one fallible heuristic to the exclusion of another, reliance on conditional testimony under normal conditions, which is a further essential part of our practice in using conditionals. The tension between the two heuristics is not confined to cases where an agent must pool conditional testimony from several sources. The tension also arises for testimony from a single source.

Here is a schematic example. You regard Trudy as a fairly trustworthy source of information. She offers you the indicative conditional testimony 'If A, C', where there is no relevant context-sensitivity in the sentences A and C, taken by themselves. Should you assume that her credence in C conditional on A is high? Even granted that she is sincere, the assumption may not be safe, for she may be like the controller

---

[3] Angelika Kratzer further complicates the Sly Pete case by suggesting that since 'No particular circumstances of evaluation were given in Zack's and Jack's notes', 'the claims made by their notes might have remained genuinely undetermined', and proposing the use of supervaluationist technology to deal with the 'unresolved vagueness' (2012: 103). On the face of it, their notes are somewhat unin-formative rather than imprecise in any relevant way: the vagueness looks like an artefact of the theory. Similar points apply to the nuclear reactor case.

[4] Moss (2018) suggests a related view of the Sly Pete case.

of the nuclear reactor in version V, basing her knowledge of (9) on pooled testimony even though, for her, (9) violates the Suppositional Rule However, for the sake of argument, we may imagine a situation where you know that Trudy's credence in C conditional on A is high. Still, despite your warranted trust in Trudy, it does not mean that you should follow her in that, by making your own credence in C conditional on A high. Without making any mistakes, she may lack evidence which you have; in such respects her conditional credences may be less well informed than yours. Perhaps you trust her not to have very low credence in any relevant unconditional truth, in particular in any truth-function of A and C. Given what you know about Trudy's credences, you can work out that her credence in 'A and not(C)' is very low. If you trust her enough, you may simply exclude the possibility of that conjunction altogether. In that case, your credence in C conditional on A should be 1. But suppose that you very slightly doubt Trudy's trustworthiness. Those conditions leave your credence in C conditional on A unconstrained.

To sharpen the problem, here is a toy example. Your credence that Trudy is trustworthy is 99 per cent. Your total credence is distributed over 1,000 possible worlds, which you regard as all equiprobable. Thus Trudy is trustworthy in 990 of them. Conditional on her being trustworthy, you exclude 'A and not(C)' altogether. More specifically, as it happens, 'A and C' is true in just one of those 990 worlds; A is false in all the other 989. Consider two possibilities for the remaining 10 worlds, in which Trudy is untrustworthy.

(i)   In all the remaining worlds, either A is false or C true. Then your credence in C conditional on A is 1. Applying the Suppositional Rule, you have credence 1 in 'If A, C'.

(ii) In all the remaining worlds, A is true and C false. Then your credence in C conditional on A is 1/11. Applying the Suppositional Rule, you have credence 1/11 in 'If A, C'.[5]

Of course, there are also intermediate cases, in which A is true and C false in some but not all of the remaining 10 worlds, and your credence in C conditional on A is strictly between 1 and 1/11. The point is that although you are very confident (but not perfectly confident) that Trudy is trustworthy, your credence in C conditional on A can still vary wildly, depending on what you think about the tiny sliver of doxastically live cases where she is untrustworthy. Thus treating indicative conditionals as vehicles for conditional credences undermines indicative conditional testimony, even in conditions of near-perfect trust, because conditional credences are too sensitive to slight differences in agents' individual circumstances to be amenable to being passed on from one agent to another. The phenomenon is not restricted to cases where one agent pools testimony from several others.

---

[5] Since every proposition in the example has a probability of $n/1000$ for some natural number n between 0 and 1000, and no such ratio equals 1/11, the conditional probability of C on A in case (ii) is not the probability of any proposition. Thus the case illustrates the difficulty of assigning propositions to indicative conditionals in accord with the Suppositional Rule for credences. However, the no-proposition view no more helps to solve the problem of conditional testimony than it does to resolve the more intrinsic inconsistencies in the Rule. For related triviality arguments see Hájek 1989.

When we trust Trudy in case (ii), as we may reasonably do, we assign high credence to 'If A, C', even though we assign low credence to C conditional on A. We thereby violate the Suppositional Rule for credences, overriding it with the secondary testimonial heuristic.

Vann McGee has an example with a similar structure. The details are omitted here because they are complicated, though amusing, but this is his summary (2000: 107):

> The counterexample [. . .] is a conditional 'If p, then q' that we are willing to assert on the basis of the testimony of someone we regard as a highly reliable authority. Learning p would, however, sufficiently undermine our confidence in the authority that we would no longer have grounds for believing either the conditional or its consequent. So even though 'If p then q' is assertable, the posterior probability that q would have once we learned that p, and hence our present conditional probability of q given p, is low.

Here is a simpler case of the same kind. An expert psychiatrist is giving a lecture. Pointing somewhere on a slide of the brain, she says: 'This area of the brain is enlarged in all congenital liars. For example, if I am a congenital liar, this area of my brain is enlarged.' We are very confident, though not perfectly confident, of the generalization and so of its instances, including (24), on her testimony:

(24)   If she is a congenital liar, that area of her brain is enlarged.

But our credence for that area of her brain being enlarged, conditional on her being a congenital liar, is not high, for we do not find the conjunction 'She is a congenital liar and that area of her brain is enlarged' more probable than 'She is a congenital liar and that area of her brain is not enlarged.'[6] We thereby violate the Suppositional Rule for credences, overriding it again with the secondary testimonial heuristic.

Slightly imperfect trust in conditional testimony makes trouble for most views on which indicative conditionals are distinctively epistemic in meaning. Sometimes, with good reason, the hearer is fairly but not fully confident of 'If A, C' even though, amongst the relevant epistemic possibilities, comparatively few A-worlds are C-worlds. On views like Stalnaker's, the truth-value of 'If A, C' is the truth-value of C at the selected A-world; but the hearer has no good reason to expect the selected A-world to be a C-world. On the restrictor view, when 'if A' restricts a covert epistemic necessity operator, 'If A, C' is false in ways which should be obvious to hearers, since their slight distrust focuses on a few relevant epistemically possible A-worlds which are not C-worlds, any one of which falsifies the restricted claim of epistemic necessity. Similar difficulties also arise for Veltman's data semantics (1986), and for update semantics and contextually shifty strict conditional analyses (Gillies 2009, Starr 2014, Willer 2010, Yalcin 2012), for they all make 'If A, C' crash in the presence of a relevant epistemic A-possibility which is not a C-possibility. *Contra* Descartes, slight doubts have no such drastic consequences.

Of course, agents can tell each other about their individual conditional probabilities and other epistemic states. That is not in doubt. The difficulty is in passing on

---

[6] See Magidor 2018 for a related example.

information *about the external world* in such epistemic forms. Yet in cases like the nuclear reactor, indicative conditionals *are* used to pass on information about the external world from speaker to hearer. Our general practice of communicating information by testimony makes no special exception of indicative conditionals. Consequently, the view of those conditionals as vehicles for conditional probabilities or other epistemic states makes little sense of their role in testimony. The very reliance on the Suppositional Rule which provides the strongest motivation for epistemic and contextualist views of indicative conditionals is also what gets such views into trouble with testimony.

Instead, the key to these examples is the interplay between primary and secondary ways of assessing conditionals. Sometimes, we have good reason to defer the assessment of an indicative conditional to someone we take to be in a better position to judge than we are, perhaps along a whole chain of testimony. We let the fallible secondary heuristic override the fallible primary heuristic. That does not make our thinking somehow inauthentic or uncritical or irrational. By normal standards we are doing well, trusting a trustworthy source. Without the testimony of others, we are all lost.

The standard case with testimony is that speaker and hearer are in different epistemic and linguistic contexts, before and after the testimony is received. When you read a news report, your epistemic and linguistic context differs in various ways from the reporter's. The commonplace phenomenon of conditional testimony is no exception. If we could not use conditionals under such conditions, they would be far less useful than they are. Any adequate theory of conditionals must be able to explain how they can be reliably passed around between different epistemic and linguistic contexts. As explained for version IV of the examples in section 5.1, the hardest case of the challenge for contextualist accounts to meet is to explain probabilistic acceptance of conditional testimony under conditions of imperfect trust.

Of course, in principle, testimony of any kind raises the question of the speaker's trustworthiness. In that respect, conditionals are no different from any other kind of sentence. To avoid gullibility, the hearer must be alert to evidence for or against the speaker's trustworthiness. Once we take account of these generic challenges, and of the fallibility of the Suppositional Rule, there is no good motivation for taking indicative conditionals to introduce any distinctively epistemic element of meaning.

What we need is a semantic account of indicative conditionals which explains how they can be both tested by the Suppositional Rule and freely communicated from one context to another by testimony and, of course, memory. That bipartite structure is commonplace in other areas of epistemology. For example, in perception we learn facts about our environment which we remember and communicate to others who were absent from the original scene. Similarly, mathematicians prove theorems and communicate their results to others, who may apply them while taking the proof on trust. Indicative conditionals work similarly, often with no need to modify their wording. Chapter 6 argues that a simple-minded semantics for such conditionals best explains the combination of primary and secondary heuristics, despite their structural tensions.

6

The Role of Conditional Propositions

6.1 Underpinned propositions

Previous chapters identified two general heuristics on which we rely in assessing indicative conditionals. The primary heuristic is the Suppositional Rule, which includes adjusting one's credence in the conditional to one's conditional credence in its consequent on its antecedent. The secondary heuristic is the passing on of conditionals from one epistemic context to another by testimony and memory, in the same linguistic form, under the usual conditions. While both heuristics are cognitively effective, the first is implicitly inconsistent and the second is obviously fallible. Sometimes they give conflicting results; each acts as a check on the other. We have seen various indications that this complex practice is best understood by assuming that an indicative conditional expresses a piece of information about the world, in short a *proposition*, detachable from any particular context in which it is expressed (except for incidental context-sensitivity in its antecedent or consequent), intelligible from, and communicable between, many different perspectives. Although some evidence seemed to tell strongly against that hypothesis, interpreting the evidence that way turned out to depend on uncritically following an implicit bias in favour of the primary heuristic against the secondary heuristic. Since the primary heuristic is inconsistent, that strategy is unlikely to end well. In this chapter, we explore the consequences of a strategy more evenly balanced between the two heuristics.

To do justice to the secondary heuristic, we assume that 'if' introduces no context-sensitivity of its own. For simplicity, we also ignore any context-sensitivity in the antecedent and the consequent themselves; factoring it back into the account is a routine exercise. Thus we assume that the sentences A, C, and the indicative conditional 'If A, C' expresses the propositions $[A]$, $[C]$, and $[if A, C]$ respectively, independently of context.

Doing justice to the primary heuristic is trickier. The probabilistic version of the Suppositional Rule (SRP) treats the conditional probability of $[C]$ on $[A]$ as a proxy for the probability of $[if A, C]$. Consider a policy of probabilistically accepting conditionals on that basis. For the policy to be *sound*, the probability of $[if A, C]$ must not be lower than the conditional probability of $[C]$ on $[A]$. For the policy to be *complete*, the probability of $[if A, C]$ must not be higher than the conditional probability of $[C]$ on $[A]$. Then a natural way of doing justice to the Suppositional Rule is by assuming that the proposition $[if A, C]$ is selected so as to make the policy sound, and as complete as is compatible with soundness. The idea is that avoiding

Suppose and Tell: The Semantics and Heuristics of Conditionals. Timothy Williamson, Oxford University Press (2020).
© Timothy Williamson.
DOI: 10.1093/oso/9780198860662.001.0001

inappropriate acceptance takes priority over avoiding inappropriate failure to accept.

We can implement that idea formally. For simplicity, we treat propositions as subsets of a domain of worlds Ω (more complicated arguments in a similar spirit could be given within a framework of more fine-grained propositions). In what follows, Φ, Ψ, and X are any propositions. The negation $\neg\Phi$ of Φ is of course the complement $\Omega\backslash\Phi$. We stipulate:

SRP on Φ for Ψ *underpins* X just in case $\text{Prob}(\Psi|\Phi) \leq \text{Prob}(X)$ for every probability distribution Prob over Ω on which the probabilities are defined.

In other words, if we treat the conditional probability of Ψ on Φ as an estimate of the probability of X, it never overestimates, but may underestimate; if it errs, it does so on the side of caution. For brevity, we can just speak of Ψ on Φ underpinning X. Thus soundness is the constraint that [C] on [A] should always underpin [if A, C]. The role of completeness will be explained later.

Clearly, soundness does *not* guarantee that only true indicative conditionals pass the suppositional test. In bad cases, one's probabilities are a poor guide to the truth: $\text{Prob}([C]|[A])$ may be high, so the test is passed, although [if A, C] is false. Epistemic agents try to apply the test in good cases, where their probabilities are a good guide to the truth. Nevertheless, even the most rational of agents have limited evidence, which can mislead them by making a false proposition highly probable. That is just the generic problem of error, which semantics by itself cannot be expected to solve. What the semantics of indicative conditionals *can* be expected to do is not *undermine* SRP, by making it unsound, as it would sometimes do if [C] on [A] did not underpin [if A, C].

What does it take for Ψ on Φ to underpin X?

In one special case, underpinning is trivial. When Φ and Ψ are incompatible—as subsets of Ω, their intersection is empty—$\text{Prob}(\Psi|\Phi)$ is zero whenever it is defined, so automatically $\text{Prob}(\Psi|\Phi) \leq \text{Prob}(X)$, whatever proposition X is. In that case, the conditional probability never recommends X. That case is comparatively uninteresting.

In the interesting case, Φ and Ψ are *compatible*: $\Phi \cap \Psi$ is non-empty. Let w be a world in $\Phi \cap \Psi$. Now suppose also that Ψ on Φ underpins X. On those two assumptions, we establish two facts:

First, $\Psi \subseteq X$. For let $v \in \Psi$. Consider any probability assignment Prob such that $\text{Prob}(\{v\}) \neq 0$; $\text{Prob}(\{w\}) \neq 0$; $\text{Prob}(\{u\}) = 0$ for every other $u \in \Omega$.[1] Thus $\text{Prob}(\Psi|\Phi) = 1$. Since Ψ on Φ underpins X, $\text{Prob}(X) = 1$, so $v \in X$.

Second, $\neg\Phi \subseteq X$. For suppose that $v \in \neg\Phi$. Consider any probability assignment Prob such that $\text{Prob}(\{v\}) \neq 0$; $\text{Prob}(\{w\}) \neq 0$; $\text{Prob}(\{u\}) = 0$ for every other $u \in \Omega$. Thus $\text{Prob}(\Psi|\Phi) = 1$. Since Ψ on Φ underpins X, $\text{Prob}(X) = 1$, so $v \in X$.

---

[1] We can extend Prob from singletons to all subsets of Ω, even when Ω is uncountably infinite, by specifying that for $\Phi \subseteq \Omega$: if $v\in\Phi$ and $w\in\Phi$, $\text{Prob}(\Phi) = 1$; if $v\in\Phi$ and $w\notin\Phi$, $\text{Prob}(\Phi) = \text{Prob}(\{v\})$; if $v\notin\Phi$ and $w\in\Phi$, $\text{Prob}(\Phi) = \text{Prob}(\{w\})$; if $v\notin\Phi$ and $w\notin\Phi$, $\text{Prob}(\Phi) = 0$.

The two facts together mean that $\neg\Phi \cup \Psi \subseteq X$. But $\neg\Phi \cup \Psi$ is just the material conditional $\Phi \supset \Psi$. Thus, whenever Φ and Ψ are compatible and Ψ on Φ underpins X, $\Phi \supset \Psi$ entails X.

Conversely, whenever $\Phi \supset \Psi$ entails X, Ψ on Φ underpins X. For, as noted in chapter 3, $\mathrm{Prob}(\Psi|\Phi) \leq \mathrm{Prob}(\Phi \supset \Psi)$ whenever the probabilities are defined. But when $\Phi \supset \Psi$ entails X, $\mathrm{Prob}(\Phi \supset \Psi) \leq \mathrm{Prob}(X)$, so $\mathrm{Prob}(\Psi|\Phi) \leq \mathrm{Prob}(X)$.

To sum up in set-theoretic terms, we have established this:

FACT. For subsets Φ, Ψ, and X of Ω with $\Phi \cap \Psi \neq \{\}$: Ψ on Φ underpins X just in case $\Phi \supset \Psi \subseteq X$.

Thus, for compatible Φ and Ψ, $\Phi \supset \Psi$ is the *strongest* proposition underpinned by Ψ on Φ. However, in the special case where Φ and Ψ are incompatible, the strongest proposition underpinned by Ψ on Φ is the contradiction $\{\}$, which is distinct from $\Phi \supset \Psi$ except when Φ is the tautology Ω and Ψ is the contradiction $\{\}$.

An application of FACT is that if X is the result of strengthening the material conditional proposition $\Phi \supset \Psi$ by some epistemic condition (for compatible Φ and Ψ), then Ψ on Φ does *not* underpin X. Such a semantics makes one vulnerable to overestimating the probability of the conditional by using the Supposition Procedure.

When the antecedent and consequent are compatible, FACT shows that the material conditional $[A] \supset [C]$ is the best candidate to be the conditional proposition [if A, C] in terms of the desiderata above. For the material interpretation of 'if' makes the probabilistic acceptance policy based on SRP sound in the sense above, because [C] on [A] underpins $[A] \supset [C]$. The material interpretation also makes that policy as complete as is compatible with soundness, in the sense above, because whenever a candidate X to be [if A, C] makes the policy sound, [C] on [A] underpins X, so $\mathrm{Prob}([A] \supset [C]) \leq \mathrm{Prob}(X)$, so X does not beat $[A] \supset [C]$ by having a probability that exceeds $\mathrm{Prob}([C]|[A])$ by less. Indeed, one can easily show that unless $X = [A] \supset [C]$, $\mathrm{Prob}([A] \supset [C]) < \mathrm{Prob}(X)$ for some probability distribution Prob, in which case X does less well than $[A] \supset [C]$ in respect of completeness. In brief, the material conditional is the strongest content fit to be verified by the probabilistic Suppositional Rule and communicable from context to context by memory and testimony.

The result would still hold even if we significantly weakened the definition of 'underpin', and so the demands of soundness. For the proof of FACT goes through even when we require only that in the special case where $P(\Psi|\Phi) = 1$, $P(X) = 1$. Indeed, the proof is robust under many further variations. For example, in the simple case where Ω is finite, one can require Prob to be *regular*, in the sense that $\mathrm{Prob}(\{w\}) > 0$ for every $w \in \Omega$, so that only a contradiction has probability zero; the proof can easily be adapted to that requirement by assigning the worlds not of interest very low positive probability rather than probability zero.

The result can also be extended to the 'locally' calculated heuristics for probabilities discussed in section 4.2, where SRP is applied separately to each cell Π_i in a partition of Ω, rather than globally to all of Ω at once. For we have:

$$\sum_{i \leq n} \mathrm{Prob}(\Pi_i)\mathrm{Prob}([C]|[A] \wedge \Pi_i) \leq \sum_{i \leq n} \mathrm{Prob}(\Pi_i)\mathrm{Prob}([A] \supset [C]|\Pi_i)$$
$$= \mathrm{Prob}([A] \supset [C])$$

Thus the partition-sensitive 'local' probabilistic connection also in effect never overestimates the probability of the material conditional, for any probability distribution Prob and any partition on which those probabilities are defined. Likewise for the result that when [A] and [C] are compatible, [A] ⊃ [C] is the *strongest* proposition to be underpinned by [C] on [A]: that result too extends to 'local' probabilistic connections, since the 'global' conditional probability Prob([C]|[A]) is just the special case of the 'local' probabilistic connection for the trivial partition with only one cell. Thus the argument that SRP tracks the material interpretation of 'if' generalizes to 'local' implementations of SRP.

We still have to consider the anomalous-looking case where the antecedent and consequent are incompatible, and SRP singles out the contradiction rather than the material conditional as the best candidate for the indicative conditional proposition. One might be tempted to take the result at face value, and endorse this hybrid interpretation of 'if'.

Of course, the hybrid interpretation defuses only a small proportion of the cases where the material interpretation looks bad, because the conditional probability of the consequent on the antecedent is so much lower than the probability of the material conditional. For example, consider (1) and (2), in the context of a fair lottery with a million tickets and only one winner; the draw has already been made, but the result is not yet known:

(1)   If my ticket won, it lost.
(2)   If my ticket won, I will live to the age of two hundred.

On the material interpretation, both (1) and (2) are highly probable, because their antecedent is almost certainly false. On the hybrid interpretation, (1) has probability zero, because its consequent is incompatible with its antecedent, so (1) is treated as a contradiction, but (2) is treated as highly probable as before, because its consequent is compatible with its antecedent, so (2) is treated as a material conditional. The hybrid interpretation is not a way of validating the Suppositional Rule; nothing is.

However, the hybrid interpretation leads to implausible discontinuities in the properties of the indicative conditional. For example, suppose that x is a point somewhere in the real closed interval $[0, 1]$, with a uniform probability distribution. Let c be a constant such that $0 \leq c \leq \frac{1}{2}$, R_c be the interval $[c, c+\frac{1}{2})$, and C_c be the indicative conditional 'If x is in R_0, x is in R_c'. For $c < \frac{1}{2}$, R_0 and R_c overlap, so the antecedent and consequent of C_c are compatible, so C_c is interpreted as the material conditional, true whenever x is either not in R_0 or in R_c, in other words whenever $c \leq x \leq 1$, so its probability is $1 - c$. Thus as c approaches $\frac{1}{2}$ from below, the probability of C_c approaches $\frac{1}{2}$ from above. However, when $c = \frac{1}{2}$, R_0 and R_c do not overlap, so the antecedent and consequent of C_c are incompatible, so C_c is suddenly interpreted as a contradiction, so its probability is 0. This is very odd behaviour, but perhaps not utterly absurd.

A better way to deal with the case of incompatibility is by considering a wider range of applications of the Suppositional Rule. In particular, there is its application to deductive consequence:

$$\text{SR}\vdash\Leftrightarrow \qquad BB \vdash \text{if } A, C \Leftrightarrow BB; A \vdash C$$

As observed in section 3.2, this rule forces 'if' to be equivalent to the material conditional. Thus, overall, the material interpretation of 'if' fits the Suppositional Rule better than does the hybrid interpretation.

By any reasonable standard, the material interpretation is also more *natural* than the hybrid interpretation. It satisfies simpler principles, by being uniformly truth-functional. Even where [A] and [C] are incompatible, [C] on [A] still underpins [A] ⊃ [C], so soundness is not violated: it is just that [A] ⊃ [C] may not be the *strongest* proposition underpinned by [C] on [A], so the interpretation is not quite as complete as is compatible with soundness. But that is a small price to pay for the advantages of the material interpretation.

This book defends a uniform material interpretation of the conditional. Such a view goes back at least to Philo the Megarian (c. 300 BCE).[2] In modern times, Paul Grice and Frank Jackson, amongst others, have given sophisticated defences of the material interpretation of indicative conditionals.[3] However, for several decades it has been very much a minority view amongst theorists of conditionals. The problem is not a shortage of arguments for the material interpretation. Quite the opposite: it is derivable from various highly plausible principles of conditional logic. For example, by normal standards the usual natural deduction rules for the conditional— conditional proof and *modus ponens*—are compelling for 'if', and they force its equivalence with ⊃. As section 3.1 explained, those principles are corollaries of the Suppositional Rule. More generally, a common motif in the literature is that attractive-looking principles for 'if' collapse it into a material conditional. The trouble is that those arguments do not enable us to explain away the wide range of equally compelling-looking counterexamples to the material interpretation. Both Grice and Jackson made ingenious attempts to do so, but it is widely agreed that they could not adequately deal with the full range of recalcitrant data.[4]

Accumulated experience strongly indicates that any defence of the material inter-pretation will have to attribute systematic errors to native speakers. Theorists of language are understandably reluctant to pay that price. Just about any view can be maintained with a heavy enough dose of error theory. However, the arguments *for* the material interpretation strongly indicate that any theory incompatible with the material interpretation will also have to attribute systematic errors to native speakers. For example, they may pre-theoretically accept the argument from the hypothesis 'Everyone in the hall wore a tie' to the conclusion 'If John was in the hall, he wore a tie' in contexts for which the theory counts the universal generalization as true and

---

[2] See Sextus Empiricus, *Outlines of Pyrrhonism*, 2.110.

[3] See Grice 1989: 58–85 (originally one of his 1967 William James lectures) and Frank Jackson 1979, 1987. Other recent sympathetic treatments include Abbott 2004, Kratzer 2020, Lewis 1986a,b, and Rieger 2006, 2013, 2015. Koralus (forthcoming) upholds a material view of the truth-conditions of indicative conditionals while treating the connection with suppositional reasoning as built into their semantics rather than as a heuristic.

[4] Section 5.1 raised a specific objection to Jackson's theory, concerning its implications for Gibbard-style stand-offs. Another divergence between my approach and Jackson's is that his is based on a distinction between truth and assertibility, a normative standard, whereas a claim generated by a heuristic, even a hardwired one, need not be assertible in any normative sense—it may just be an insidious fallacy. Section 6.3 will discuss some general difficulties for Grice's approach. For a sample of further objections to Grice and Jackson see Edgington 1995 and Bennett 2003.

the conditional as untrue. The Suppositional Conjecture predicts that not all our judgements of conditionals are correct, because they are not mutually consistent. In the end, any theory of conditionals may have to attribute errors to native speakers somewhere. Those errors are not all just random. Pre-theoretic reactions to examples are rather predictable and similar across speakers, though of course not perfectly so. They manifest patterns. Errors cannot simply be attributed to 'confusions' which happen to afflict some speakers but not others, or a given speaker at some times but not others. A more systematic form of explanation is needed. Postulating the Suppositional Rule as a primary heuristic meets that need. The underpinning result then demonstrates a key respect in which the material interpretation provides the best semantic fit to our practice of using the heuristics we do.

For the rest of Part I, the main aim is to understand the divergences between the semantics and the heuristics, in what look like the hardest cases for the material interpretation of indicative conditionals. Part II extends that account to so-called *subjunctive* or *counterfactual* conditionals. The present account of 'if' is a very general version of the material interpretation, because it applies uniformly to its occurrences in *both* classes of sentence, explaining the differences compositionally by reference to the meaning of other constituents of 'counterfactual' conditionals. By contrast, most proponents of the material interpretation have limited it to indicative conditionals, leaving counterfactual conditionals for separate treatment, as Grice, Lewis, and Jackson did. But a uniform semantic treatment of 'if' is preferable, if it can be achieved: and it *can* be achieved.

All that is to anticipate. First, we must consider an issue raised but not settled by the underpinning result.

6.2 Overpinned propositions

The underpinning constraint is, in an obvious way, one-sided. It is aimed primarily at avoiding the inappropriate acceptance of indicative conditionals rather than their inappropriate rejection. It excludes overestimates of the probability of the target proposition, but not underestimates. For the sake of balance, we should also consider a dual constraint, aimed primarily at avoiding the inappropriate *rejection* of indicative conditionals rather than their inappropriate acceptance. It excludes underestimates of the probability of the target proposition, but not overestimates:

SRP for Ψ on Φ *overpins* X just in case $\mathrm{Prob}(\Psi|\Phi) \geq \mathrm{Prob}(X)$ for every probability distribution Prob over Ω on which the probabilities are defined.

But $\mathrm{Prob}(\Psi|\Phi) \geq \mathrm{Prob}(X)$ just in case $\mathrm{Prob}(\neg\Psi|\Phi) = 1 - \mathrm{Prob}(\Psi|\Phi) \leq 1 - \mathrm{Prob}(X) = \mathrm{Prob}(\neg X)$, and $\mathrm{Prob}(\Psi|\Phi)$ is defined just in case $\mathrm{Prob}(\neg\Psi|\Phi)$ is defined. Thus Ψ on Φ overpins X just in case $\neg\Psi$ on Φ underpins $\neg X$. Consequently, FACT means that whenever $\neg\Psi$ is compatible with Φ, then Ψ on Φ overpins X just in case $(\Phi \supset \neg\Psi) \subseteq \neg X$. The latter inclusion holds just in case $X \subseteq \neg(\Phi \supset \neg\Psi) = \Phi \wedge \Psi$. In other words, unless Φ entails Ψ, Ψ on Φ overpins X just in case X entails $\Phi \cap \Psi$. Consequently, the conjunction $\Phi \cap \Psi$ is the weakest proposition overpinned by Ψ on Φ.

If we impose the constraint on the indicative conditional that [C] on [A] should overpin [if A, C], the result is that unless the antecedent entails the consequent, their conjunction is the weakest proposition to meet the constraint. It is the weakest content fit to be probabilistically rejected by the Suppositional Rule whose falsity can be communicated from context to context by memory and testimony.

Just as requiring soundness and as much completeness as is compatible with soundness for SRP as a probabilistic *acceptance* rule singles out a material reading of 'if', so requiring soundness and as much completeness as is compatible with soundness for SRP as a corresponding probabilistic *rejection* rule singles out a conjunctive reading of 'if'. In section 3.2, we already noticed a similar pattern for the application of the Suppositional Rule to *deductive* acceptance and rejection. Just as the equivalence SR$\vdash\Leftrightarrow$ (above) for deductive acceptance singles out a material reading of 'if', so the corresponding equivalence SR$\vdash^-\Leftrightarrow$ for deductive rejection singles out the conjunctive reading of 'if':

$$\text{SR}\vdash^-\Leftrightarrow BB \vdash^- \text{ if } A, C \Leftrightarrow BB; A \vdash^- C$$

At first sight, these results suggest a severe tension between the verification and falsification procedures for indicative conditionals: they require utterly different contents. But consider: in probabilistic terms, when the Suppositional Procedure rejects 'if A, C', it does so because the conditional probability of $[C]$ on $[A]$ is low, which is so just in case the conditional probability of $[\text{not}(C)]$ on $[A]$ is high, which is exactly the condition for the procedure to accept 'if A, not(C)'. Thus a negative result of the test on a conditional can be communicated as a positive result with respect to the opposite conditional. That is in fact what we tend to do. When the conditional probability of $[C]$ on $[A]$ is low, it is far more natural and more easily intelligible to say 'if A, not(C)' than 'not(if A, C)': for example, 'If the shutters are down, she is not at home' rather than 'It is not the case that if the shutters are down, she is at home.' Since our practice with conditionals is skewed towards communicating acceptance rather than rejection, underpinning is a more appropriate constraint than overpinning. That fits the fact that the material conditional has traditionally been considered a much more serious candidate for the content of the indicative conditional than has conjunction—despite occasional examples suggestive of a conjunctive content, such as 'If she was strict, she was also fair.'

The preference for negating the consequent over negating the conditional is an important but little-remarked feature of mathematical practice. Although one may negate the universal generalization of a conditional, directly negating a conditional itself plays very little role in mathematics, especially in proofs, where mathematicians are at their most careful. Correspondingly, mathematicians do not treat deductive *rejection* of a conditional, as in SR$\vdash^-\Leftrightarrow$, on a par with deductive acceptance, as in SR$\vdash\Leftrightarrow$. Mathematics accumulates theorems, not anti-theorems. Pre-theoretic unclarity as to the consequences of a negated conditional may be part of the explanation, though negating a conditional may not even be an option on the standard menu of linguistic moves for articulating a mathematical proof. That is all just as well, given the inconsistency implicit in combining SR$\vdash\Leftrightarrow$ with SR$\vdash^-\Leftrightarrow$. Somehow, with admirably sound instincts, mathematicians have steered away from trouble without even thinking about it.

The asymmetry between asserting and denying conditionals also tells against a strategy which tries to minimize overall distances between the probability of the conditional and the conditional probability while remaining neutral between excess of the former and excess of the latter. Such a middle way between the underpinning and overpinning strategies has other disadvantages too: it is much harder and less natural to characterize formally; the results are liable not to be robust under small variations in how distances in probability are measured and aggregated; the implications for the logic of conditionals may be unclear. Since our practice in effect favours selecting conditionals suitable for acceptance, rather than rejection, underpinning is the key virtue in fitting a semantics to the Suppositional Rule. Underpinning in turn favours the material semantics for the plain conditional. In what follows, we therefore focus on challenges to the material interpretation.

6.3 Illusions of truth-value

The emerging picture predicts that when we are forced to assess a conditional 'If A, C', and deprived of the option of using the opposite conditional 'If A, not(C)' instead, our assessments will often come apart from a clear-eyed assessment of an overt material conditional. In the relevant cases, $\text{Prob}([A] \supset [C])$ is high while $\text{Prob}([C]|[A])$ is low. Equivalently, we can compare the corresponding negations: $\text{Prob}(\neg([A] \supset [C]))$ = $\text{Prob}([A] \wedge \neg[C])$ is low while $\text{Prob}(\neg[C]|[A])$ = $\text{Prob}([[A] \wedge \neg[C])/\text{Prob}([A])]$ is high. The difference is greatest when $\text{Prob}([A])$ is low, making the multiplier $1/[A]$ large, and $[A]$ is incompatible with $[C]$, given background information, giving $\text{Prob}([[A] \wedge \neg[C])$ its maximum value (for fixed A) of $\text{Prob}([A])$. In that case, $\text{Prob}([C]|[A])$ = o but $\text{Prob}([A] \supset [C])$ = $\text{Prob}(\neg[A])$, which is high. Similar points can be made for $[C]$ barely compatible with $[A]$.

An extreme example is (1) above ('If my ticket won, it lost'). The probability of the material conditional is just the probability that my ticket did not win, which is almost 1.

The probability of my ticket losing, conditional on it winning, is zero. In response to the forced-choice question 'Is (1) true or false?', the answer feels pre-theoretically like a no-brainer: 'False', not 'Almost certainly true'.

The impression that (1) is false cannot easily be explained away, following Grice, as somehow caused by a false conversational implicature. Such an implicature should be cancellable, but how is the speaker to do that? To add 'But I don't mean to imply that winning is compatible with losing' would merely add to the hearer's confusion. Of course, there is no obvious conversational point to uttering (1). Moreover, the speaker is arguably not in an epistemic position to *assert* (1), for he does not *know* that if his ticket won, it lost, since that would amount to knowing that it did not win, which he presumably does not know (Williamson 2000). Even if he does know that his ticket did not win, why not assert that directly rather than the needlessly complex (1)?

However, considerations about assertibility do not get to the heart of the matter. For we can assess propositions as unasserted hypotheses, presented by a list of declarative sentences in a context. When presented with (1) in such a list, the natural judgement is still that it is false. At first sight, (1) looks like a perfectly good counterexample to the truth-functional interpretation of indicative conditionals.

But not all is as it seems. In examples such as (20) ('If the detector is working, the core is overheating and the core is not overheating') in section 5.1, we saw how testimony can provide good evidence for an indicative conditional with a consistent antecedent and inconsistent consequent, even though the conditional probability of the latter on the former is zero. Indeed, we can have the evidence of the strongest kind for an indicative conditional whose consequent is inconsistent with its antecedent: mathematical proof. Consider this example:

(3)   If p is the largest prime, $p! + 1$ is a prime larger than p.

At first sight, (3) looks just as bad as (1), and for the same reason: its consequent is inconsistent with its antecedent. Our initial reaction to both is to dismiss them. But to prove by *reductio ad absurdum* that there is no largest prime number, a mathematician may suppose that p is the largest prime, and from that supposition prove that $p! + 1$ is a prime larger than p. By standard mathematical reasoning, in particular the rule of conditional proof, she can then discharge the assumption and assert (3). She can conclude that there is no largest prime, but that does not undermine (3), which remains a legitimate step in the proof. This involves no special mathematical meaning of 'if'. Mathematicians use 'if' with the same meaning as everyone else. They are simply more careful in their reasoning with it. They hold themselves to the unusually high standard of deductive validity: (3) was established by that high standard. When one acquires skill in mathematical reasoning, one learns to resist the temptation to dismiss indicative conditionals such as (3). To insist on still dismissing (1) as false is to ignore the lessons of mathematical experience.

Of course, the hybrid interpretation discussed in section 6.1 makes (1) and (3) contradictions. However, we have already found ample reason to reject it, not least because it still leaves us with the task of explaining why many indicative conditionals whose consequent is barely compatible with the antecedent (such as (2) above) strike us as almost certainly false when they are in fact almost certainly true. We now have another reason to reject the hybrid interpretation: it makes informal arithmetic *inconsistent*, where informal arithmetic is simply arithmetic as standardly done by mathematicians in English. For (3) is a lemma of informal arithmetic, and the hybrid interpretation makes (3) a contradiction. On the uniformly material interpretation of indicative conditionals, we can still explain the apparent falsity of (1) and (2), and even of (3) as naïvely considered, by our reliance on the Suppositional Rule as our primary heuristic for assessing indicative conditionals.

Just as the Suppositional Rule creates illusions of falsity, it also creates illusions of truth. When a true conditional appears false, its false negation appears true. In particular, on the material reading, conditionals have surprisingly weak truth-conditions, so their negations have surprisingly strong truth-conditions. For example, (4) is truth-conditionally equivalent to (5):

(4)   It is not the case that if there is a god, all your prayers will be answered.
(5)   There is a god and not all your prayers will be answered.

An atheist might confidently assert (4), but deny (5) for its first conjunct: the truth-conditional equivalence of (4) and (5) is obviously not obvious. Presumably, the atheist has applied the Suppositional Procedure to (6):

(6)   If there is a god, all your prayers will be answered.

Even on the supposition that there is a god, the atheist rejects the consequent that all the addressee's prayers will be answered; after all, different theists pray for mutually incompatible futures. By the Suppositional Procedure, the atheist rejects (6), and so asserts its negation (4).

Admittedly, there is a gap between *rejecting* something and *denying* it, in the sense of asserting its negation. On the supposition that there is a god, is the atheist really willing to assert that *not* all the addressee's prayers will be answered? Indeed, the theist might object to the denial of (6) by saying 'They might be!' (on the supposition that there is a god). At this point, the atheist might reasonably retreat from denying (6) to just rejecting it on epistemic grounds, which was really what the Suppositional Procedure supported all along. Such confusion about negated conditionals may explain why they are so rare in mathematical practice. In any case, even this more circumspect atheist will confidently reject (5) without confidently rejecting (4): this gap between the assessments of (4) and (5) remains, and the Suppositional Rule explains it.

Chapters 7 and 8 use the Suppositional Rule to explain many more examples in which the material reading seems at odds with our pre-reflective use of 'if'.

6.4 The practice of using plain conditionals

A fruitful reaction to the material interpretation of indicative conditionals is to ask why we should have a practice of relying on the Suppositional Rule when it is so liable to give incorrect results. Has an account on which the indicative conditional is truth-functional the resources to explain our practice? It is not very plausible that our ordinary use of indicative conditionals simply happens to have a major dysfunctional feature. The practice is too central to our cognitive lives for that. After all, our use of 'and', 'or', and 'not' as more or less sentential operators fits their interpretation as truth-functional conjunction, disjunction, and negation quite well (though with some subtle complications). Why should it be so much harder with the conditional?

Much of our knowledge of the world is knowledge of connections. I know that tigers are dangerous. I hear a rustle in the undergrowth. I have no idea whether it is a tiger. I have very little idea even how likely it is to be a tiger. But I do know that *if* it is a tiger, it is dangerous. Again, a characteristic feature of decision-making is that one does not yet know what one will do. One must think through the options. Typically, that involves assessing what will happen *if* one takes a given option. Sometimes one may need to use subjunctive conditionals, but often we use indicative conditionals.[5] Shall I leave this shelter and head home? If I do, I shall get wet, but I shall also get home before nightfall. In such cases, we need a way of testing a conditional, using our background knowledge, *without* testing its antecedent or consequent independently of the other. The Suppositional Rule is such a way. It is well adapted to cases where we are not (yet) in a position to test the antecedent and consequent independently. In

[5] Causal decision theory uses subjunctive rather than indicative conditionals to deal adequately with Newcomb problems.

probabilistic terms, our conditional probabilities are sometimes better informed than our unconditional probabilities; the Suppositional Procedure is designed for such cases.

As Edgington writes (1995: 266–7):

> Humans are not endowed with complete belief-distributions over the finest partitions they need to consider. They need to work out some degrees of belief (as the need arises) in terms of others which are more readily accessible. $b(B|A)$ can be accessible en route to $b(A \ \& \ B)$, and can be accessible when $b(A)$ is not (for humans, fortunately, are capable of supposing).

For example, when the conditional probability is low, the probability of the conjunction is also low, no matter how high the probability of the first conjunct.

As already emphasized, the Suppositional Rule is as vital to the most sophisticated reasoning as it is to more instinctive and apparently naïve thinking. In order to test a scientific theory by eliciting predictions from it, one reasons on the supposition of the theory. In rough schematic terms, the upshot is a conditional of the form 'If theory T is true and experiment E is done, the result will be R.' Obviously, when the conditional was derived, scientists were not yet in a position to know the truth-value of the antecedent or of the consequent: the point of the derivation and then the experiment is to help them find out. The use of variables in reasoning typically involves conditionals whose antecedent and consequent cannot be separately assessed. For example, in a mathematical proof, one can assert 'If $x^2 < 25$, $x < 5$' when the value of 'x' may be any real number, so it would make no sense to try to find out whether '$x^2 < 25$' really is true. Pronouns in natural language can produce similar effects, as in the generic advice 'If it is a tiger, it is dangerous', uttered before any particular animal is at issue.

Although much remains to be understood about how suppositional reasoning is implemented in the human imagination, we can grasp in principle the power and efficiency of a system that somehow allows many of the cognitive capacities we use online to assess propositions non-hypothetically to be reused offline to assess propositions hypothetically. The Suppositional Procedure relies on that system. It is powerful and efficient, but it has a cost. Especially under pressure to classify conditionals as true or false, the procedure can lead us to misclassify false conditionals as true, as with (1), even when we neither rely on false background beliefs nor execute the procedure improperly. The structure of conditional probability makes that virtually inevitable. Most powerful and efficient heuristics are fallible, because they take shortcuts. What is so mysterious about that?

Here is a possible response. It is not mysterious that we should use a fallible heuristic to assess conditionals. What is mysterious is that, even when its verdicts get us into trouble, we should still find them, on reflection, compelling. After all, when we apply the availability heuristic, and estimate the frequency of names by their familiarity, we have no deep difficulty in understanding how statistics may prove our estimates wrong. After we judge that the double-fletched line is longer than the double-tipped line in the Müller-Lyer illusion, we learn by measurement that they are the same length, and we correct our judgement accordingly, even though one line continues to *look* longer than the other. Why does nothing similar happen with the Suppositional Procedure, if it is really leading us into error?

That we use a fallible heuristic does not by itself guarantee that some way of correcting its errors is available to us. After all, brains in vats have no way of correcting their belief that they are not brains in vats. But, the objector may insist, if native speakers never correct the errors supposedly induced by the Suppositional Procedure, what reason have we as theorists to think that there are such errors? Would it not be methodologically better to give 'if' a more charitable interpretation on which systematic errors are not imputed?

The trouble is that there is no such interpretation. Our primary heuristic for assessing indicative conditionals, the Suppositional Rule, is implicitly inconsistent both with itself and with our secondary heuristic, our cognitively efficient practice of freely passing such conditionals from one context to another, by memory and testimony. Our practice with indicative conditionals is deeply paradoxical. In that respect, it is like our practice with the terms 'true' and 'false', which is deeply paradoxical in the presence of the Liar and other semantic paradoxes. Perfect charity is not an option.

Furthermore, native speakers sometimes *do* correct errors induced by the Suppositional Procedure. As observed in section 5.1, they can inhibit the dispositions to judgement it induces in favour of other epistemic sources treated as more authoritative, such as expert testimony and mathematical proof. Rather, what native speakers as such lack is any systematic or theoretically justified way of arbitrating such disputes. They have to fly by the seat of their pants.

Given the deep paradoxicality of human practice with conditionals, it is no surprise that their semantics has been the subject of controversy since ancient times. In the third century BCE, the poet and scholar Callimachus wrote 'Even the crows on the rooftops are cawing about the question "Which conditionals are true?"'. By then, Philo had already defended the truth-functional interpretation of conditionals, while Diodorus Cronus proposed a stricter interpretation, on which a conditional is true only if the conjunction of its antecedent and the negation of its consequent was always impossible: a strict conditional interpretation. Both ancient and modern debates make clear that *no* semantic account is accessible to ordinary reflection. What such reflection can provide are various considerations and bits of evidence, some of which might seem decisive when taken in isolation: but taken all together they are deeply puzzling. We need a theory to make sense of this ragbag of data. Rival theories are judged against each other by their comparative explanatory power. Given the internal tensions in our practice, different theories will locate our errors at different points. Since that is the situation, it would be quite unrealistic to expect ordinary speakers to be able to tell just where they are going wrong.

As we saw in section 3.2, applying the Suppositional Rule to deductive attitudes provides what, in isolation, looks like a conclusive case for the material reading of 'if', since it delivers standard introduction and elimination rules for the material conditional, in effect *modus ponens* and conditional proof, which together enforce equivalence to the material reading. Of course, once confronted with apparent counterexamples to the material reading, we need to go much deeper. Mere appeals to *modus ponens* and conditional proof themselves will not explain what is going on in such cases. Of course, the point cuts both ways: semantic accounts of 'if' that invalidate conditional proof or *modus ponens* must explain why they look so compelling, even on reflection.

A more general methodological concern with the postulation of basic fallible heuristics for assessing sentences, in particular the Suppositional Rule, is what they do to the empirical status of truth-conditional semantic theories for natural languages. Such theories are intended to explain the data of language use, and thereby to be confirmed or disconfirmed. The fear is that the postulated fallible heuristics will take over the role of explaining the data, leaving the truth-conditional semantics with no explanatory task to perform. That would leave the truth-conditional semantics not just empirically unconstrained but redundant, an idle wheel.

That fear is an overreaction. It depends on a simplistic understanding of the relation between theory and evidence in semantics. Admittedly, the recognition of such fallible heuristics contributes to the indirectness of the relation. But it is in any case quite indirect. Even in the less complex case of 'and', the raw data are extremely messy. For example, in the well-documented conjunction fallacy, many people judge a conjunction more probable than one of its conjuncts, in defiance of an elementary principle of probability theory. Is that evidence against the standard truth-functional interpretation of the relevant uses of 'and'? It cannot simply be dismissed as irrelevant, but in general we should not envisage the cognitive assessment of a sentence as anything like isomorphic to its truth-conditional meaning. Even in the case of conjunction, we sometimes reject a conjunction without rejecting any conjunct, even though the conjunction is false if and only if at least one conjunct is false. For example, if you have no idea how tall Maria is, you can still reject the conjunction 'Maria is at least six foot tall and she is less than six foot tall', because it is inconsistent, although you reject neither conjunct. The gap is even greater for universal quantification. The universal generalization 'No grain of sand is larger than a football' is true if and only if no member of the relevant domain as a value of the variable 'x' makes 'x is a grain of sand' true and 'x is larger than a football' false. But we do not assess the generalization by going through all its instances in the domain and assessing each of them separately.

For cognitive and computational reasons, we need epistemic methods to mediate between the bare semantics and our judgements of truth-value about particular examples. That is so not only when the sample sentences are encountered 'in the wild', but also when they are embedded in stipulated fictions. In the latter case, we are carrying out a miniature thought experiment (Williamson 2007a). That involves applying some of the same cognitive capacities offline, in imagination, that we might apply online if we encountered a similar sentence in a real-life analogue of the situation in which we imagined the sample sentence uttered. In effect, as theorists we are applying the Suppositional Procedure to assess what status the sentence would have were it uttered in the imagined circumstances. The issue is not whether we need methods to do that—of course we do—but whether they are fallible or infallible. Given the general nature of human cognition, it would be astonishing if they all turned out to be infallible. We have no realistic hope of dissolving the methodological problems posed by language users' reliance on fallible heuristics. Instead, we should confront those problems, and do our best to solve them.

So what role is left for the semantics to play, since it is ill-suited to play that of the heuristics themselves? A key feature of the heuristics is that they form a mixed bunch. Specifically, for conditionals, there is the Suppositional Rule, but also reliance on trusted testimony. As we saw, their relations are sometimes tense, but achieving a proper balance between them is crucial to the cognitive value of our use of conditionals. Thus a natural question arises: what holds the mixed bag of heuristics together? We cannot expect a mere random assortment to have much cognitive value. This is the natural point for the truth-conditional semantics to enter, and explain how the mixed bag is unified as a practice of using given kinds of sentence to express, retain, and communicate given kinds of information. For indicative conditionals, exactly that was done by the argument above that the material conditional [A] ⊃ [C] is almost always the strongest proposition to underpin [C] on [A] probabilistically. That shows how the Suppositional Rule and reliance on conditional testimony are held together by their complementary roles in the acquisition, retention, and communication of material conditional information. Without such a truth-condition, the practice falls apart.

On this view, the truth-conditional semantics plays a central role in explaining linguistic practice, but at a high level. Although it is not a component of the mechanics, we need it to understand what the mechanics are designed *for*. Thus the semantics is by no means immune to testing by the data of natural language use. Its support comes from those data, such as the evidence that estimates of the probability of indicative conditionals tend to correlate with estimates of the conditional probability of the consequent on the antecedent. It is just that the testing does not take a naïve falsificationist form, because native speakers are not assumed infallible. It takes a more sophisticated abductive form typical of developed sciences with a large gap between theory and observation.

6.5 Semantic equivalence and cognitive equivalence

A case has been built for the simple material interpretation of the conditional. Quite generally, 'If A, C' is true just in case either A is false or C true, and it is false just in case A is true and C false. Thus 'If A, C' has the same *extension* as 'Either not(A) or C'. Indeed, they have the same *intension*, since the sameness in extension holds with respect to any world and time. Thus, in David Kaplan's terminology, 'If A, C' and 'Either not(A) or C' have the same *content* (at least by a coarse-grained standard). They also have the same *character*, since the sameness in content holds with respect to any context of utterance. Consequently, if linguistic meaning is equated with Kaplan character, 'If A, C' and 'Either not(A) or C' have the same linguistic meaning.

But 'If A, C' and 'Either not(A) or C' do not have the same *cognitive significance*. Their sameness of content or character is not cognitively available to be known on straightforward reflection by a normal, intelligent, native speaker of the language. Such a speaker might easily be baffled by the question, or come down against the material interpretation on the basis of serious reasons. The case for the material interpretation, as developed in this book, is highly theoretical. It is the outcome of a century-long tradition of research on conditionals, and the central dispute it addresses goes back more than two millennia. It applies the mathematical techniques of probability theory and formal semantics, and it is influenced by ideas from cognitive science and cognitive psychology. The argument is fundamentally abductive: the material interpretation is integral to the best explanation of our complex, messy, and inconsistent practice of using the conditional. Given our limited cognitive resources, our practice does a good job of acquiring, storing, and communicating the sort of information expressed with the material conditional—a better job than it does of fulfilling that function on any other hypothesis as to the information at issue. Of course, any rival hypothesis as to the semantics of 'if' would also have to be defended in the same highly theoretical way.

The main cognitive differences between 'If A, C' and 'Either not(A), or C' are interpersonal, systematic, and broadly predictable. On the present view, they originate in the fact that the Suppositional Rule is central to our use of 'if', and marginal to our use of 'or'.

The difference cannot be explained in terms of Grice's theory of implicatures, conventional or conversational. Differences of implicature are normally far more accessible to native speaker reflection. Moreover, implicatures are supposed to be pragmatic phenomena, arising at the level of language as a medium of communication, whereas the key cognitive difference between 'If A, C' and 'Either not(A), or C' is already in full force at the level of thought, as Edgington (1995) has rightly emphasized. Indeed, the basic Suppositional Rule may well be a heuristic for conditional *thought*, to which the word 'if' subsequently becomes attached.

For such reasons, the difference is not one of conventional implicature. A normal native speaker of English in possession of a lottery ticket (after the draw, before the announcement) will confidently reject as absurd the thought 'If my ticket won, it lost', while being almost certain of the (redundant) thought 'Either my ticket didn't win or it lost.' It is nothing like one's position with respect to 'Bertrand Russell was a philosopher but English', which one accepts as historically accurate, because he was both a philosopher and English, whatever one thinks of the contrast, conventionally implied by the choice of the word 'but', between being a philosopher and being English. The cognitive difference between 'If A, C' and 'Either not(A) or C' is not a difference in conventional implicatures.

Nor is the cognitive difference between 'If A, C' and 'Either not(A) or C' a difference in conversational implicatures. They are normally *detachable*, in the sense of being shared between truth-conditionally equivalent statements. For example, saying 'The professor was sober today' conversationally implicates that the professor is often drunk, and a logically equivalent statement will typically conversationally implicate the same thing. But 'If A, C' and 'Either not(A) or C' are truth-conditionally equivalent by hypothesis, so they have the same detachable conversational implicatures. Admittedly, some conversational implicatures are generated not by *what* one says but by *how* one says it, and so are non-detachable: for example, one can speak pompously in reporting someone's views to implicate that he is pompous; a non-pompous but truth-conditionally equivalent report of the same views might lack that implicature. However, the cognitive difference between 'If A, C' and 'Either not(A) or C' is not like that either (this point is reinforced below). Thus, despite its many achievements, Grice's theory of implicatures is not what is needed here.

In general, what primarily differentiates conditionals is not *pragmatic* but *cognitive*. But is it also *semantic*?

Many philosophers and linguists interpret the large cognitive differences between 'If *A*, *C*' and 'Either not(*A*) or *C*' as amounting to a difference in linguistic meaning. Since the two sentences have the same Kaplan character, the obvious conclusion is then that there is more to linguistic meaning than Kaplan character.

A cheap move is to appeal to the compositional semantic difference between 'If *A*, *C*' and 'Either not(*A*) or *C*': the latter has a constituent with the meaning of 'not(*A*)'; the former has not. Similarly, if one seeks a semantic difference between 'If *A*, *C*' and another of its truth-conditional equivalents, 'Not(*A* and not(*C*))', one can point to the fact that the latter has a constituent with the meaning of 'Not(*C*)'. That move is cheap for at least two reasons.

First, it does not explain the cognitive inaccessibility of the identity in truth-conditions. After all, 'Either not(*A*) or *C*' and 'Not(*A* and not(*C*))' have the same truth-conditions as each other while differing from each other in a compositional semantic way, but in their case the identity in truth-conditions is cognitively far more easily accessible, to a little reflection by an intelligent native speaker.

Second, the move depends on a non-mandatory choice of truth-conditional equivalent. A sentence can present the material truth-conditions of 'If *A*, *C*' more perspicuously without need of complex compositional structure. For one can explain the symbol ⊃ by giving its truth-table, and then simply use '*A* ⊃ *C*'. The truth-conditional equivalence of 'If *A*, *C*' to '*A* ⊃ *C*' is as inaccessible to ordinary speakers as its truth-conditional equivalence to 'Either not(*A*) or *C*', without a semantic difference in their constituents. 'If *A*, *C*' and '*A* ⊃ *C*', so explained, still differ cognitively, since the Suppositional Rule is central to the use of the former and need play no role in the use of the latter (when *A* and *C* themselves are conditional-free).

The temptation is to redescribe the cognitive difference between 'If *A*, *C*' and '*A* ⊃ *C*' as constitutive of a semantic difference. Doing that in any systematic way would require a framework of some sort of epistemically inflected semantics, perhaps in terms of verification conditions, falsification conditions, assertibility conditions, inferential rules, or whatever. It is notoriously unclear how to get beyond programmatic remarks and the odd example to a credible working semantics in epistemic terms for a significant fragment of natural language. In particular, real-life evidential relations tend to be holistic in a way inimical to attempts to craft a systematically compositional verificationist semantics, determining verification-friendly semantic values for complex expressions in terms of their structure and the verification-friendly semantic values of their constituents.

At first sight, the Suppositional Rule might look like the beginnings of a solution for the case of conditional sentences. On closer examination, however, things are less promising. It is not just that the Rule is inconsistent in some cases. Even when there is no inconsistency, it often leaves vast discretion to the individual's imagination in how the initial supposition of the antecedent is developed. It is also often up to the individual's discretion whether to apply the Rule or instead to rely on memory or testimony; as we have seen, the results can be quite different. Moreover, as we have also seen, the reliability of the Rule depends on our tendency to avoid direct negations of conditionals, even though such negations are by no means forbidden:

'It is not the case that if it rains, the match will be cancelled' is good English, though somewhat unclear as to what it entails. Heuristics are not semantic rules, and we have no reason to expect to find semantic rules hidden inside them.

It might be replied that when the Suppositional Procedure is central to the use of 'if' and absent from that of ⊃, the difference is too closely tied to the understanding of the two connectives *not* to be semantic. If you were teaching English as a second language, and you introduced your pupils to 'if' only by giving them its truth-table, and they were not bright enough to connect it with the conditional in their native language, but applied it solely on the basis of labour-intensive truth-tabular computations, would they really have mastered English 'if'?

Such pupils would indeed perform very poorly with 'if' in ordinary conversation. They would appear slow-witted and sometimes perverse. But are those specifically linguistic faults, or more general cognitive ones?

Imagine a native speaker of English who often takes several minutes to make inferential connections which ordinary speakers make in a fraction of a second. As a result, normal conversation with him may be impossible. One spends most of the time waiting in silence for the penny to drop. But that is no reason to postulate something in the semantics of English which he lacks. Speed of computation is not a dimension of meaning.

Similarly, consider an exceptionally unimaginative native speaker of English, the sort of person whose response to the comment 'If she doesn't win first prize, she'll be very disappointed' is 'Why?' As we have seen, in many situations the normal human way of assessing a conditional is by using one's imagination in implementing the Suppositional Procedure. By contrast, in such situations, this unimaginative speaker tries to assess the conditional in some other way, sometimes successfully, often not. As a result, he tends to perform poorly in conversations about hypothetical questions. Anecdotally, speakers vary significantly in their capacity to handle such questions effectively. I have witnessed native speakers of English failing to engage with overtly counterfactual questions. When their interlocutor (not me) explained 'I know that it is not so, but what if it *had been* so?', they could only repeat 'But it *isn't* so', with uncomprehending obstinacy. Their tunnel vision blocked the normal development of the conversation. Yet they did not seem to lack part of the semantics of English. Rather, the problem seemed to be their limited imaginative intelligence.

Return to the pupils, laboriously working out truth-tables to assess ordinary 'if' sentences. Their problem is not ignorance of the semantics of English. Rather, it is that they are attempting to apply the semantics by conscious inference, which is exorbitantly expensive of scarce resources such as time and attention, rather than relying on quick and easy, largely unconscious processing, like normal speakers. After all, *whatever* the correct semantic theory of 'if', the attempt to use 'if' in ordinary conversation by consciously drawing inferences from that theory will lead to social disaster. Humans did not evolve to speak that way. Complicating the semantics cannot solve the problem; it only makes it worse. Unless the pupils are inordinately slow on the uptake, as they become more familiar with using 'if' they will start taking shortcuts, and hook 'if' up with their general cognitive capacity for offline thinking, which they have independently of their knowledge of English as a second language, and which supports the capacity to apply the Suppositional Procedure.

The contrast between the normal use of 'if' and the use of ⊃ based on truth-tables is quite compatible with the material interpretation of 'if', on which they determine the same truth-conditions. The difference is cognitive rather than semantic. To say that is not at all to downgrade the cognitive aspect of language use. Rather, it is a precondition for doing that aspect justice. For an unintended consequence of the attempt to encode central features of the cognitive aspect of language use in the semantics is a distorted, diminished, and impoverished view of the cognitive aspect. That happens because conformity with the semantics is supposed to be a condition for minimal linguistic competence. Thus when an eccentric speaker displays at least minimal linguistic competence although their use lacks a cognitive feature, that feature gets excluded from the semantics, even if it is central to the practice of most other speakers with the relevant terms. The hypothetical eccentric speaker need not even exist; the mere possibility of one is enough to disqualify the cognitive feature. Consequently, what is left in the semantics is only a faint trace of the original rich cognitive practice, a trace so weak as to be cognitively almost useless.

A sobering case in point is the fate of descriptive theories of reference. A candidate description is eliminated as soon as a minimally competent speaker can consistently entertain a hyperbolical doubt whether the description might be false of the referent of the term. The ordinary, informative descriptions offered by the original proponents of descriptive theories were easily shown to fail that test. Later descriptive theorists fell back on uninformative metalinguistic descriptions like 'whatever whoever I got the term from was referring to' or intractably complex descriptions conjectured to live in the shaky framework of cognitively interpreted two-dimensional semantics. The effect of such moves was to distract attention from the most cognitively significant features of the terms at issue. For example, recognitional capacities for the referent play a central role in the practice of using ordinary proper names and natural kind terms, even though many speakers use such terms without possessing those capacities, and even in more expert hands they are not 'analytically' guaranteed to pick out the right thing or kind. There is an important subject for investigation here: cognitive aspects of our practices of using proper names and natural kind terms. Hilary Putnam made a promising start, with his remarks on the role of experts and the division of linguistic labour. But the general tendency was to marginalize cognitively central phenomena once it became clear that they were not purely semantic. A less imperialistic conception of semantics is needed to permit inquiry into cognitive aspects of linguistic practices to flourish in its own right.

Similar distortions can be observed in the study of logical constants. Both in proof theory and in less formal investigations of the epistemology of logic, the focus has been too much on 'analytic' rules of inference, even when those get restricted to a weak fragment of the logic we actually and successfully use. To achieve a more faithful understanding of the cognitive aspects of our ordinary practice of using the logical constants, we need to stop concentrating on 'analytic' rules. That applies in particular to 'if', and its associated heuristics.

The Suppositional Procedure is central to the normal practice of using a conditional, but it does not belong in a systematic compositional semantics for the language. The relation between the meaning of 'if' and its associated heuristics is less direct than that. As theorists, we arguably have to ascribe the standard truth-table to 'if' in order to best understand how the mix of heuristics used for 'if' hang together as a coherent and pointful cognitive practice. The truth-conditional semantics and the heuristics are at different explanatory levels. With respect to the truth-conditional semantics, 'If A, C' and 'Either not(A) or C' are synonymous, even though they are composed from operators associated with entirely different heuristics. As a result, their synonymy is not transparent to native speakers. Similarly, coreferential proper names and natural kind terms may be synonymous with respect to the truth-conditional semantics, even though their synonymy is not transparent to native speakers, because they are associated with different heuristics. For example, 'Hesperus' and 'Phosphorus' are associated with different recognitional capacities.

6.6 Synonymy and vagueness

Arguably, *vagueness* also generates cases of non-transparent synonymy. On an epistemicist view, the vague adjective 'small', as applied to natural numbers, has the same truth-conditional meaning as the precise expression 'at most n', where 'n' stands in for the appropriate numeral (at least for a given context). But we are in no position to know exactly which numeral is the right one, because vague and precise expressions are associated with different cognitive practices. 'At most n' belongs to the language of pure arithmetic, and so is associated with the practice of applying various algorithms for numerical calculation, whereas the vagueness of 'small' prevents us from applying those algorithms to it. Instead, 'small' may be associated with a capacity which enables us to recognize some natural numbers (for instance, 0) as small, and others (for instance, in some contexts, a thousand) as not small, but fails to determine some intermediate cases either way.

As discussed in section 3.8, many theorists have sought to understand vagueness and its tendency to generate sorites paradoxes in terms of *tolerance principles*, to the effect that if two things are similar enough in the respects relevant to a vague predicate, then the predicate applies to one of them only if it also applies to the other. Unfortunately, a finite sorites series starts with something to which the predicate recognizably applies and ends with something to which the predicate recognizably fails to apply, although any two neighbouring members of the series are as similar as the tolerance principle requires.

Despite the contradictions, tolerance principles are often claimed to be somehow built into the meanings of vague expressions, making those meanings inherently paradoxical. Competence with a vague expression may even be held to require a disposition to accept instances of the relevant tolerance principle. But even if speakers are so disposed, why regard the tolerance principles as anything more than fallible heuristics generally associated with vague expressions? As section 3.8 explained, in any given context, almost all instances of the tolerance principle are truth-preserving inferences. The point is easiest to appreciate on a classical, bivalent semantics, which epistemicism about vagueness supports, but it applies at least as strongly on non-classical semantics for vague languages, since they are designed for the express purpose of making the tolerance principle fail less drastically than it does on the classical approach (at the cut-off point). Thus the tolerance principle is a *highly* though not *perfectly* reliable heuristic. That it gets us into trouble in a sorites paradox merely shows that with sufficient ingenuity one can exploit its limitations to make trouble, which is just what one would expect of a fallible heuristic.

Since tolerance principles are so simple and reliable for one-off applications, it is no wonder that we use them as heuristics for casual judgement. It would be surprising if we did not. A one-off applications of a tolerance principle does not draw attention to the limits of its validity. Its fallibility is not transparent to native speakers, just as the fallibility of the Suppositional Rule is not transparent to them. We need paradoxes to warn us that something is wrong, and even then they do not tell us *what* is wrong. In using such heuristics pre-theoretically, we are not conscious of them as mere rules of thumb.

There is no reason why a tolerance principle for a vague term should be any more semantic than the Suppositional Rule for 'if'. From the present perspective, tolerance principles are at the wrong level to figure in a compositional semantics for a vague language. Consequently, epistemicism about vagueness has no difficulty in giving tolerance principles their due at the level at which it is due. If speakers are disposed to use them, that is no evidence at all against epistemicism.

Indeed, making tolerance principles semantic add-ons undermines their role in explaining vagueness, because such add-ons are unnecessary for vagueness. For there could be a language just like English except that no tolerance principles had been added on. Removing the add-ons would not magically make the language precise. Tolerance principles are not special semantic rules. They are just manifestations of a more general and quite efficient cognitive tendency to treat apparently similar cases alike.

7

More Challenges

7.1 Dorothy Edgington and the Queen

Our heuristics for assessing indicative conditionals are inconsistent, but the truth-functional semantics arguably does the best job of rationalizing our overall practice. That view obviously needs to be tested further. How powerful are its explanatory resources? A good way to explore that issue is by working through various proposed counterexamples and challenges to the material interpretation. This chapter takes a few steps in that direction.

We start with an example from Dorothy Edgington (1995: 243), which she uses to highlight the implausibility of the material reading:

> '(i) I think that my husband isn't home yet. But if he is, he'll be worried about where I am. So I should try to phone.' Compare (ii): 'I think that the Queen isn't home yet (at Buckingham Palace, that is). But if she is, she'll be worrying about where I am.' The first thoughts are sane enough, the second a sign of madness.

In the third person, the conditionals at issue are these:

(1) If DE's husband is home, he is worried about where DE is.
(2) If the Queen is home, she is worried about where DE is.

As Edgington observes, she may be rationally confident of both (1) and (2) on the material interpretation—in the case of (2), because it follows from the Queen's not being home, and 'having read in the newspaper of her day's engagements', she is 'about 90 oer cent certain that the Queen isn't at home yet'. She rightly emphasizes that the issue is not confined to assertive *utterances* of (1) and (2): more fundamentally, it already arises for unspoken *thoughts* with those contents. Thus appeals to the conversational or conventional implicatures of (1) and (2) miss the heart of the problem.

A salient contrast between the two conditionals is that, on DE's evidence, the conditional probability of the consequent of (1) on its antecedent is reasonably high, while the conditional probability of the consequent of (2) on its antecedent is very low. Correspondingly, the Suppositional Procedure will accept (1) and reject (2). That explains the initial cognitive asymmetry between (1) and (2).

Applying the Suppositional Procedure, DE accepts that the Queen is not at home (with about 90 per cent confidence) while rejecting (2) (with virtually 100 per cent confidence), even though the Queen's not being home entails (2) on the material interpretation. Thus the material account appears to impute what Edgington calls

Suppose and Tell: The Semantics and Heuristics of Conditionals. Timothy Williamson, Oxford University Press (2020).
© Timothy Williamson.
DOI: 10.1093/oso/9780198860662.001.0001

'an Incredibly Gross Logical Error' (1995: 244) to normal, indeed highly rational, native speakers of English. But the case looks quite different once one sees that our ordinary understanding of our native language may not always represent the truth-conditions of our sentences to us in a perspicuous form—just as it is not pre-theoretically perspicuous to us whether 'All *Fs* are *Gs*' entails 'Some *Fs* are *Gs*'. After all, *no* semantic account of indicative conditionals is obviously correct. When our understanding of them is mediated by an inconsistent heuristic such as the Suppositional Procedure, no wonder their semantic nature is not transparent to us. Thus the logical error is one which normal, highly rational native speakers of English may very easily and intelligibly commit.

Edgington has a larger worry: if the semantics of the indicative conditional obliterates the significance of conditional probabilities, it may thereby undermine the value for us of our practice with such conditionals. Her example is a case in point, since the envisaged cognitive asymmetry between (1) and (2) has practical conse-quences. For (1) is meant to motivate phoning DE's home, while (2) is meant not to motivate phoning the Queen's home. For the sake of the example, we should also assume that, in the circumstances, the desirability of trying to phone depends only on the probability that the person is home and worried about where DE is, and not also on their status and relation to DE, otherwise the practical asymmetry between (1) and (2) could be explained away on the latter grounds. Thus what finally matters for the prospective action motivated by the conditional (1) or (2) is the probability of the conjunctions of its antecedent and consequent, (3) for (1) and (4) for (2):

(3)   DE's husband is home and worried about where DE is.
(4)   The Queen is home and worried about where DE is.

Of course, if the probability that DE's husband is home is zero, then the probability of (3) is also zero, and if the probability that the Queen is home is zero, then the probability of (4) is also zero. But in the present case, the antecedents of both conditionals have positive probability. The role of the conditional is then to leverage a significantly positive probability for the antecedent into a significantly positive probability for the conjunction of the antecedent and the consequent. The standard equation $\text{Prob}([A] \cap [C]]) = \text{Prob}([A])\text{Prob}([C]|[A])$ confirms the key role of the conditional probability in that leveraging.

We also need the probabilities of (1) and (2), on the material interpretation, to be less than one. For if $\text{Prob}([A] \supset [C]) = 1$ and $\text{Prob}([A]) > 0$ then $\text{Prob}([C]|[A]) = 1$ too, so in that special case the difference between the probability of the material conditional and the conditional probability, crucial for Edgington's argument, dis-appears. That is why we should envisage DE's evidence as warranting high but not perfect confidence in the two material conditionals. That fits what Edgington says about (2): its high probability on the material interpretation is supposed to come entirely from the low probability of its antecedent; we have just seen that the latter probability must not fall as low as zero. Hence the probability of (2) on the material interpretation must be less than one. The example works best when the probability of (1) on the material interpretation is also less than one. For when (1) is certain and (2) less than certain, that asymmetry might help explain the practical differences between them.

Since the probabilities of the material conditionals are less than one in the cases of interest, so are the conditional probabilities, since a conditional probability never exceeds the probability of the corresponding material conditional. Thus, by the equation $\text{Prob}([A] \cap [C]) = \text{Prob}([A])\text{Prob}([C]|[A])$, the probabilities of the conjunctions (3) and (4) must be at least slightly lower than the probabilities of the respective antecedents, which are non-zero by hypothesis. That still leaves room for a wide divergence in probability between (3) and (4).

In the simplest case, the probability of the Queen being worried about where DE is conditional on the Queen being home is zero. In that case, the probability of (4) is automatically zero, irrespective of the probability of her being home, so the practical question of trying to phone Buckingham Palace does not arise. By contrast, the probability of DE's husband being worried about where DE is, conditional on DE's being home, is fairly high, and the probability of his being home is also positive, so the probability of (3) is not zero. Thus the practical question of trying to phone home does at least arise. That practical asymmetry turns on the difference in conditional probability. Is that not enough to vindicate Edgington's argument?

The problem is that, as we have already seen, independently of the material interpretation, rational acceptance of an indicative conditional does *not* always correlate with the conditional probability of its consequent on its antecedent. In the example from section 5.1, the controller rationally accepts (5) on East's testimony and, simultaneously, (6) on West's testimony:

(5)   If the detector is working, the core is overheating.
(6)   If the detector is working, the core is not overheating.

The conditional probabilities corresponding to (5) and (6) cannot both be high. The controller does not apply the Suppositional Procedure to (5) and (6) herself. Instead, she rationally trusts her trustworthy informants, and assigns high probability (perhaps just less than one) to both (5) and (6). As a consequence, she also assigns high probability to the hypothesis that the detector is not working.

The original example was symmetrical between East and West, and so between (5) and (6): the two engineers were equally trustworthy, and equally trusted. We can tweak the example so that the controller rationally trusts East slightly more than West. She is 100 per cent confident of (5), but only 99 per cent confident of (6) (the same point can be made with a less extreme value for (5), but this stipulated case is more vivid). Her probability for the detector not working is less than 100 per cent, and her probability for the core overheating conditional on the detector working is 100 per cent, so her probability for the core *not* overheating conditional on the detector working is 0 per cent. Thus she rationally has very high confidence in (6) while assigning its consequent a probability of zero conditional on its antecedent.

Similarly, DE's rational confidence in (1) does not guarantee that her conditional probability for the consequent on the antecedent is high. Of course, one strength of Edgington's example is that we do not naturally imagine DE basing her beliefs about her husband's hypothetical feelings on others' testimony. For such conditionals about people we know well, we naturally use the Suppositional Procedure.

However, we can science-fictionally reimagine the example as concerning *geeks*, who have a technologically advanced culture but unfortunately lack a mindreading faculty. However, at home DE keeps two amazing, elaborate, and highly reliable machines. When DE's husband is home and worried about where DE is, machine 1 detects it (on the basis of subtle neurophysiological cues). When DE's husband is home and *not* worried about where DE is, machine 2 detects it (on the basis of subtle neurophysiological cues). Both machines send DE monitoring reports. On good evidence, DE has perfect confidence in machine 1's reports and almost perfect confidence in machine 2's reports. In the present case, machine 1 sends DE this report:

(7)   If DE's husband is home, he is not worried about where DE is.

Meanwhile, machine 2 sends DE the report (1), the opposite conditional to (7). Since DE rationally has perfect confidence in machine 1, her rational credence in its report (7) is 100 per cent. Since she rationally has almost perfect confidence in machine 2, her rational credence in its report (1) is 99 per cent. As a consequence, her rational credence that her husband is not home yet is at least 99 per cent, but less than 100 per cent. Since her credence in (7) is 100 per cent, her credence in (3) ('DE's husband is home and worried about where DE is') is 0 per cent. But since she has a small positive credence in the hypothesis that her husband is home, she also has a small positive credence in the conjunction (8):

(8)   DE's husband is home and not worried about where DE is.

These credences require a credence of zero for DE's husband being worried about where DE is conditional on his being home. In these circumstances, DE rationally has very high credence in (1) while rationally assigning its consequent a probability of zero conditional on its antecedent.

Needless to say, that variant of the example is wildly fanciful. But it shows how the cognitive role of the conditional (1) in Edgington's case depends on background conditions which go far beyond its semantics: in particular, on the implicit assumption that the agent assessed (1) by means of the Suppositional Procedure. When those conditions were removed, it no longer played that role. Since the present account explains why those conditions typically hold in such cases, using the resources already developed, it is not under threat from the example.

7.2 Conditional commands and questions

'If' occurs naturally in both commands and questions. Such occurrences have been used against the material interpretation. We will briefly consider the problems here. Of course, a full discussion of the issues would have to situate them with respect to general theories of the semantics and pragmatics of imperatives and interrogatives. More specifically, the semantics of conditional imperatives should be derived from the semantics of the conditional together with the semantics of the imperative, and the pragmatics of conditional imperatives should then be derived from their semantics together with general principles of pragmatics. Similarly, the semantics of conditional interrogatives should be derived from the semantics of the conditional together with the semantics of the interrogative, and the pragmatics of conditional interrogatives should then be derived from their semantics together with the general principles of pragmatics. Those tasks are far too large to fit into this book. This section has the more limited aim of rebutting some specific objections to the material interpretation involving these forms of speech. Since the explanatory resources of the material interpretation have been drastically underestimated for conditional declaratives, one may suspect that they have been drastically underestimated for conditional imperatives and interrogatives too.

We start with imperatives. When you are told 'Pay cash', you comply, in the sense of doing as you were told to do, just in case you pay cash. Similarly, when you are told 'Pay cash if you have it', you comply, in the same sense, just in case you pay cash if you have it. In other words, on the material interpretation of the indicative conditional, you complied just in case either you pay cash or you do not have (enough) cash. More generally, when one is told 'Do φ if C', one complies just in case either one does φ or it is not the case that C. Thus the material interpretation has a natural and plausible generalization to conditional commands, treating them simply as commands with a conditional content.

Michael Dummett endorses such an account for conditional commands whose antecedent is within the agent's power (1959; 1973: 340). His example is the dutiful child told 'If you go out, wear your coat', who cannot find his coat and, in order not to disobey, stays in. The restriction to cases where the antecedent is within the agent's power is unnecessary. Whether one has enough cash may not be within one's power.

Dorothy Edgington has argued against the material interpretation of conditional commands. Her first example is this injunction:

(9)   If you write the article, submit it to *Mind*.

She denies that (9) amounts to the material reading (10):

(10)   Either don't write the article or submit it to *Mind*.

Her reason is that 'you could easily make that true in ways which would please me least of all' (1995: 288). Presumably, she means by not writing the article at all. But when you tell me to do something, you do not commit yourself to indifference as to *how* I do it. There are almost always tacit constraints and rankings. For instance, one could comply with the letter but not the intended spirit of (10) by writing the article with some deliberate but well-hidden mistakes, and submitting it to *Mind* under the name of one's hated rival. Such tacit constraints and rankings matter for simple commands too. The coach may tell a player:

(11)   Take the opposition by surprise!

The player could comply with (11) by scoring a gratuitous own goal, but that is no objection to the literal reading of (11).

Admittedly, the grammatical symmetry in (10) between the two disjuncts might be taken to conversationally imply the speaker's indifference as to which clause is realized, while the grammatical asymmetry in (9) between the main and subordinate clauses might be taken to conversationally imply the speaker's preference for the main clause to be realized. But such Gricean implicatures can easily be cancelled by the conversational context. The speaker of (10) may already have made it very clear that her overwhelming priority is for the procrastinating addressee to write the article. Equally, the grudging speaker of (9) may already have made it very clear that her overwhelming priority is for the headstrong addressee *not* to write the article, since its bizarre conclusion may do his reputation lasting damage, which might be partially alleviated by its publication in a high-prestige journal. Edgington's considerations do not establish any difference in semantic content between (9) and (10).

Edgington's other main example against the material interpretation of conditional imperatives is similar in form:

> If, in the emergency ward, you're told 'If the patient is still alive in the morning, change the drip', and you smother the patient, you can hardly claim to have merely carried out an order. (1995: 290)

> On this [material] analysis I command 'Make it the case that either the patient is not alive in the morning, or you change the drip'. You obey my command if you kill the patient. (1995: 301)

These comments ignore the background constraint that nurses are not to kill or harm patients. After all, when several patients are in question, the command is likely to take a plural form such as (12):

(12)   Change the drip for every patient who is still alive in the morning.

Now (12) is conditional-free. The homicidal nurse could comply with the letter of (12) by smothering all the patients overnight, but that is no objection to the literal reading of (12). Nurses are assumed to know and respect the background constraints. Thus Edgington's second example is equally impotent against the material interpretation.

On Edgington's alternative account, 'Do φ if C' is a conditional command in the strong sense that when it is not the case that C, nothing is commanded; when C, doing φ is commanded. Thus there is no command to comply with when the condition fails. That is not the natural view. The child who stays in because he cannot find his coat is naturally understood to have complied with the instruction 'If you go out, wear your coat.' In Edgington's main examples, the addressee is naturally understood to have been instructed what to do. That is so irrespective of whether the article is actually written or the patient is actually still alive in the morning. 'Instruct' here is just an appropriate, less military-sounding variant of 'command'. Even more idiomatically, the addressee has been *told* what to do. As a result, even when the condition is not actually met, the addressee still *knows* what to do.

In thinking through cases, one complicating factor is variation in how loosely or tightly the verb 'do' is being used. In a tight sense of 'do', a 'No Smoking' sign is not an instruction to do anything; it is only an instruction *not* to do something: smoke. By contrast, in a loose sense of 'do', the sign is an instruction to do something: not smoke. Correspondingly, loose action types are closed under negation; tight action types are not. Similarly, loose action types are closed under conditioning while tight action types are not. For example, *evacuating the building* is an action type in both the tight and loose senses. *Evacuating the building if the fire alarm sounds* is an action type in the loose sense, but not in the tight one. Thus, in saying 'Do φ if C', the instructor unconditionally commands a loose action type. In that sense, even when it is not actually the case that C, the addressee has still been instructed what to do.

The material interpretation is well able to explain the phenomena of conditional imperatives. There is no need to invoke the complications of conditional commands in Edgington's special sense.

Conditional interrogatives such as (13)–(16) raise similar questions to conditional imperatives:

(13)   Who, if anyone, married Edward Heath?
(14)   If I write the article, where should I submit it?
(15)   Will the match be cancelled if it rains?
(16)   Are there angels, if the Bible is to be believed?

On a conditional speech act account of conditional questions similar to Edgington's conditional speech act account of conditional commands, no question is asked unless the antecedent condition is met. Thus the speaker asks no question with (13) if Heath did not marry, with (14) if the article is never written, with (15) if it does not rain, and with (16) if the Bible is not to be believed. That is not a very natural view of such cases. Let us take them in turn.

Compare (13) with (13–):

(13–)   Who married Edward Heath?

In some sense, (13–) presupposes that someone married Heath. Since no one did, (13–) suffers presupposition failure. One might say that the question 'does not arise'. The obvious reason for inserting 'if anyone', as in (13), is to guard against such presupposition failure: that is, to ensure that the question *does* arise even if no one married Heath. Thus conditioning the question has the reverse effect to what the conditional speech act theory would lead one to expect. When the condition fails, the question with the conditional clause still flies, while the question without that clause crashes.

There is a similar contrast between (13) and its unconditional variant (13–):

(14–)   Where should I submit the article?

In some sense, (14–) presupposes that the article will be written. If it is not, (14–) suffers presupposition failure. One might say that the question 'does not arise'. The obvious reason for inserting 'if I write the article', as in (14), is to guard against such presupposition failure: that is, to prevent the question from crashing if in fact the article is never written. Again, conditioning the question has the reverse effect to what the conditional speech act theory would lead one to expect.

Issues of presupposition failure are less salient for (7) and its unconditional variant:

(15–)   Will the match be cancelled?

However, unless the qualification 'if it rains' is irrelevant in (15), the answer to (15–) may depend on the weather. A respondent who knows that it will probably not rain may answer 'Yes' to (15) and 'Probably not' to (15–). In any case, whether or not it rains, (15) expects an answer, so presumably a question has been asked.

The right answer to (16) is clearly 'Yes': according to the Bible, there are angels. You can give that answer whether or not you hold that the Bible is trustworthy. In my case, I am quite confident that the Bible is not in general to be believed. Nevertheless, by an application of the Suppositional Procedure of a sort for which it is reliable, I am also confident that *if* the Bible is to be believed, there are angels. I give that answer to (16) without prejudice to my answer to the unconditional question:

(16–)   Are there angels?

Of course, I also hold that if the Bible is to be believed, there are no angels, because I am confident that the antecedent is false and I accept the material interpretation of the conditional. But it would be conversationally misleading to give the latter conditional as an answer to (16), for in line with the Suppositional Rule, the questioner presumably wants to know whether there are angels *according to* the Bible, a question left open by the falsity of the antecedent.

There are significant differences between the expected forms of the answers to the conditional interrogatives (13)–(16). For (13), the respondent is presumably expected to know (or guess) whether Heath was married, and to give an *unconditional* answer—either that he was not, or who married him. For (14) and (15), by contrast, the respondent is expected to be unable to determine whether the condition holds, and so to be confined to a *conditional* answer, with the same antecedent as the question. For (16) too, a conditional answer is expected, though in that case the respondent's ability or inability to determine whether the condition holds is irrelevant.

Surely much remains to be said about conditional interrogatives. But the examples above are inconsistent with the conditional speech act view, on which the antecedent states a precondition for a question to have been asked. As with conditional imperatives, the resources available to the material interpretation are quite sufficient for a simpler explanation of the phenomena.

The idea that imperatives and interrogatives with 'if' are used to perform conditional speech acts, in the strong sense that nothing has been commanded or asked unless the antecedent is true, goes naturally with the corresponding idea about declaratives with 'if': that they too are used to perform conditional speech acts of assertion, in the strong sense that nothing has been asserted unless the antecedent is true.[1] Philosophers who take that line may appeal to an analogy between indicative conditional statements and conditional bets.[2] Consider (17) and (18):

(17)   If the race is not cancelled, Whistlejacket will win.
(18)   I bet that if the race is not cancelled, Whistlejacket will win.

Normally, if the race *is* cancelled, the bet in (18) is off: I neither win nor lose a bet. The analogy may suggest that if the race is cancelled, (17) is neither true nor false, because no statement has been made, no proposition expressed. But if the race is *not* cancelled, then the bet is on: if Whistlejacket wins the race, I win my bet; if

---

[1] Belnap 1970 implements a version of this idea in a kind of three-valued logic. Such approaches are discussed in section 8.2.

[2] See Dummett 1959 for discussion.

Whistlejacket loses, I lose. On that analogy, if the race is run then if Whistlejacket wins, (17) is true; if Whistlejacket loses, (17) is false. Either way, if the race is not cancelled, a statement has been made, and a proposition expressed. More generally, if the antecedent of an indicative conditional is false, then the conditional as a whole expresses no proposition; if the antecedent is true, then the conditional as a whole expresses whatever proposition its consequent expresses. To make the analogy between conditional bets and conditional assertions: when one states 'If A, C', if A is false, one has made no assertion; if A is true, one has asserted C.

On the conditional speech act account, classical logic ensures both CEM (conditional excluded middle) and CNC (conditional non-contradiction) hold whenever a proposition is expressed, for then those principles reduce to the unconditional principles of excluded middle and non-contradiction respectively. In that respect, the conditional speech act account is in line with the Suppositional Rule. However, its endorsement of CNC leads it into trouble with the argument of section 5.1. In the example there, the controller accepts a pair of opposite conditionals, even though on the conditional speech act account either they fail to express propositions or they express mutually contradictory propositions.

An amusing consequence of the conditional assertion proposal is that one can assert an epistemically risky proposition by stating what looks like a trivial logic truth, such as (19):

(19)   If there is life in other galaxies, there is life in other galaxies.

I utter (19). Suppose that there is no life in other galaxies. Then, on the proposal, I have asserted nothing. But suppose instead that there *is* life in other galaxies. Then, on the proposal, I have asserted that there is life in other galaxies. But I do not know that there is life in other galaxies. I have no evidence either way. The proposition then expressed is epistemically risky. Similarly, I might state (20) before the die is cast:

(20)   If the die comes up six, the die comes up six.

If the die does come up six, then, on the proposal, I asserted that the die would come up six. Yet, on my evidence at the time I made the assertion, the probability that the die would come up six was only 1/6. In such cases, what I asserted was true, but I did not know that it was true. Of course, friends of the proposal may take themselves to know that the conditional assertion account is correct, and so to know that in uttering sentences such as (19) and (20) they make assertions only when they make true assertions, although they do not know the truth of what they then assert. There is no risk of making a *false* assertion by uttering (19) or (20). Nevertheless, it is like putting random propositions in a message for posterity, knowing that my guardian angel will delete all the false ones at my death—although in that case posterity will reap the benefit of the remaining true ones.

As we shall see in section 8.2, such a view of conditional assertions has serious difficulty in explaining their interaction with universal quantification. In any case, we have already found the conditional speech act approach doing less well than the straightforward material interpretation in handling non-declaratives. It is an inadequate solution to an illusory problem.

7.3 Interactions with epistemic modals

Epistemic modals such as 'must' and 'may' (or 'might') pose another serious challenge to the material reading of 'if'. They interact with 'if' in ways hard to understand on that reading. Anthony Gillies isolates two plausible-sounding principles about such interactions both of which the material reading violates, as do many other readings (2010: 15–16). We can formulate them thus, without committing ourselves to the relative scope of 'if' and the epistemic modals:[3]

(P1)   'if A, must C' has the same truth-conditions as 'if A, C'.

(P2)   'if A, may C' has the same truth-conditions as 'may(A and C)'.

For instance, there is no very obvious potential reason for accepting (21) while rejecting (22), or for accepting (22) while rejecting (21) (with all occurrences of 'he' coreferential):

(21)   If he is in Moscow, he is guilty.

(22)   If he is in Moscow, he must be guilty.

Similarly, there is no very obvious potential reason for accepting (23) while rejecting (24), or for accepting (24) while rejecting (23):

(23)   If he is in Paris, he may be innocent.

(24)   He may be in Paris and innocent.

To assess P1 and P2 on the material reading of 'if', we first have to parse the sentences quoted on their left-hand sides: is 'if' to scope over or under the epistemic modal?

Suppose that 'if' takes narrow scope. Then the material reading interprets P1 and P2 as P1N and P2N respectively:

(P1N)   '$A \supset \text{must}(C)$' has the same truth-conditions as '$A \supset C$'.

(P2N)   '$A \supset \text{may}(C)$' has the same truth-conditions as 'may(A and C)'.

Since 'must' has a strengthening effect, presumably 'must(C)' is sometimes false when 'C' is true; for example, 'He must be guilty' can be false even when 'He is guilty' is true. In any such case, '$A \supset \text{must}(C)$' is true and '$A \supset C$' false whenever A is true. Thus P1N is invalid. As for P2N, it is invalid for several reasons. When A is false, '$A \supset \text{may}(C)$' is automatically true, but 'may(A and C)' can still be false, because A is incompatible with C given the relevant knowledge base. But even when A is true, and 'may(C)' is also true, so '$A \supset \text{may}(C)$' is true, 'may(A and C)' can still be false, again because A is incompatible with C given the relevant knowledge base. For instance, when the conversational participants have no idea where the man is, but he is in fact in Paris, 'He is in Paris $\supset$ he may be in Moscow' is true while 'He may be (in Paris and in Moscow)' is false.

---

[3] In line with much of the literature, Gillies takes 'might' rather than 'may' as an epistemic modal. However, 'might' is the past tense of 'may', and so may introduce an additional modal dimension by the use of a 'fake past' tense (see chapters 9 and 11 for more discussion). Epistemic 'may' avoids that complication. Gillies also routinely uses consequents with 'then'. My arguments do not depend on those differences.

Alternatively, suppose that 'if' takes wide scope. Then the material reading interprets P1 and P2 as P1W and P2W respectively:

(P1W)   'must($A \supset C$)' has the same truth-conditions as '$A \supset C$'.

(P2W)   'may($A \supset C$)' has the same truth-conditions as 'may(A and C)'.

Here P1W is invalid because a material conditional may be true without being known. To invalidate P2W, just consider any case where some epistemic possibility falsifies A (and so verifies $A \supset C$) but no epistemic possibility verifies both A and C.

Thus, however scopes are assigned, the material reading of 'if' invalidates both P1 and P2. By contrast, on Angelika Kratzer's view that 'if' is not an operator but instead a restrictor on operators, combined with suitable auxiliary assumptions, P1 and P2 are valid (Kratzer 1981, 1986, 2012). In particular, it is assumed that in indicative conditionals with no overt operator for the 'if' clause to restrict, 'if' restricts a covert 'must' operator. That trivially explains P1. Furthermore, given that 'may' is in effect an existential quantifier over possibilities, 'if A may C' is trivially equivalent to 'may(A and C)', which explains P2. A dynamic semantics can achieve a somewhat similar effect by interpreting 'if' as a strict conditional operator over contextually relevant epistemic possibilities which also updates the context by excluding worlds where the antecedent is false as irrelevant (Gillies 2010).

Clearly, given that P1 and P2 are valid, any theory which predicts their validity has the advantage in that respect over any theory which does not, and any theory which predicts their *in*validity is false. But, equally, if P1 and P2 are in fact invalid, any theory which predicts their validity is false. So are P1 and P2 indeed valid?

Here is a reason for doubting P1. Imagine a team of detectives being encouraged to brainstorm about the whereabouts of the suspect. In this context, what they say will be treated as informed guesswork, not as an assertion. One detective says:

(25)   If he's in France, he's in Paris.

That may be a shrewd guess, and a good thing to say in the context, even though it is common ground in the conversation that it is epistemically possible that the suspect is in France but not in Paris. In this context, saying (25) is quite different from saying (26):

(26)   If he's in France, he must be in Paris.

For (26) is a bad thing to say in this context. By uttering (26), the detective would be interpreted as speculating that it is epistemically *im*possible that the suspect is in France but not in Paris, contrary to what is common ground in the conversation. Thus (26) differs truth-conditionally from (25); its truth-conditions are significantly more demanding. Therefore, P1 is invalid.

Here is a reason for doubting P2 too. The conspiracy theorist and former professional footballer David Icke maintains that the world is controlled by a network of reptiles, including many well-known public figures. They are not just metaphorically reptiles; they are literally reptiles. Thus one can truly assert (27) in a normal context:

(27)   If David Icke is to be believed, the Pope may be a reptile.

But, given P2, (27) is truth-conditionally equivalent to (28):

(28)   It may be that David Icke is to be believed and the Pope is a reptile.

But one could not truly assert (28) in a normal context. By non-sceptical standards, it is not epistemically possible that the Pope is literally a reptile. Thus (28) differs truth-conditionally from (27); its truth-conditions are significantly more demanding. Therefore, P2 is invalid.

Why do P1 and P2 look so plausible, when they are in fact invalid? To understand what is going on, we need to identify the key features of the examples.

We start with the counterexample to P1. A key feature is that uttering (25) in the context is treated as speculating rather than asserting. Had someone *asserted* (25), and the assertion been accepted into the common ground, asserting (26) would then have been treated as acceptable, because epistemic possibilities in which the suspect is in France but not in Paris would have been excluded. Thus when the utterance of (25) is treated as mere guesswork, (25) is *not* interpreted as involving an epistemic necessity operator. That is where the attempts to explain the alleged validity of P1 break down.

The counterexample to P2 is different, because the speaker asserts (27) flat out. A key feature is that the antecedent of (27) is not presupposed to be epistemically possible. That may be unusual, but it is not anomalous. The assertion of (27) is felicitous. With no epistemic possibility in which David Icke is to be believed, we have no epistemic possibility in which David Icke is to be believed and the Pope is a reptile. Often, the epistemic impossibility of an antecedent causes cognitive difficulty in assessing a conditional; that does not happen here because the speaker accepts (27) on the basis of quite shallow processing, simply given knowledge of David Icke's theories. That is unsurprising given the Suppositional Conjecture.

In both cases, what triggers the counterexample is pragmatic, not semantic. In one case, the speaker is not making a flat-out assertion; in the other, the speaker is not presupposing the epistemic possibility of the antecedent. Thus the natural hypothesis is that P1 and P2 look so plausible because they involve pairs of sentences between which some rough pragmatic equivalence holds under a wide range of ordinary circumstances, but sometimes fails quite badly outside that range.

The most economical hypothesis is that P1 and P2 are just the special cases for indicative conditionals of more general patterns in the pragmatics of epistemic modals.

For P1, the natural starting point is the generally close correlation between A and 'must A'. What was said above about (21) and (22) can also be said about (29) and (30), which are conditional-free: there is no very obvious potential reason for accepting one of them while rejecting the other:

(29) He is guilty.
(30) He must be guilty.

Arguably, (30) entails (29) because what is known is true, so what follows from what is known is also true (see von Fintel and Gillies 2010 for discussion). The converse entailment from (29) to (30) does not hold: people are sometimes guilty of crimes despite the indecisiveness of the evidence. Still, if one is in a position to *assert* (29), then—arguably (Williamson 2000)—one knows (29), so (30) is true with respect to the relevant knowledge base. Arguably, it does not quite follow that one is in a position to *assert* (30), since one might know that he is guilty without knowing that one knows that he is guilty and so without knowing that he *must* be guilty (Williamson 2000). Even so, *usually*, when one is in a position to assert (29), one is also in a position to assert (30). A special case of this rough correlation in assertibility-conditions between A and 'must A' is a rough correlation in assertibility-conditions between 'if A, C' and 'must(if A, C)', which is the wide-scope reading of 'if A must C'. That holds both for the material reading of 'if' and for other operator readings.

However, there is more to the connection between A and 'must A' than just the approximate correlation in assertibility-conditions. For imagine a conversation about Portia's whereabouts. We do not know which country she is in. This exchange occurs:

(31)   Speaker 1: Suppose that Portia is in a Portuguese-speaking country in South America.
       Speaker 2: OK, she must be in Brazil.

Speaker 2's contribution is felicitous, given that Brazil is well known to be the only Portuguese-speaking country in South America. Yet speaker 2 does not know that Portia is in Brazil. Rather, in going along with speaker 1's supposition ('OK'), speaker 2 tacitly restricts the range of epistemic possibilities contextually relevant for 'must' to those in which the supposition holds. The same restriction applies to 'may'. That is why speaker 2 is naturally understand as saying something false in this variant on the exchange, given that Chile is well known not to be Portuguese-speaking:

(32)   Speaker 1: Suppose that Portia is in a Portuguese-speaking country in South America.
       Speaker 2: OK, she may be in Chile.

These restrictions on epistemic modals do not involve any special semantic or syntactic mechanism. The dialogues (31) and (32) are conditional-free. Nor is the restriction mandatory. The conversation could easily have gone as in (33), where speaker 2 rejects the supposition:

(33)   Speaker 1: Suppose that Portia is in a Portuguese-speaking country in South America.
       Speaker 2: No, she only visits English-speaking countries—she must be in one of them.

The restriction on 'must' and 'may' in (31) and (32) is pragmatically generated.

This pattern tightens the correlation between A and 'must(A)'. For even when speakers merely suppose A, without purporting to have knowledge, they typically restrict the range of epistemic possibilities contextually relevant for 'must' to those in which the supposition A holds, thereby trivially verifying 'must(A)'.

The act of supposing the premise in the argument 'A, therefore must(A)' thus typically creates a context in which the premise is true. Moreover, the converse argument 'must(A), therefore A' is straightforwardly valid. But none of that makes A and 'must(A)' truth-conditionally equivalent. For when one *speculates* (rather than supposes or asserts) A, one typically does *not* restrict the range of epistemic possibilities contextually relevant for 'must' to those in which A holds. Thus speculating 'He is guilty' does not amount to speculating 'He must be guilty.'

We can now understand the pragmatic regularity underlying the illusion that P1 is valid, on the wide scope reading of 'if A must C' as 'must(if A, C)'. For, as a special case of the regularity just observed for epistemic modals, the act of supposing 'if A, C' typically creates a context in which 'must(if A, C)' is true, and the converse entailment from 'must(if A, C)' to 'if A, C' is straightforwardly valid. Nevertheless, the two forms are not truth-conditionally equivalent, as the counterexample (25)/(26) showed, because speculating 'if A, C' does not amount to speculating 'must(if A, C)'. These considerations apply to both the material reading of 'if' and many other operator readings.

We now turn to the illusion that P2 is valid, again on the wide scope reading. We focus on the material reading of 'if', so 'if A may C' is understood as 'may($A \supset C$)'. The question is why it should seem so close in truth-conditions to 'may(A and C)', despite the vast gap in truth-conditions between conjunction and the material conditional, which differ in truth-value whenever the antecedent is false. Of course, 'A and C' truth-functionally entails '$A \supset C$', so 'may(A and C)' entails 'may($A \supset C$)', since epistemic possibility is closed under single-premise truth-functional entailment. The problem is the converse.

A promising suggestion comes from recent work by Kratzer (2020): that the antecedent of a conditional can be reused as a tacit contextual restriction on other operators.[4] For, when that happens, 'may($A \supset C$)' is understood as equivalent to 'may(A and ($A \supset C$))'; since 'A and ($A \supset C$)' is truth-functionally equivalent to 'A and C', 'may(A and($A \supset C$))' is equivalent to 'may(A and C)', as required, again given that 'may' is closed under single-premise truth-functional entailment.[5]

But why should the antecedent of a conditional be reused as a tacit contextual restriction on other operators? The Suppositional Conjecture points towards an explanation. For in applying the Suppositional Procedure to the conditional embedded in 'may(if A, C)', one supposes A: as just observed, epistemic modals tend to be restricted by a supposition in play, for pragmatic rather than semantic reasons. More specifically, when 'if A, C' is assessed by the Procedure, all non-A epistemic possibilities are excluded from consideration, so it is natural to exclude them from consideration in the subsequent assessment of 'may(if A, C)' too. But that pattern also has a natural exception, when A is not presupposed to be epistemically possible, for then restricting epistemic modals by A risks making them all vacuous. Thus the counterexample (27)/(28) to P2 is unsurprising.

In short, P1 and P2 are not good evidence against the material reading of 'if', because there is independent evidence that they are invalid. The material reading predicts both their invalidity and their appearance of validity. Instead, their invalidity poses a threat to any semantic or syntactic theory designed to explain their validity. In particular, the difference between speculating 'if A, C' and speculating 'if A, must C' is a difficulty for any theory treating unembedded occurrences of 'if A, C' as expressing claims of epistemic necessity.

---

[4] See section 8.5 for applications of Kratzer's proposal to interactions between quantifiers and 'if'.
[5] The corresponding restriction for 'must($A \supset C$)' makes no difference, since ($A \supset (A \supset C)$) is truth-functionally equivalent to $A \supset C$.

8

Interactions between Plain Conditionals and Quantifiers

8.1 Universal instantiation by conditionals

The material interpretation of indicative conditionals faces further challenges from their interactions with quantifiers. In this chapter, we start by amplifying deductive connections suggested in section 1.2 between universal generalizations and indicative conditionals. The material interpretation is ideally placed to explain these connections. However, in many other cases, interactions between quantifiers and indicative conditionals look much less favourable to it. Such cases are a good test of theories of conditionals, because our understanding of how quantifiers work is comparatively firm, so there is much less wiggle room on that side. Nevertheless, once we take the Suppositional Rule into account, we can predict the challenging appearances, in ways consistent with the material interpretation. Indeed, on further probing, the cases can be seen to favour that interpretation.

The logical power of universal generalizations depends on the rule of inference which allows us to unpack their implications for individual instances. Consider (1):

(1)   Every spy is clever.

Suppose that Kim is in the contextually relevant domain. What does (1) imply about him? The obvious answer is (2):

(2)   If Kim is a spy, Kim is clever.

The move from (1) to (2) depends on a suitable rule of universal instantiation. Schematically, the move is from 'Every N Vs' to 'If t is an N, t Vs', where t is a singular term denoting a member of the domain. This is the obvious candidate for implementing in English the rule of universal instantiation, without which, the universal generalization would be impotent.

Section 1.2 explained the problems facing alternative candidates for the natural language instances of a universal generalization, formulated in terms of negation and disjunction or conjunction. Those arguments need not be repeated here. The effect of a non-material semantics is to leave us with no natural way of expressing instances of universal generalizations in natural language. Given the cognitive significance for us of universal generalizations, the idea that there is no natural way of expressing their instances is quite implausible. One great advantage of the material reading of indicative conditionals is that it explains how we can simply express instances of universal generalizations in natural language.

Suppose and Tell: The Semantics and Heuristics of Conditionals. Timothy Williamson, Oxford University Press (2020).
© Timothy Williamson.
DOI: 10.1093/oso/9780198860662.001.0001

In particular, (2) is an indicative conditional. Thus we want universal generalizations to entail the corresponding indicative conditionals. For that, it does not suffice that when we can assert (1), we can also assert (2). For we may be treating (1) as a mere hypothesis, not a known fact. To assess the hypothesis, we still need the ability to draw consequences like (2) from it. Mathematical proofs routinely involve the step from a claim of the form 'Every N Vs', resting on various hypothetical suppositions, to a claim of the form 'If *t* is an N, *t* Vs'. Unless it is valid, standard mathematical proofs are riddled with fallacies.

How does (1) entail (2)? For our purposes, we can ignore members of the domain other than Kim. Occasionally, instances of a generalization for other members of the domain have implications for the member we started with, most dramatically when they render the generalization inconsistent, but the reasoning from (1) to (2) is much easier and more direct than that. If it works at all, it does so simply by plugging in the name 'Kim' for the implicit bound variable in (1). What does the truth of (1) require of Kim, as a member of the domain? Given that Kim is a spy, the truth of (1) requires Kim to be clever. Given that Kim is not a spy, the truth of (1) requires nothing more of Kim. Those two conditions are met just in case the material conditional holds:

(3)   Kim is a spy $\supset$ Kim is clever.

All is well if (2) has the material reading, and so is equivalent to (3). But if (2) has a stronger reading, and so is not entailed by (3), there is trouble. For what stops (2) being false while (1) is true, thereby making the entailment fail?

Imagine that every spy happens to be clever, so both (1) and (3) are true. Imagine further that Kim is not a spy. Take a context in which it is not known whether Kim is a spy, and (1) and (3) are also not known, but the available evidence decisively excludes Kim's being a spy who is clever, so (4) is known:

(4)   Kim is a spy $\supset$ Kim is not clever.

Thus, in some relevant epistemic possibilities, including the actual one, Kim is not a spy. In other relevant epistemic possibilities, Kim is a spy but (1) and (3) are false because Kim is not clever. Most semantic accounts of indicative conditionals make (4) true in such a context (perhaps with some suitable filling-in of the example):

(5)   If Kim is a spy, Kim is not clever.

Most non-material semantic accounts of the indicative conditional makes opposite conditionals, such as (2) and (5), mutually incompatible, at least when their shared antecedent is epistemically possible, as it is here. An example is Stalnaker's account (1968, 1975, 1984). On his view, in this context, since (5) is true, (2) is false. Similarly, on any view for which the truth of an indicative conditional implies the epistemic necessity of the corresponding material conditional, (2) is not true in this context, for (3) is not epistemically necessary. Thus (1) does not entail (2); the argument from (1) to (2) is invalid, because it is not truth-preserving.

The usual fall-back from the validity of an argument is its pragmatic *reasonableness*. Even when the premises do not entail the conclusion, the argument may still be reasonable in the sense that in any context in which it is appropriate to suppose or assert the premises, one cannot do so without committing oneself to the conclusion. Stalnaker (1975, 1984) elaborates this idea, and shows that some attractive arguments involving indicative conditionals are invalid but still reasonable on his semantics. But the argument from (1) to (2) is not even pragmatically reasonable in a robust sense, given that it is inappropriate to accept (2) and (5) together. For in the context at issue, one may tentatively accept (1), in order to work out its implications. By assumption, (1) is in fact true, though not known—but there may be plenty of non-conclusive evidence for (1). When tentatively accepting (1), one still knows (5), which is much more secure epistemically than (1), so it is still appropriate to accept (5): to test a hypothesis properly, one needs to *use* one's background knowledge, not suspend it. But then, in that context, it is *not* appropriate to accept (2), given its incompatibility with (5). On such a view, the argument from (1) to (2) does not preserve appropriate tentative acceptance. Even if it is reasonable on some narrower understanding of Stalnaker's notion, that is insufficient for the needs of much ordinary good reasoning, such as the testing of universal generalizations.

Similar problems face any attempt to make opposite indicative conditionals, such as (2) and (5), semantically mutually incompatible, at least given the epistemic possibility of the antecedent.

By contrast, the material account makes opposite indicative conditionals compatible, except when the shared antecedent is a logical truth. We even saw in section 5.1 that speakers can sometimes appropriately assert opposite conditionals together, when they are inhibiting the Suppositional Procedure, for instance because they accept the conditionals on the testimony of different witnesses. The present case need not involve testimony: the speaker might achieve a similar effect by compartmentalization, accepting (2) simply by deriving it from (1), without reference to the evidence for (5). Of course, once the Suppositional Procedure is given free rein, it tends to destabilize joint acceptance of pairs such as (2) and (5)— as we saw, that explains the allure of the principle of conditional non-contradiction. At this point, as ordinary reasoners, we may simply deduce 'Kim is not a spy', the negation of the shared antecedent, from (2) and (5) together, and work with it, ignoring the problematic pair of conditionals thereafter. In practice, we may not worry about the legitimacy of our proceedings. Nevertheless, they had better *be* legitimate. Accounts on which (2) and (5) are not true together are ill placed to explain how we can soundly reason from them together to the negation of their antecedent. The material account faces no such difficulty in explaining the legitimacy of the practice.

8.2 Gappy conditionals?

The need for indicative conditionals to instantiate universal generalizations makes trouble for another proposal sometimes made: that indicative conditionals with a false antecedent are 'undefined' or 'gappy', neither true nor false, perhaps because they fail to express propositions—an idea closely related to the 'conditional speech act' view of indicative conditionals mentioned in section 7.2. Bruno de Finetti (1935, 1995) and

Nuel Belnap (1970, 1973) developed such a view, which has been revived by Janneke Huitink (2009, 2010).[1]

The proposal is this: when A is true, 'if A, C' has the same value as C (true, false, or gappy); when A is not true, 'if A, C' is gappy. Thus when every spy is clever but Kim is not a spy, (1) is true but (2) is gappy rather than true.[2] That view makes the argument from (1) to (2) invalid by ordinary standards of validity, which require truth to be preserved from the premises to the conclusion of a valid argument.

The natural response for proponents of the gappy view of conditionals is to revise the standard of validity to permit valid arguments with true premises and a gappy conclusion.

Huitink (2009) appeals to Strawson-entailment: some premises Strawson-entail a conclusion just in case the semantics guarantees that the argument preserves truth whenever all the sentences are non-gappy. The Strawson-entailments are those which never lead from true premises to a false conclusion. Since (2) is never false when (1) is true, (1) Strawson-entails (2).

Huitink justifies the choice of Strawson-entailment by claiming that when we judge the validity of an argument, we assume that its premises and conclusion are true or false. But that claim does not fit her trivalent semantics. For it implies that we should judge any argument from 'if A, C' to A valid, since whenever the premise is true or false, the conclusion is true. But although 'If he is guilty, he deserves punishment' Strawson-entails 'He is guilty', few people would judge the argument valid.

A related problem for Strawson-entailment is that it is structurally unsuited to be a consequence relation. In particular, it is non-transitive. For example, on the trivalent semantics, 'not(A)' always Strawson-entails 'if A, C' (since the conclusion is gappy whenever the premise is true) and 'if A, C' always Strawson-entails A (as just explained), but 'not(A)' does not always Strawson-entail A. A consequence relation should be transitive to allow us to chain short valid arguments together into a long valid argument.[3]

A more promising alternative to Strawson-entailment is nf-entailment ('nf' for 'non-falsity'): some premises nf-entail a conclusion just in case the semantics guarantees that the argument preserves non-falsity. The nf-entailments are those which never lead from non-false premises to a false conclusion. Nf-entailment has the right

---

[1] See Bradley 2002 for a related view.

[2] Belnap (1970) interprets 'if A, C' as expressing the same proposition as C when A is true, and no proposition otherwise. He also interprets the unary universal quantifier 'for each thing' as expressing the conjunction of those propositions expressed by its instances, if any are, and no proposition otherwise. He then parses 'all Fs are Gs' as 'for each thing, if it is an F, then it is a G', whose semantics he treats compositionally. Thus whenever there is at least one F, 'all Fs are Gs' expresses the conjunction of the propositions expressed by 'x is a G' for values of 'x' for which 'x is an F' is true. Hence, on Belnap's original view, since there is at least one spy, 'All spies are spies' expresses a contingent proposition attributing spyhood to various people (those who happen to be spies). If there were no spies, it would not express a proposition at all. Huitink's treatment avoids these radical consequences, in order to avoid a different problem for Belnap's approach pointed out by Edgington (1995).

[3] Huitink (2009) uses her approach to explain why opposite conditionals in Gibbard stand-off cases are mutually inconsistent: they Strawson-entail a contradiction. However, chapter 5 observed that opposite conditionals *can* be consistently maintained in such cases, so that is a cost, not benefit, of her approach.

structural properties to be a consequence relation; in particular, it is transitive. Appropriately, 'if A, C' does not generally entail A, since the premise is gappy and the conclusion false when A is false. But (1) does nf-entail (2), as required.

Nf-entailment is not obviously an appropriate standard for reasoning in science, where 'if' is used ubiquitously. That some consequences of a theory are 'undefined', rather than true, sounds like an objection to the theory. Presumably, one does not want to be asserting those untrue consequences.

At a more everyday level, on the trivalent semantics, nf-entailment has strange consequences. Consider (6)–(9) with respect to a small domain of farm animals, where (6) and (7) are in fact true:

(6)   Every cow is brown.
(7)   Toro is not a cow.
(8)   If Toro is a cow, Toro is brown.
(9)   It is not the case that if Toro is a cow, Toro is brown.

Here (6) nf-entails (8); the point of invoking nf-entailment was to give the thumbs up to such arguments. Moreover, (7) nf-entails (9), for (7) guarantees that (8) is undefined, and the negation of an undefined sentence is itself undefined. Since nf-entailment is monotonic—it persists under the addition of premises—two very ordinary sentences, (6) and (7), together true in very ordinary circumstances, together nf-entail a contradiction, (8) and (9) together.

Of course, gappy theorists will not deny the true conjunction of (6) and (7). Rather, they must reject the standard rule of *reductio ad absurdum*, because it does not preserve nf-entailment. They must reject it even in mathematical contexts, since one can easily construct mathematical analogues of (6)–(9):

(6m)   Every rational number is algebraic.
(7m) $\sqrt{2}$ is not a rational number.
(8m)   If $\sqrt{2}$ is a rational number, $\sqrt{2}$ is algebraic.
(9m)   It is not the case that if $\sqrt{2}$ is a rational number, $\sqrt{2}$ is algebraic.

Both (6m) and (7m) are mathematical truths. Just as (6) and (7) respectively nf-entail (8) and (9), so (6m) and (7m) respectively nf-entail (8m) and (9m). Thus two commonplace mathematical truths entail a contradiction under the favoured standard of nf-entailment. That requires a more revisionary view of mathematics than one might have guessed from the gappy account as originally presented.

Could gappy theorists go further, and revise the semantics of negation too, counting the negation of a gappy sentence as false rather than gappy? Then, given (7), (9) would be false, so (7) would no longer nf-entail (9); similarly, (7m) would no longer nf-entail (9m). However, the revised semantics for negation undermines the non-classical approach. Gappy status, neither true nor false, is intended for cases when reality is somehow neutral between a sentence and its negation; the question 'Which?' does not properly arise. That is how it is meant to work on approaches attributing gappy status to the Liar and other semantically paradoxical sentences, or to vague sentences in borderline cases. It is meant to work similarly on approaches treating the utterance of an indicative conditional as making a conditional assertion, so that nothing is asserted when the antecedent is false. But on the just-revised semantics, when (6) and (7) are true, reality is not at all neutral between (8) and (9). For (8) is not only not false, it follows from the true premise (6) by the operative standard of nf-entailment. By contrast, its negation (9) is counted simply false. Far from neutrality, that is a misuse of gappy status to mark in effect true indicative conditionals with a false antecedent.

Non-classical approaches to indicative conditionals differ markedly from non-classical approaches to vagueness, semantic paradoxes, and the like. Proponents of the latter are usually at pains to argue that their accounts disrupt classical reasoning only in the presence of the pathological phenomena, and pose no threat to it in normal circumstances. By contrast, non-classical approaches to indicative conditionals predict widespread breakdowns of ordinary reasoning in ordinary circumstances—for example, the failure of standard arguments by *reductio ad absurdum*. That is more evidence that conditionals must be studied in a wider setting, paying more attention to the implications of their ubiquitous role for our cognitive lives.

8.3 Applying the Suppositional Rule to generalizations

The links between generalizations and indicative conditionals are not only deductive. We often use analogues of the Suppositional Procedure to assess generalizations non-deductively as well as deductively. This section discusses a range of examples. In considering them, one must keep in mind that neither the Suppositional Procedure nor the Suppositional Rule is part of the semantics proper, for reasons explained in earlier chapters. Instead, we apply the Procedure or Rule in cognitively assessing sentences which are, by and large, already interpreted, just as we apply other cognitive capacities for reasoning and pattern recognition. Thus one should not be asking where they fit into the compositional semantic analysis of the sentence; they do not. Of course, we expect the semantics and the heuristics for a natural language to cohere into a reasonable overall story about its use: the reliability of the heuristics for assessing sentences depends on their truth-conditions, as determined by the compositional semantics. But for present purposes, we need not go far into the fine-grained semantic structure of quantifier constructions, given that we know the truth-conditional upshot.

Here is a case where a universal generalization is assessed by an analogue of the Suppositional Procedure:

(10)   Every grain of sand is smaller than a football.

We cannot assess (10) by checking each grain separately. Right now, I cannot see or bring to mind any one particular grain of sand. But I can still assess (10). I *imagine* a grain of sand, though there is no particular grain of sand that I imagine. I am doing the informal equivalent of supposing that x is a grain of sand (where 'x' is a variable). On that supposition, I am very confident that it is smaller than a football. I involuntarily imagine it as smaller than a football. In casual mode, I may immediately assent to (10). In a more conscientious mode, I may play variations on the imaginative theme. For example, I imagine a lump of sand the size of a football and, within that imaginative exercise, conclude that it contains many grains of sand and is not itself such a grain. The method may not be ultra-reliable, but we still use it; often, we have no realistic alternative.

In mathematics, the Suppositional Procedure is the normal way of proving universal generalizations. Consider (11):

(11)   Every set has fewer members than subsets.

To prove (11), one supposes that *s* is a set. On that supposition, one proves that *s* has fewer members than subsets, by Cantor's diagonalization argument. One thereby reaches the conclusion (11). Indeed, a mathematician may formulate the same generalization as an implicitly quantified conditional:

(12)   If *s* is a set, *s* has fewer members than subsets.

Similarly, one may express the same generalization as (10) with a conditional:

(13)   If something is a grain of sand, it is smaller than a football.

In less formal settings, the natural conclusion of the Suppositional Procedure is often more cautious: a generic generalization such as (14) or (15):

(14)   Llamas have four legs.
(15)   A llama has four legs.

There may well be three-legged llamas, but occasional exceptions are insufficient to falsify a generic statement. One may also express a similar generic generalization with an implicitly quantified conditional:

(16)   If it's a llama, it has four legs.

Indeed, even in assessing (10), a cautious person may worry about the physical possibility of anomalously large grains of sand, and suspect that the proper conclusion to draw from applying the Suppositional Procedure imaginatively to that case is not a strictly universal generalization but a generic one, such as (17), (18), or (19):

(17)   Grains of sand are smaller than footballs.
(18)   A grain of sand is smaller than a football.
(19)   If it's a grain of sand, it's smaller than a football.

One may even use something like the Suppositional Procedure to check statements with 'many', such as (20):

(20)   Many books have a green cover.

When I first considered (20), no particular book with a green cover immediately came to mind. Rather I imagined a book, though there was no particular book that I imagined; since I very easily imagined a book with a green cover, I concluded that, of all the books in the world, many must have a green cover (admittedly, the final reflection involves a move beyond the Suppositional Procedure). It is not a very scientific procedure, but it is what we often do. It is a form of what psychologists call the *availability heuristic*. When we lack statistics, as we usually do, we use psychological availability as a proxy for frequency.

In the cases just described, generalizations were verified rather than falsified. A two-way test for a proposition Φ verifies Φ if the result is positive and falsifies Φ if the result is negative. Since falsifying Φ is equivalent to verifying ¬Φ, and verifying Φ to falsifying ¬Φ (in classical logic), a two-way test for Φ also constitutes a two-way test for ¬Φ. The Suppositional Procedure is set up as a two-way test for conditionals, since it licenses the transfer of *all* assessments, negative as well as positive, from the consequent on the supposition of the antecedent to the whole conditional with the supposition discharged. However, as noted in section 6.2, we tend to treat a negative result for 'If *A*, *C*' as a positive result for 'If *A*, not(*C*)' rather than for 'Not(if *A*, *C*)', and the result FACT about underpinning proved in section 6.1 shows that the Suppositional Procedure is more reliable in that role, on the material reading of indicative conditionals. The case of generalizations works analogously.

For example, I suppose that *x* is a golden mountain. Under that supposition, I ask whether *x* is smaller than a football, and answer 'No'. The natural conclusion to draw is a negative universal generalization with 'no':

(21)   No golden mountain is smaller than a football.

Equivalently but less naturally:

(22)   Every golden mountain is not smaller than a football.

Those generalizations correspond to a conditional with a negated consequent:

(23)   If it is a golden mountain, it is not smaller than a football.

Of course, neither (21) nor (22) is the negation of the universal generalization corresponding to the answer 'Yes':

(24)   Every golden mountain is smaller than a football.

Nor is (23) the negation of the corresponding conditional:

(25)   If it is a golden mountain, it is smaller than a football.

The negation of (24) is simply:

(26)   Not every golden mountain is smaller than a football.

For good reason, (26) is a far less natural conclusion than (23) to draw from the negative result of the Suppositional Procedure. After all, (26) entails (27):

(27)   Some golden mountain is not smaller than a football.

In turn, (27) entails (28):

(28)   There is a golden mountain.

But (28) is false, at least in a context with a terrestrial domain. Thus (26) is a terrible conclusion. Moreover, the Suppositional Procedure provides one with no good reason to accept (28). This is especially clear in mathematics, where one would never draw a conclusion of the form 'Not every *F* is a *G*' simply from having proved '*x* is not a *G*' on the supposition '*x* is an *F*'; one would routinely require a proof of 'There is an *F*' too.

However, even with quantifiers, we are not completely clear of confusion. Assuming that there are no golden mountains, we deny (28). By contraposition, we assert the negation of (27):

(29) It is not the case that some golden mountain is not smaller than a football.

Standard reasoning takes us from (29) to (24). But that is troubling, for now we find ourselves committed to both (21) and (24), which make a pair of the form 'No F is G' and 'Every F is a G'. Such pairs tend to strike us as inconsistent. That is not just because such pairs tend to be uttered under the conversational presupposition that F is non-empty. For that presupposition has already been cancelled by the explicit denial of (28): we are aware that there is no golden mountain.

The trouble is rather that when the Suppositional Procedure is applied, 'Every F is a G' and 'No F is G' correspond to a pair of explicit contraries on the supposition 'x is F'. In this case the inconsistency is between 'It is smaller than a football' and 'It is not smaller than a football' on the supposition 'It is a golden mountain.' Since the Suppositional Procedure involves exporting assessments from inside the scope of the supposition to outside, with respect to the relevant complex sentences, while discharging the supposition, it tells us to export the assessment of 'It is not smaller than a football' and 'It is smaller than a football' as contraries inside the supposition 'It is a golden mountain' to the assessment of the corresponding negative and positive universal generalizations, (21) (equivalently, (22)) and (24), as contraries, outside the scope of the supposition. We can put the generalizations in conditional form, which amounts to taking (23) and (25) to be contraries. Schematically, the temptation to regard 'Every F is a G' and 'No F is a G' as contraries has the same origin as the temptation to regard 'If it is an F, it is a G' and 'If it is an F, it is not a G', and more generally 'If A, C' and 'If A, not C', as contraries (in other words, CNC, the principle of conditional non-contradiction). The origin is the Suppositional Procedure. These conflicting pressures may help explain the long history of uncertainty and dispute over the alleged 'existential import' of universal generalizations.

The role of the Suppositional Procedure in assessing universal generalizations motivates the prediction that they will be involved in analogues of the nuclear reactor and Sly Pete cases. That prediction is borne out. Imagine a team of zoologists surveying the fauna on a small island just off the coast of Australia. They provisionally expect to find kangaroos. One group is sent to survey the forest in the centre of the island, while another group is sent to survey the rest of the island, which is unforested. Having carefully surveyed the forest, the first group makes its truthful report to the expedition leader:

(30) No kangaroo is in the forest.

Meanwhile, having carefully surveyed the unforested remainder, the second group makes its own truthful report to the leader:

(31) Every kangaroo is in the forest.

Superficially, (30) and (31) look mutually inconsistent. But the expedition leader is not fazed. She rightly trusts both groups, and draws the correct conclusion: there are no kangaroos on the island.

As already noted, the Suppositional Procedure is not suited to verifying 'existential' generalizations. Thus, in situations oriented more towards verifying than falsifying statements made, universal generalizations are more likely than existential ones to evoke the Suppositional Procedure. Consequences of that asymmetry will be discussed later.

The tendencies just described apply to generics as well as to universal generalizations. For example, we tend to treat the mutual inconsistency of the predicates of generics as making the generics themselves mutually inconsistent, as with (32) and (33):

(32)   Talking donkeys are white.
(33)   Talking donkeys are black.

It is not just that one cannot felicitously *assert* both generics in the same context; (32) and (33) seem mutually inconsistent even as suppositions. The corresponding conditionals are (34) and (35):

(34)   If it's a talking donkey, it's white.
(35)   If it's a talking donkey, it's black.

We tend to treat (34) and (35) as mutually inconsistent too. Again, it is not just that one cannot felicitously *assert* both conditionals in the same context; (34) and (35) look mutually inconsistent even as suppositions. Even acknowledging the possibility of the empty case does not make it clear to us that (32) and (33), or (34) and (35)) are mutually consistent:

(36)   There may be no talking donkeys.

Of course, the truth-conditions of generics are so unclear that the truth of both (32) and (33) in the absence of talking donkeys cannot be taken for granted, but one also cannot take for granted that generics *do* have existential import. The central role of the Suppositional Procedure in the acceptance of generics, including with respect to fully specified hypothetical scenarios, may help explain the unclarity in their truth-conditions.

8.4 Quantifying conditionals

The interaction between indicative conditionals and quantified sentences is not confined to their logical relations as separate premises or conclusions of arguments. The constructions can be combined in a single sentence. In particular, indicative conditionals can be quantified—this is one of many kinds of evidence against the no-proposition view. It has also been used as a source of evidence against the material interpretation. We will consider a sample of the data.

Here is a typical example (from Higginbotham 1986):

(37)   Every student will succeed if he works hard.

In the envisaged all-male college, we imagine a student, Fred, who will clearly never succeed, however hard he works. We have not completely ruled out the possibility that he will work hard, but we regard it as very unlikely. Presumably, the truth of (37) requires the truth of (38), by some form of universal instantiation:

(38) Fred will succeed if he works hard.

We reject (38). Since we estimate the probability that Fred will succeed, even conditional on his working hard, as zero, our rejection of (38) conforms to the probabilistic form of the Suppositional Rule, SRP. Yet, on the material reading, (38) is very likely to be true, since its antecedent is very likely to be false. Since we reject (38), we reject (37) too. By contrast, we do not treat Fred as a counterexample to the corresponding conditional-free universal generalization:

(39) Every student who works hard will succeed.

For any counterexample to (39) is a student who works hard but will not succeed, and we regard Fred as very unlikely to work hard. The material reading of 'if' makes (37) equivalent to (39), but we are inclined to treat Fred as a counterexample to (37) but not to (39). Thus we implicitly treat (39) as not entailing (38), contrary to the material interpretation. We may call that the *orthodox view* of the example.

The assessment of (38) as false is a straightforward application of the Suppositional Procedure, in a case of exactly the sort which manifests its fallibility as a heuristic, on the account developed in previous chapters. We should take a closer look.

On the orthodox account, (39) does not entail (38), otherwise Fred would be just as much a counterexample to (39) as to (37). As section 8.1 noted, that already recommends suspicion of the orthodox account, since (38) provides the most natural way of expressing what (39) implies about Fred, given that he is a student.

Furthermore, if (37) was false when made as a prediction about the future, then the corresponding past tense statement should be false in retrospect, decades later:

(40) Every student succeeded if he worked hard.

Imagine looking at the college files on alumni, which we know to be scrupulously accurate. For the relevant group of (former) students, there are two lists: one of those who worked hard, another of those who succeeded. We notice that everyone on the former list is also on the latter. Thus any student is on the latter list if he is on the former; in effect, any student succeeded if he worked hard. We can therefore verify (40) by enumeration, without using the Suppositional Procedure. Of course, an orthodox theorist may claim that (40) as uttered with the benefit of hindsight has different truth-conditions from (37) as uttered in the earlier epistemic setting. But how plausible is that claim? A more natural view of (37), with the benefit of hindsight, is that, as uttered decades earlier, it did in fact make a true prediction, whether or not anyone at the time was in a position to know or assert its truth.

Of course, just as (37) implies (38), so (40) implies (41), in both cases by universal instantiation, given that Fred was one of the students:

(41) Fred succeeded if he worked hard.

Consider someone who regarded (38) as false, and so Fred as a counterexample to (37), from the earlier epistemic perspective already described, and then learnt nothing more about the students for decades. Will she now regard (41) as false, and so Fred as a counterexample to (40)? Assume that she knew all along of the two lists of alumni, those who worked hard and those who succeeded, but knows only that they exist. It seems rather odd to be confident, in that epistemic situation, that Fred is a counterexample to (40), as uttered in her own context. What seems more reasonable is to take those who know the contents of the two lists to be in a better position to assess (40) than she is, and to hold that Fred *may* be a counterexample to (40), rather than to insist that he *is*. That attitude depends on treating what proposition (40) expresses as *not* ultra-sensitive to context, but as constant over a range of contexts.

On one view, (38)—and so (37) too—is false if Fred has stopped working because he knows that he will fail no matter how hard he works (von Fintel and Iatridou 2002, Higginbotham 2003, Huitink 2010). On that basis, 'if' is taken to import an intensional element into (37) and (38), making (37) inequivalent to (39). Such claims are much less plausible for the past tense (40) and (41). This suggests that any modal element in (37) and (38) comes from 'will', not from 'if': the future is being treated as *open*, a realm of many possibilities rather than one actuality for 'will' to quantify over, whereas the past is treated as closed. Then (37) and (39) may be inequivalent because 'works hard' is outside the scope of the modally interpreted 'will' in (39) but may be inside it in (37) and (38). In that case, the interference from 'will' makes (37) and (38) bad examples for studying the interaction of 'if' with quantifiers. We therefore continue to work with past tense cases.[4]

The quantifier 'no' provides more drastic challenges to the material interpretation. Consider (42):

(42)   No student failed if he worked hard.

When the conditional is read materially, the truth of (42) requires the falsity of (43) for each student as a value of the variable pronoun 'he' in both occurrences:

(43)   He failed if he worked hard.

But (43) is false on the material reading just in case the student assigned to 'he' worked hard and did not fail, in other words, he succeeded (for simplicity, we treat 'succeed' as contradictory to 'fail'; readers who wish to do so can instead substitute 'did not fail' for 'succeeded'). Thus (42) is true on the material reading just in case (44) is true:

(44)   Every student worked hard and succeeded.

Needless to say, we normally do not assess sentences like (42) as even roughly equivalent to sentences like (44). Instead, we may use the Suppositional Procedure. On the supposition that an unspecified student worked hard, we ask whether he failed; in effect, we assess the open sentence 'x failed' on the foreground supposition 'x worked hard' and the background supposition 'x was a student'. If the verdict is negative, our attitude to (42) is correspondingly positive. Consequently, if one seeks a

---

[4] See also the comments in Klinedinst 2011 on the difference between past and future conditionals in response to Leslie 2009.

conditional-free sentence similar in truth-conditions to (42), the obvious candidate is not (44) but (45):

(45)   No student who worked hard failed.

Thus the material reading diverges more dramatically from what seems natural for 'no' than for 'every'. Whereas the material reading of (37) makes it equivalent to (39), which is at least also its natural-seeming conditional-free approximation, the material reading of (42) makes it equivalent to (44), which is truth-conditionally quite different from its natural-seeming conditional-free approximation (45).

At this point, one might invoke Kratzer's account of 'if' as a restrictor, and propose that the role of the phrase 'if he worked hard' in both (37) and (42) is simply to restrict the quantifier, 'every student' in (37) and 'no student' in (42).[5] On that analysis, (37) is equivalent to (39), and (42) to (45); indeed, the conditional-free versions display the underlying semantic structure more perspicuously. Of course, by itself, this account does not explain any appearance of intensional behaviour in (37) or (42), since such behaviour would involve a semantic difference between (37) and (39), and between (42) and (45). The Suppositional Procedure would still have a role to play in explaining away such apparent differences.

However, the restrictor view gives less promising results for 'if' in the scope of other quantifiers (von Fintel and Iatridou 2002; compare von Fintel 1998b). Consider (46)–(48):

(46)   Some student succeeded if he worked hard.
(47)   At least three students succeeded if they worked hard.
(48)   The cleverest student succeeded if he worked hard.

For 'if' to be restricting the quantifiers in (46)–(48), those sentences must be equivalent to (49)–(51) respectively:

(49)   Some student who worked hard succeeded.
(50)   At least three students who worked hard succeeded.
(51)   The cleverest student who worked hard succeeded.

But the inferences from (46) to (49), from (47) to (50), and from (48) to (51) all seem invalid. For consider (52):

(52)   Ted succeeded if he worked hard.

Given that Ted was one of the relevant students, the truth of (52) presumably suffices for the truth of (46), by straightforward existential generalization. But, under those circumstances, the truth of (52) does not suffice for the truth of (49). Perhaps Ted was a gifted but lazy student who did not work hard and did not succeed. Thus (46) does not entail (49). By a similar argument, with three instances instead of one, (47) does not entail (50). Nor does (48) entail (51), for perhaps Ted was the cleverest student, and succeeded if he worked hard, but did not work hard, and so was not the

---

[5] See von Fintel 1998b in response to Barker 1997 and Geach 1962.

cleverest student who worked hard. These cases tell against using the restrictor view of 'if' to analyse (37) and (42), as well as (46)–(48).

On further reflection, the truth-conditional equivalence of (42) ('No student failed if he worked hard') and (44) ('Every student worked hard and succeeded') is surprisingly more defensible than it first sounds. Let Ned be any one of the students. Imagine that you have carefully checked a definitive list of all those students who both worked hard and succeeded, and noted that Ned is not on it. You thereby came to know that Ned did not both work hard and succeed. You have no further relevant knowledge. You are now in a position to assert authoritatively:

(53)   Ned failed if he worked hard.

Knowing your excellent epistemic position and general trustworthiness, your hearers should take your word for (53), even though their epistemic situation may differ from yours in other respects. But (53) is presumably inconsistent with (42), given that Ned is one of the students: the truth of (53) makes it a clear counterexample to the negative universal generalization (42). Thus any case of a student who did not both work hard and succeed falsifies (42). In other words, any counterexample to (44) is also a counterexample to (42). Since (42) and (44) are true just in case they have no counterexamples, (42) entails (44), contrary to appearances. Conversely, (44) entails (42), for 'He failed if he worked hard' is false of any student who worked hard and succeeded, so (44) ensures that (42) has no counterexamples.

A hardcore contextualist about conditionals might object that the utterance of (53) changes the context, by introducing new information: although (53) may show that (42) is false as uttered in the new context, that does not mean that it was false as uttered in the original context. But that objection fails to respect the natural dynamics of the conversation. When you authoritatively assert (53), participants will interpret you as having given a counterexample to what the first speaker said, in other words, to (42) in its original context. Consider those participants who previously found Ned's having failed improbable conditional on his having worked hard. Those who fully trust your assertion of (53) raise their conditional probability to 1, although they may now find his having worked hard very improbable. Those who almost fully trust your assertion of (53), but think it just possible that you missed Ned's name on the list, now assess (53) as almost certain, but may still find Ned's having failed improbable conditional on his having worked hard, while also finding his having worked hard very improbable. The latter participants assess (53) deferentially, rather than by the probabilistic version of the Suppositional Procedure. As chapter 5 explained, in appropriate circumstances conditional testimony can trump the Suppositional Procedure.

Similar considerations apply to other apparently recalcitrant examples, such as (54), which would usually be treated as equivalent to (55), both uttered in a context where it is quite clear which cohort of students is at issue:

(54)   Most students failed if they worked hard.
(55)   Most students who worked hard failed.

Suppose that most students failed because they did not work hard; of the few who worked hard, all succeeded. The standard judgement is that both (54) and (55) are false in those circumstances. That is uncontroversial for (55), and it is the verdict of the Suppositional Procedure on (54). Now imagine that, as before, your information is limited to a definitive list of all those students who both worked hard and succeeded. For each student not on the list, you make an authoritative assertion of the same form as (53) ('NN failed if he worked hard'). As it turns out, you make such an assertion for most of the students in the cohort. Thus (54) is verified after all. Those who rejected (54) can still insist that what they really cared about all along was the proposition expressed by (55), but it is a natural illusion that (54) strictly and literally expresses that proposition—an illusion caused by our reliance on the Suppositional Procedure.

In more complex cases, 'if' combines with both a quantifier and an epistemic operator. Here is one:[6]

(56) Very few of the 100 boys are likely to have passed the exam if they took it.

Background: an elective examination in quantum physics was given last year; we do not know who (if anyone) took it. The intended overall parsing of (56) is along the lines of (57):

(57) (Very few of the 100 boys)$_x$ (if x took the exam, x is likely to have passed the exam).

On the material reading of the embedded conditional, it is verified by anyone who did not take the exam; thus (56)/(57) can be falsified by many of the boys electing not to take the exam (perhaps on the grounds that they would surely fail). Obviously, that is not the natural pre-theoretic view of its truth-conditions.

How can the Suppositional Procedure be applied to (56)/(57)? The natural place to start is with the embedded conditional, to which the procedure directly applies. Thus one assesses the open sentence 'x is likely to have passed the exam' on the foreground supposition 'x took the exam', with the background supposition 'x is one of the 100 boys', using any other relevant information one may happen to have. That should issue in some sort of (vague, tentative) probability for 'x is likely to have passed the exam' conditional on the two suppositions. Applying the Suppositional Procedure, one equates that probability with the probability of the conditional in (56)/(57). One can then use that probability that an unspecified boy satisfies the conditional as an estimate of the proportion of the 100 boys who satisfy the conditional, and on that basis finally assess the truth-value of (56)/(57). Of course, the whole process will be very rough-and-ready, but that is psychologically realistic. Such an assessment of (56)/(57) feels like a reasonable approximation to what one might pre-theoretically do. Crucially, nothing in it treats boys who elect not to take the exam as weighing against the truth of (56)/(57).

Now imagine someone whose relevant information is just an authoritative list of the 100 boys with red dots by the names of all and only those who both took the exam and are *not* likely to have passed it, by whatever standard of likelihood was intended

---

[6] Thanks to an anonymous referee for suggesting this example.

in (56). For each boy without a red dot, she can authoritatively assert (58), on the intended reading of 'likely':

(58)   If he took the exam, he is likely to have passed it.

Each boy who elected not to take the exam lacks a red dot by his name, and so gives rise to an utterance of (58). Thus, contrary to pre-theoretic expectations and the Suppositional Rule, each such boy counts against the truth of (56)/(57), in line with the straightforward compositional semantics for (56)/(57) on the material interpretation of 'if'.

The discussion of (37)–(58) has this tentative moral. As fluent speakers of a natural language, we are subject to natural illusions about the truth-conditions of many quantified conditionals, because we assess them by the Suppositional Procedure, which is (predictably) unreliable in such cases. The compositional semantics for our language determines the truth-conditions for sentences of the language, but in some cases we systematically misjudge their truth-values, even with respect to fully described hypothetical examples. Of course, if the only motivation for positing such illusions were loyalty to the material interpretation of 'if', the defence would look desperate indeed. But that is not the position. Rather, in such cases, we have independently seen the natural first impressions of truth-value turn out to be unstable, collapsing under pressure into second thoughts quite consistent with the material interpretation.

8.5 More quantified conditionals

The partial role of the Suppositional Procedure also casts light on some examples of indicative conditionals which sound odd even though they appear to be making sensible claims. Here are two examples (from von Fintel and Iatridou 2002 and Dorr 2018):

(59)   Every book happened to be on the table if I needed it.
(60)   Every coin is silver if it's in my pocket.

As one would expect, they seem to be equivalent to the perfectly reasonable (61) and (62) respectively:

(61)   Every book I needed happened to be on the table.
(62)   Every coin in my pocket is silver.

But then why do (59) and (60) sound odd, when (61) and (62) do not?

Both (61) and (62) are paradigms of sentences to be verified by enumeration rather than by the Suppositional Procedure. If true, they are true by chance. That is explicit in (59) and (61) with the use of 'happened to be' rather than 'was'. As for (61), it sounds more natural as soon as we imagine the speaker as someone who regards only silver coins as worthy of *his* pockets. Similarly, (59) sounds more idiomatic once we imagine the speaker insinuating that his guardian angel was at work. In both cases, the envisaged general trend would give more scope to the Suppositional Procedure. The use of 'if' in (59) and (60) invites, although it does not require, general assessment by the Suppositional Procedure. The respective suppositions would be

'It is a book I needed' and 'It is a coin in my pocket', with no further specification of the pronoun. Such a method of assessment would be inappropriate in the absence of any underlying general trend.

The picture is complicated further by the capacity of various parts of a sentence to supply implicit restrictions on quantifiers. First consider this sentence (Geurts 2004):

(63)   Most people change their diet if they survive a cardiac arrest.

The material reading looks bad here, because few people satisfy 'x survives a cardiac arrest' as values of x, so most people satisfy the material conditional 'x survives a cardiac arrest $\supset x$ changes x's diet' whether or not they change their diet. The natural assumption is that (63) has the same truth-conditions as the quantifier-free generalization (64):

(64)   Most people who survive a cardiac arrest change their diet.

Such examples might be taken as evidence for a restrictor account of 'if'. But, as before, that would be too quick. For consider this variant of (63):

(65)   Most people change their diet after they survive a cardiac arrest.

On one natural reading, (65) is quite similar in meaning to (63), despite the large difference in meaning between (66) and (67):

(66)   He changed his diet if he survived a cardiac arrest.
(67)   He changed his diet after he survived a cardiac arrest.

Although 'A after B' seems to entail 'A and B', it would be strange to infer (68) or (69) from (65):

(68)   Most people change their diet.
(69)   Most people survive a cardiac arrest.

Rather, (65) seems to be ambiguous between a reading on which it is elliptical for (70) and one on which it is elliptical for (71):

(70)   Most people who change their diet do so after they survive a cardiac arrest.
(71)   Most people who survive a cardiac arrest change their diet after they do so.

Both (70) and (71) have something like 'x changes x's diet after x survives a cardiac arrest' as the matrix over which the quantifier generalizes, but some of the same material is implicitly reused to restrict the quantifier 'most people', in order to make the overall claim reasonable, given the common ground of the conversation that surviving a cardiac arrest is an uncommon experience. This has nothing to do with a restrictor account of 'after'. The same points can be made with 'before', 'because', and other such connectives in place of 'after'.

Someone might object to this analogy between 'if' and 'after' on the grounds that 'C after A' presupposes A, and such a presupposition triggers a mandatory quantifier domain restriction. By contrast, 'C if A' presupposes neither A nor C, so the explanation of any quantifier domain restriction for 'if' must be different. However, the behaviour of 'after' does not sustain this objection. For a start, the allegedly mandatory quantifier restriction would be by the presupposed element, but that explains only the reading of (65) as (71), but not the equally good reading of (65) as (70), where the restriction is by what comes after (not presupposed), not by what it comes after (perhaps presupposed). Furthermore, 'after' under a quantifier need not trigger any domain restriction. Consider (72) and (73), uttered about a boarding school:

(72)   John went to bed after he brushed his teeth.
(73)   Every boy went to bed after he brushed his teeth.

Perhaps (72) presupposes that John brushed his teeth. But (73) has a perfectly good reading on which its truth requires that every boy in the school brushed his teeth and then went to bed: a boy who did neither falsifies (73) and so is not excluded from the domain. Thus 'after' is far more flexible in its restricting effects than the objection requires. The analogy with 'if' stands.

Once we see how easily material from the matrix is implicitly reused to restrict the quantifier in these more straightforward examples, we should not be surprised when the same thing happens with quantified conditionals too. Hence we can expect a reading of (61) on which it is elliptical for (74):

(74)   Most people who survive a cardiac arrest change their diet if they survive a cardiac arrest.

But, under the restriction 'x survives a cardiac arrest', on the material reading the conditional 'x changes x's diet if x survives a cardiac arrest' is logically equivalent to 'x changes x's diet', so (74) reduces to (64), which was already identified as a natural paraphrase of (63). Thus, given independently attested contextual effects, (63) poses no problem for the material reading of the indicative conditional (see Kratzer 2020 for a similar approach).

For completeness, we should also try reading 'most people' in (63) as implicitly restricted to those who change their diet, rather than to those who survive a cardiac arrest. Thus (63) is read as elliptical for (75):

(75)   Most people who change their diet do so if they survive a cardiac arrest.

On one reading, (75) is trivial, because the quantifier restriction 'x changes x's diet' guarantees the conditional 'x changes x's diet if x survives a cardiac arrest'; that is quite consistent with the material interpretation. That is not a natural reading of (63), for it is not worth saying.

We can read (75) as non-trivial by taking it to involve implicit quantification over situations in which the agent changes their diet. That gives a reading something like (76):

(76)   For most people x such that in some situation x changes x's diet, for every situation s, if x survives a cardiac arrest in s, x changes x's diet in s.

The material reading of 'if' in (76) makes good sense. It continues to do so even when the quantifier restriction is removed:

(77)   For most people x, for every situation s, if x survives a cardiac arrest in s, x changes x's diet in s.

Neither (76) nor (77) is trivial, though (77) may be true just because most people never survive a cardiac arrest, and (76) may be true because most people who change their diet never survive a cardiac arrest. Neither point is a serious objection to the material interpretation of 'if'.

We might also understand the implicit quantification over situations in (63) as generic rather than strictly universal. That would fit naturally with a tendency to assess the conditional by the Suppositional Procedure. Clearly, many more variations can be played on these themes. There is no threat to the material interpretation here.

The implicit reuse of explicit material to restrict quantifiers also gives us reason to reconsider the role of 'if' under adverbs of quantification. After David Lewis's seminal 1975 paper, it became standard to treat 'if' clauses as restricting the quantifier. For example, (78) is analysed as something like (79):

(78) Beryl usually sneezes if she is nervous.
(79) In most situations in which Beryl is nervous, she sneezes.

Here, and in the cases discussed below, the same pattern applies to 'always/often/sometimes/rarely/never', analysed as 'in all/many/some/few/no situations'. Kratzer then generalized this treatment to other occurrences of 'if'. But the phenomenon is not specific to 'if'. Thus (80) is naturally understood as ambiguous between two readings, one paraphrased by (81) and one paraphrased by (82):

(80) Beryl usually sneezes after she sniffs flowers.
(81) In most situations in which Beryl sniffs flowers, she sneezes after she does so.
(82) In most situations in which Beryl sneezes, she does so after sniffing flowers.

The cases of 'before', 'because', and other such connectives in place of 'after' pattern similarly.

It is thus no surprise that (78) may be understood as something like (83):

(83) In most situations in which Beryl is nervous, she sneezes if she is nervous.

The material reading of 'if' makes (83) equivalent to the desired paraphrase (79).

Can we also read (78) as paraphrased by (84)?

(84) In most situations in which Beryl sneezes, she sneezes if she is nervous.

Such a reading is unnatural, because the quantifier restriction trivializes the conditional. Again, that is quite consistent with the material interpretation of 'if'.

Adverbs of quantification behave similarly when applied to simpler sentences. Thus (85) can typically be paraphrased by something like (86):

(85) He usually swims breaststroke.
(86) In most situations in which he swims, he swims breaststroke.

Here too, in order to produce a reasonable claim, some of the explicit material in the matrix of (85) is implicitly reused to restrict the quantifier, without being removed from the matrix. On the same pattern, (87) can typically be paraphrased by something like (88):

(87) He usually takes the train to London.
(88) In most situations in which he travels to London, he takes the train to London.

More generally, the explicit material is implicitly reused to restrict the quantifier to situations appropriate to it, as with (89) and (90):

(89) She usually has breakfast.
(90) She usually celebrates her birthday.

Thus, as they would normally be understood, (89) implies that she has breakfast most mornings, while (90) implies that she celebrates her birthday most years.

Normally, such quantifier restriction effects are not mandatory. For example, when (87) is uttered as an answer to the question 'What does he do on Sundays?', its natural reading can be paraphrased by (91), with no quantifier restriction like that in (88):

(91) On most Sundays, he takes the train to London at least once.

What will block the restriction is highly sensitive to background information. For example, still in answer to the question 'What does he do on Sundays?', (92) is more naturally understood along the lines of (93) rather than (94):

(92) He usually whispers.
(93) On most Sundays, in most situations in which he speaks, he whispers.
(94) On most Sundays, he whispers at least once.

For creatures who travel to and from London thousands of times a day, but rarely speak more than once a day, it might be the other way round.

This flexibility in how adverbs of quantification are restricted raises a concern about their interaction with 'if' clauses, where the restrictions may seem mandatory. Then a special semantic or syntactic mechanism would presumably be needed to mandate them, undermining the analogies proposed above. The Suppositional Rule may indeed put strong pressure to make the antecedent of a conditional restrict adverbs of quantification. Nevertheless, the restriction is not mandatory. For example, I am asked to describe someone's temperament. I know that she very rarely shows anger, but I have no idea how often she is angry without showing it. I may say:

(95) Usually, if she is angry, she doesn't show it.

This is better paraphrased as the unrestricted (96) than as the restricted (97), in both cases with the material reading of 'if':

(96) In most situations, if she is angry she does not show that she is angry.
(97) In most situations in which she is angry, if she is angry she does not show that she is angry.

For I know that (96) is true, whereas for all I know (97) is false, because she is rarely angry but always shows it when she is. Thus, even for sentences with conditionals, the antecedent does not mandatorily restrict adverbs of quantification. No special semantic or syntactic mechanisms are needed. The proposed analogies stand.

Huitink (2009, following von Fintel 1994) objects to treating the 'if' clause as covertly restricting an adverbial quantifier. She considers a dialogue like this:

(98)   A: When do you play soccer?
       B: We usually play if the sun shines.

Here B's utterance is to be paraphrased as: in most relevant situations in which the sun shines, they play soccer (redundantly: if it shines). However, Huitink claims that this effect must be achieved by the overt role of the 'the sun shines' as an input to a connective 'if', not by any covert role restricting 'usually'.[7] She argues that 'if' clauses covertly restrict quantifiers only when they constitute the topic, but that 'if the sun shines' does not constitute the topic of B's utterance, even if stress on 'sun' puts it in focus. But consider this variation on (98):

(99)   A: When do you play soccer?
       B: We usually play after the rain stops.

Here B can naturally be heard as meaning that in most relevant situations in which the rain stops, they play soccer after it stops. If B's 'if' clause does not constitute the topic of her utterance in (96), her 'after' clause does not constitute the topic of her utterance in (99). That restriction is not mandatory: we could alternatively hear B as meaning that in most relevant situations in which they play soccer, they do so after the rain stops. Huitink's argument seems as applicable to the first reading of (99) as to (98). But the restricting effect in (99) cannot be achieved by the overt role of 'the rain stops' as an input to the connective 'after'. Thus Huitink's argument depends on too narrow a view of the resources available for quantifier restriction.

In general, the restriction effects occur more or less unreflectively, as part of the normal process of charitable interpretation in context. They do not require any overt lexical item dedicated to the role of introducing the quantifier restriction. In the case of indicative conditionals, they can easily be predicted on general grounds, given the material reading of 'if'. Therefore, they provide no evidence against that reading. In Part II of the book, we shall observe a similar pattern when 'if' interacts with modal operators, as in subjunctive conditionals.

The option of reusing the antecedent of a conditional in the scope of a quantifier to restrict the quantifier might suggest a less drastic treatment for some earlier cases, such as (42) ('No student failed if he worked hard'). We might try analysing it as having the truth-conditions of (100):

(100)   No student who worked hard failed if he worked hard.

On the material interpretation, (100) is equivalent to the apparent paraphrase (45) ('No student who worked hard failed'). But that does not fit the envisaged context. For (53) ('Ned failed if he worked hard') was treated, very naturally, as a direct counterexample to (42), where Ned was any one of the students. But (53) does not provide a counterexample to (100), for that would require a student who worked

---

[7] Huitink intends this as an argument for her Belnap-style trivalent semantics. Section 8.1 already noted some problems for that approach.

hard, and (53) does not specify whether Ned worked hard. It only tells us that if Ned worked hard, he was a counterexample to (100).

Tacit quantifier restrictions in any case fail to give promising treatments of the problems discussed in section 8.4, for the problems still arise when we block the restrictions by putting 'every one of the 100 students' for 'every student', 'no one of the 100 students' for 'no student', and so on.

By contrast, consider (101):

(101)   Beryl never sneezed if she was nervous.

Imagine someone objecting:

(102)   At her swearing-in, Beryl sneezed if she was nervous.

It is hard to hear (102) as a direct counterexample to (101), since it does not imply that Beryl was nervous at her swearing-in. It only tells us that if she was nervous at her swearing-in, there was a counterexample to (101).

Such differences need not mark any deep structural contrast between adverbs of quantification such as 'never' and nominal quantifiers such as 'No student'. Perhaps that cohort of students constitutes a more salient domain than does the collection of all situations Beryl was in during that period of her life, making it easier to hear implicit restrictions on the latter than on the former. The reliability of the Suppositional Procedure may be sensitive to such subtle cognitive differences, since contextual restrictions can work analogously to suppositions.

At first sight, the interaction of conditionals with quantifiers looked like a rich source of devastating counterexamples to the material interpretation. But once we examine the data more critically, we find that they do not point in that direction. Their complexities can be understood as the product of the complexities of quantification and the complexities of the conditional already analysed in previous chapters. In particular, without a proper account of the fallible heuristics we use in assessing conditionals, we cannot understand the competing pressures which sometimes drive us to inconsistency in our assessments of fairly simple sentences of our native language.

PART II
Would If

9
Conditionals and Abduction

This book has emphasized the need for a theory of conditionals to make sense of their cognitive role, not least in scientific reasoning. Here is another example. Scientists are testing two theories T and T*, which make different predictions about the result of an experiment. The scientists do the experiment. The observed result is exactly what T predicted, and inconsistent with what T* predicted. The observation itself is beyond serious doubt. The scientists assert both (1) and (1*):

(1)   If T were true, the experiment would have exactly this result.
(1*)   If T* were true, the experiment would not have exactly this result.

Thus the experiment confirms T and disconfirms T*. If they are the only available competitors, the scientists may reason non-deductively to T, by inference to the best explanation of the evidence, or *abduction*. The conjunction of (1) and (1*) captures the relevant difference between T and T*.

Try simplifying the account by eliminating 'would' from the conditionals:

(2)   If T is true, the experiment has exactly this result.
(2*)   If T* is true, the experiment does not have exactly this result.

The trouble is that, unlike (1*), (2*) is indefensible. The result of the experiment is known independently of T and T*. The scientists assume that if T* is true, one of the auxiliary assumptions on which the falsified prediction depended was false, although they currently regard the possibility of such an error as wild. The scientists assert (2) but deny (2*). Thus (2) and (2*) are no substitute for (1) and (1*) as premises for the abductive argument.

Conditionals grammatically like (1) and (1*) are often described as 'counterfactual', to contrast them with conditionals grammatically like (2) and (2*). However, the term is quite misleading. Both the antecedent and the consequent of (1) may well be true, and the scientists do not assume otherwise when using (1) as a premise of their argument for the antecedent of (1) (the point goes back to Anderson 1951, with reference to a similar example).

Another traditional way of distinguishing between conditionals like (1)/(1*) and conditionals like (2)/(2*) is by calling the latter 'indicative' and the former 'subjunctive'. For convenience, the term 'indicative' was sometimes used like that in Part I of this book. Grammatically, however, it is not clear that English has a genuine subjunctive mood, let alone that such a mood is present in (2)/(2*). Even amongst languages with a clearly identifiable subjunctive mood, it is not properly correlated with the intended semantic distinction (Iatridou 2000).

Suppose and Tell: The Semantics and Heuristics of Conditionals. Timothy Williamson, Oxford University Press (2020).
© Timothy Williamson.
DOI: 10.1093/oso/9780198860662.001.0001

The large difference in cognitive role between (1)/(1*) and (2)/(2*) needs to be explained, preferably in terms of some difference in their compositional semantic structure which fits them to play different cognitive roles.

The scientists may assess both (1)/(1*) and (2)/(2*) by versions of the Suppositional Procedure. However, the difference in their verdicts on the two pairs is evidence of some crucial difference in how the suppositions are handled. Identifying the difference will deepen our understanding of reasoning under suppositions. Somehow, the scientists' knowledge of the result of the experiment is available to reasoning from suppositions as to which theory is true for sentences like (2)/(2*) but not for sentences like (1)/(1*).

Chapter 10 discusses the difference in semantic structure between the two classes of sentence in detail. For now, we can rely on our pre-theoretic understanding of the sentences to make a preliminary exploration of the cognitive role of sentences like (1)/(1*) in abduction.

Here is another example. A tracker comes across some trampled mud, broken twigs, and the like. After careful examination, he comes to the conclusion that two stags had a fight here. That hypothesis explains the evidence better than any other does. His reasoning involves a conditional like this:

(3)   If two stags had fought here, there would now be marks like these.

Clearly, (3) is closer to (1)/(1*) than to (2)/(2*).

Morphologically, 'had fought' looks like the past perfect tense of 'fight', a further step back into the past from the ordinary past tense 'fought', and 'would' looks like the ordinary past tense of 'will'. Thus (3) looks like the result of putting (4) into the past:

(4)   If two stags fought here, there will now be marks like these.

However, the semantic function of the past morphology in (3) cannot be understood simply in terms of time reference, for the relevant time when there are marks like these is the time of utterance itself. Thus, if time reference were all that mattered, (3) would reduce to (5):

(5)   If two stags fought here, there are now marks like these.

But (5) is relevantly like (2)/(2*). The tracker can trivially verify (5), with no need to consider whether a stag fight would leave marks like these. For, independently of whether two stags fought here, he can see that there *are* marks like these. Updating online or offline on the supposition that two stags fought here will not make him withdraw the claim that there are marks like these. Nor will updating on the supposition that no two stags fought here make him withdraw the claim. He will assent equally to (6):

(6)   If no two stags fought here, there are now marks like these.

Thus (5) is insensitive to the connection between antecedent and consequent crucial to the abductive reasoning to which (3) is crucial.

Of course, (5) and (6) may be misleading and conversationally inappropriate in the absence of a connection between antecedent and consequent, but the tracker should not deny them on that account. For he can clearly assert the following two conditionals, given his independent knowledge that there are marks like these:

(7)   If two stags fought here, there are now marks like these after two stags fought here.

(8)   If no two stags fought here, there are now marks like these after no two stags fought here.

But (7) clearly entails (5), and (8) clearly entails (6), just by eliminating a conjunct from the consequent, a trivial move licensed by the Suppositional Procedure.

The tracker cannot assert (3), unlike (5), merely on the basis of knowledge of its consequent, independent of its antecedent. For then he could equally assert the corresponding conditional with the antecedent negated:

(9)   If no two stags had fought here, there would now be marks like these.

In the envisaged circumstances, the tracker has no basis for asserting (9). Indeed, he may reasonably deny (9).

We might provisionally envisage the difference between (3) and (5) like this. The past tense morphology in (3) is being recruited to constrain what background knowledge may be used in the assessment of its consequent on the supposition of its antecedent. The assessment is to be made as of the time of the stag fight, rather than as of the time of the marks. In a sense, the point is to avoid giving the development of the antecedent the benefit of hindsight, although, more precisely, the constraint is not on the times when the speaker acquired the knowledge, but on the times of any specific events the knowledge is about. If the tracker acquired some purely general information about stags after what was in fact the time of the fight, he can legitimately use it in the assessment of (3). By contrast, if he knew before the time of the fight that there would be marks like these, because a reliable oracle told him, he cannot verify (3) on that basis. Such constraints on the role of background knowledge enable the assessment process to focus on the underlying causal connections. Although these remarks constitute only a crude preliminary sketch, they give some feel for the difference between conditionals like (3) and conditionals like (5).

One might put the contrast like this: in applying the Suppositional Procedure to conditionals with 'would' like (1)/(1*) and (3), one can make sensible comparisons with actuality *from inside* the development of the supposition; in applying the Procedure to plain conditionals like (2)/(2*) and (5), one cannot. Thus (10) is perfectly sensible in appropriate circumstances:

(10)   If two stags had fought here, there would now not be marks like these.

One can conclude from (10) that no two stags fought here. By contrast, the consequent of (11) is self-defeating, on its intended reading:

(11)   If two stags fought here, there are now not marks like these.

On the development of the supposition that two stags fought here for the plain conditional, one does not deny that there are the very marks one is pointing out. One does not even conclude that one must be hallucinating them, though one might say:

(12)   If two stags fought here, they left surprisingly few marks.

The contrasts suggest that 'would' introduces some sort of modal element.

To vary the example, imagine that a friend is complaining of a hangover. One might say, self-righteously:

(13)   If you had drunk less last night, your head would not hurt as much.

Less elliptically, more pedantically:

(14)   If you had drunk less last night than you did drink last night, your head would not hurt as much now as it does hurt now.

It would be absurd to use the corresponding sentence without 'would', saying:

(15)   If you drank less last night, your head does not hurt as much.

For here (15) is understood as elliptical for (16):

(16)   If you drank less last night than you did drink last night, your head does not hurt as much now as it does hurt now.

The antecedent of (15) and (16) is inconsistent, since it says that something is less than itself. Similarly, the consequent of (15) and (16) is inconsistent, since it says that something is not as much as itself. Thus (15) and (16) have no lesson for one's friend.

In another context, (15) might be understood as elliptical for (17) instead of (16):

(17)   If you drank less last night than you drank the night before, your head does not hurt as much this morning as it hurt yesterday morning.

That makes good sense. In developing the supposition for a plain conditional, one can compare different times. One can even compare different possibilities. For in a suitable context (15) might be understood as elliptical for (18):

(18)   If you drank less last night than you hoped you would, your head does not hurt as much as you feared it would.

Here 'hoped' and 'feared' introduce comparisons between possibilities. But simply within the supposition itself, no contrast with actuality is available.

The contrast between plain conditionals and 'would' conditionals persists even when they concern future events. For example, the future analogue of the subjunctive (14) is (19):

(19)   If you were to drink less tonight than you will drink tonight, your head would not hurt as much tomorrow morning as it will hurt tomorrow morning.

That makes sense as a defeatist warning. In a suitable context, one can say the same thing less verbosely, in a form corresponding to (13):

(20)   If you were to drink less tonight, your head would not hurt as much tomorrow morning.

By contrast, the plain analogue of (16) is just as unhelpful as (16), and for exactly the same reason:

(21)   If you drink less tonight than you will drink tonight, your head will not hurt as much tomorrow morning as it will hurt tomorrow morning.

The elliptical form standing to (21) as (20) stands to (19), and as (15) stands to (16), is (22):

(22)   If you drink less tonight, your head will not hurt as much tomorrow morning.

It is hard to hear (22) as elliptical for (21) simply because more sensible readings are so salient, for example (23):

(23)   If you drink less tonight than you usually drink, your head will not hurt as much tomorrow morning as it usually hurts the morning after.

The future tense examples show that one cannot understand the point of the past tense morphology of 'would' conditionals as being merely to pull the assessment of the connection between antecedent and consequent back to a perspective before the events at issue, when they are still in the potential future. As imaginarily uttered in the past, (21) is no less absurd. The function of 'would' in setting up a comparison with actuality from within the development of the supposition is irreducible to considerations of time and tense. Semantically, the past tense in which (2) differs from (1) is a 'fake past' tense, even though the use of such a 'fake past' is common to many languages (Iatridou 2000). It may have its roots in the architecture of human cognition.[1]

A further sign that the function of the past tense in 'would' conditionals is not fundamentally about time is that the function can be served even when the antecedent concerns an alternative course of events that shares no initial segment with the actual course of events, so there is no time before the divergence from actuality to go back to. Here is an example:

(24)   If space had always been two-dimensional, thinking would never have occurred.

That is a sensible claim, quite likely true. It is very doubtful that two-dimensional space would permit the sort of complexity necessary for thought. By contrast, the plain analogue of (24) is this:

(25)   If space has always been two-dimensional, thinking has never occurred.

By normal standards, (25) is crazy, or at best like the example 'If he's a qualified doctor, I'm the Pope' (sections 2.2 and 5.2). We know for sure that thinking *does*

---

[1] For an older interpretation of counterfactuals which takes their tense more literally see Woods 1997: 78–92, criticized by Edgington 1997: 127–30. For a recent account of the distinction between indicative and subjunctive conditionals which also give more weight to the temporal reading of the tense morphology see Khoo 2015. For more discussion of modal versus temporal readings of the tense morphology see Schulz 2014, 2017 and Mackay 2015.

occur. On the development of the supposition that space is two-dimensional for (25), one does not conclude that thinking must be an illusion. Rather, if one is willing to take the supposition seriously in the first place, one assumes that two-dimensional space must, somehow or other, allow thinking to occur.

Such contrasts are pertinent to how we assess scientific theories. When we suppose a theory true, in order to work out and test its consequences, we want to make comparisons with actuality from within the development of the supposition. That requires the development to be like that for the 'would' conditionals (1)/(1*) rather than like that for the plain conditionals (2)/(2*). The theory may be about all of spacetime: there is no going back to a time before the hypothetical universe diverged from the actual universe.

A more general characterization of this cognitive role of 'would' conditionals is that they allow us to test hypotheses more fully by making our development of the hypothesis being supposed depend more on the resources of the hypothesis itself and less on our antecedent background knowledge and belief. Of course, we must almost always use *some* of our background knowledge or belief, for we should get nowhere from a perfectly isolated hypothesis, but the aim is to limit the assumptions we use to those not at issue between the hypothesis under test and its competitors. How far we live up to that ideal is another matter. The 'would' conditional typically gives us our best shot.

10

The Interaction of 'If' and 'Would': Semantics and Logic

10.1 Conditionals with and without 'would'

Traditionally, conditionals have been divided into two sorts: 'indicative' and 'subjunctive'. To use a traditional example, (1) and (2) can easily differ in truth-value in the same context:

(1) If Oswald did not shoot Kennedy, someone else did.
(2) If Oswald had not shot Kennedy, someone else would have.

The grammatical distinction can correlate with deep cognitive differences: chapter 9 sketched an aspect of that phenomenon.

'Indicative'/'subjunctive' minimal pairs such as (1) and (2) are standardly formalized with the same antecedent and the same consequent. Typically, they are represented with different connectives, for example as $A \rightarrow C$ and (in Lewis's notation) $A \,\square\!\!\rightarrow C$. Although Lewis intended the symbol $\square\!\!\rightarrow$ to suggest a combination of the necessity operator $\square$ with the indicative conditional $\rightarrow$, his semantics does not work that way; it treats $\square\!\!\rightarrow$ non-compositionally as a primitive connective. Alternatively, such 'indicative'/'subjunctive' pairs may receive exactly the same formalization, for example (in Stalnaker's notation) $A > C$, which does not yet explain how they can differ in truth-value in the very same context. Either way, in effect, two readings of 'if' are postulated: the 'indicative' conditional and the 'subjunctive' conditional. In practice, they are often studied quite separately from each other, sometimes as though they were completely independent.

Methodologically, it is odd to explain the truth-conditional difference between two sentences by postulating something like ambiguity in a part where they look exactly the same, while marginalizing the visible differences elsewhere. The natural default hypothesis is that 'if' means exactly the same in (1) and (2), while the semantic difference between them comes from the overt difference in verb forms. On a simple compositional approach, the latter difference expresses a semantic difference in parts of the sentences other than 'if', which then interacts with a uniform semantics for 'if' to generate the difference in truth-conditions between (1) and (2). Although lack of progress with that approach might eventually force one to try something different, that should be a desperate resort, not the starting point.

Suppose and Tell: The Semantics and Heuristics of Conditionals. Timothy Williamson, Oxford University Press (2020).
© Timothy Williamson.
DOI: 10.1093/oso/9780198860662.001.0001

Independent treatment for (1) and (2) is anyway implausible: the role of 'if' in them is too similar. Moreover, the binary division sits uneasily with intermediate examples such as (3), uttered in 1962:

(3)   If Oswald does not shoot Kennedy, someone else will.

On one hand, (3) seems to differ grammatically from (1) only in tense: past morphology for (1), future morphology for (3). On the other hand, (3) seems to differ cognitively from (2) only in temporal perspective: assessing (2) seems very close to simulating a past assessment of (3). One cannot study such patterns if one treats (1) and (2) in isolation from each other. A binary taxonomy is too crude. This chapter will confirm that initial judgement.

While the material reading for 'if' has traditionally been treated as at least a candidate for the interpretation of 'indicative' conditionals, albeit one in current disfavour, it has generally been assumed to be a non-starter for the interpretation of 'subjunctive' conditionals. By contrast, this chapter will explain how a uniformly material semantics for 'if' interacts compositionally with a modal treatment of 'would' to generate plausible truth-conditions for sentences like (2). The semantic poverty of 'if' acts as a fruitful methodological constraint on theorizing about such conditionals, reducing the number of moving parts, and avoiding ultimately unhelpful complications.

Another unfortunate side effect of standard approaches is that 'would' is not treated semantically in isolation from 'if', as though 'would-if' were a primitive. Yet 'would' can occur without 'if', meaning exactly what it does in (2). To understand its semantics, we must first examine such simpler 'would' sentences, to see how the word works independently of the conditional. We shall see how the modal semantics of 'would' is rich enough to interact with the austere truth-functional semantics of 'if' to generate an adequately complex semantics of sentences like (2).

10.2 The semantics of 'would'

Here are some comparatively simple sentences with 'would' but no 'if':

(4)   Loretta would not betray a friend.
(5)   Marmaduke would have married someone.
(6)   The wall would have collapsed sooner or later.

None of these need be understood as elliptical for a conditional (contrast Kasper 1992). For example, (4) reports in effect on Loretta's loyal character: that she has never betrayed a friend is no coincidence; it is modally robust. Likewise with (5): however unsuitable Marmaduke is to be a husband, it is not bad luck that he has married; it was always going to happen. Again, with (6): don't bother looking for proximal causes of this collapse; they are not what matter.

The natural hypothesis is that 'would' in (4)–(6) is a modal operator, introducing generalizations over possibilities. Thus (4)–(6) have approximately the overall form of (4a)–(6a):

(4a)   Would(Loretta does not betray a friend).
(5a)   Would(Marmaduke married someone).
(6a)   Would(the wall collapses sooner or later).

For simplicity, (4a)–(6a) treat the subject term of (4)–(6) ('Loretta', 'Marmaduke', 'the wall') as within the scope of 'would'. That is natural enough. Consider this comment on a hypothetical constitutional crisis:

(7) The Prime Minister would be consulted.

This is ambiguous between a *de re* reading on which 'the Prime Minister' is outside the scope of 'would' and denotes the *actual* Prime Minister, and a *de dicto* reading on which it is inside the scope of 'would' and denotes whoever *would* be Prime Minister in the hypothetical circumstances. Which reading is more natural depends on the conversational context. In what follows, sample sentences are to be read *de dicto*.

What sort of modality does 'would' express? It is not epistemic; (4) may in fact be true even though no relevant body of knowledge entails that Loretta does not betray a friend; similarly for (5) and (6). Nor is the modality deontic; (5) may be true even though it is a thoroughly bad thing that Marmaduke married someone; similarly for (6), and it is no part even of what (4) says that Loretta *ought* not to betray a friend. Rather, 'would' in (4)–(7) is *circumstantial*; it ranges over objective possibilities, determined by how things are. That is not to say that (4)–(6) attribute metaphysical necessity to the complement of 'would' in (4a)–(6a). Even though it is metaphysically possible for Loretta to betray a friend, (4) may still be true. 'Would' typically expresses a restricted or local circumstantial modality (more on this below).

'Would' is naturally read as a restricted *necessity* operator, in effect a restricted universal quantifier over worlds. That explains the validity of deductions like that from (8) and (9) to (10) (in a fixed context), by contrast with a rival candidate such as a restricted possibility operator or one equivalent to 'in most relevant worlds'. The conversation concerns a potential volcanic eruption:

(8) It would last more than a day.
(9) It would last less than a week.
(10) It would last between a day and a week.

On an interpretation of 'would' as 'in some relevant worlds', (8) and (9) are true while (10) is false in a context where the eruption lasts just one hour in some relevant worlds, and a month in others, but does not last between a day and a week in any relevant world. Similarly, on an interpretation of 'would' as 'in most relevant worlds', with 'most' understood as 'more than 50 per cent', (8) and (9) are true while (10) is false in a context where the eruption lasts less than a day in one third of the worlds, between a day and a week in another third of them, and more than a week in the remaining third.

The argument (8)–(10) is analogous to a valid argument about a future eruption, with 'will', of which 'would' is grammatically the past tense.

(8*) It will last more than a day.
(9*) It will last less than a week.
(10*) It will last between a day and a week.

However, 'will' seems to concern just *one* open future history. For claims such as (11*) sound valid:

(11*) Either it will last less than a week or it will last at least a week.

If 'will' is a universal quantifier over several open future histories, (11^*) is false when the eruption lasts at least a week in some open future histories and less than a week in others, as may easily be, since then each disjunct is false. Thus one may hold that 'will' is restricted to just one open future history (Cariani and Santorio 2018). For a domain with only one member, the differences between 'all', 'some', and 'most' disappear.

Analogous claims with 'would' may also sound valid:

(11)   Either it would last less than a week or it would last at least a week.

If 'would' is a universal quantifier over several relevant worlds, (11) is false when the eruption lasts at least a week in some relevant worlds and less than a week in others, as may easily be, since then each disjunct is false. Thus one may be tempted to hold that 'would' is restricted to just one relevant world. Again, for a domain with only one member, the differences between 'all', 'some', and 'most' disappear.

However, the singular interpretation of 'would' looks implausible for (4)–(6), which a speaker may use to emphasize the modal robustness of a phenomenon, its persistence across a wide range of contextually relevant worlds. For now, this issue will be left unresolved. It will be revisited later in the chapter, where reason will emerge to regard the singular interpretation of 'would' as an illusion, generated by an analogue of the Suppositional Procedure. We will provisionally treat 'would' as a universal quantifier over many contextually relevant worlds, while postponing the full justification for this approach.

As already noted, in an ordinary context (4) does not claim that it is metaphysically impossible for Loretta to betray a friend: she could have been brought up in very different circumstances and acquired a different character. The implicit universal quantifier in (4) is implicitly restricted to contextually relevant worlds, in some sense. Presumably, they are worlds in which Loretta's character is roughly the same as in the actual world and other circumstances are not radically different. But they should include some such worlds in which she has an opportunity and incentive to betray a friend, otherwise the truth of (4) comes too cheap to be relevantly informative. The contextual restriction has been left utterly schematic, but that is as it should be. Just which worlds will be relevant in a given context is a messy, complicated business. Pragmatics may cast some light on it, but for present purposes the schema suffices.

The actual world has *not* been assumed to satisfy the contextual restriction, although it plausibly does so in the contexts envisaged for (4)–(6). But when we are developing a scenario in which a tiger enters the room, we may truly utter (12), even though in the actual world Francis is not frightened:

(12)   Francis would be frightened.

'Would be' does not in general imply 'is'.

We can now start combining such sentences with 'if', as in (13)–(15):

(13)   If Loretta is loyal, she would not betray a friend.
(14)   If Loretta would not betray a friend, she is loyal.
(15)   If Loretta would not betray a colleague, she would not betray a friend.

These are all 'would' conditionals in the minimal sense that they are all conditionals with 'would' in the antecedent or consequent, or both.

However, on their most natural readings, (13)–(15) work differently from the usual kind of so-called 'subjunctive' conditionals, such as (2) and (16):

(16) If Loretta were loyal, she would not betray a friend.

'Were' in the subordinate clause of (16) is anaphoric on 'would' in the main clause: the idea is that Loretta does not betray a friend in a world in which she is loyal. Thus the natural analysis of (16) according to the proposed schema assigns 'would' wide scope with respect to 'if':

(16a) Would(if Loretta is loyal, Loretta does not betray a friend).

By contrast, 'is' in the subordinate clause of (13) is *not* anaphoric on 'would' in the main clause; the idea is more demanding than in (16): that if Loretta is loyal in the actual world, then she does not betray a friend in any contextually relevant world. Thus the natural analysis of (13) according to the schema assigns 'would' narrow scope with respect to 'if':

(13a) If Loretta is loyal, would(Loretta does not betray a friend).

The natural analysis of (14) is just the converse of (13b):

(14a) If would(Loretta does not betray a friend), Loretta is loyal.

Similarly for (15):

(15a) If would(Loretta does not betray a colleague), would(Loretta does not betray a friend).

We may assume that the contextual restriction remains constant from antecedent to consequent in (15a), but there is still no world-by-world coordination between the antecedent and consequent.

The differences become even clearer on a material interpretation of 'if':

(13b) Loretta is loyal ⊃ would(Loretta does not betray a friend)

(14b) Would(Loretta does not betray a friend) ⊃ Loretta is loyal

(15b) Would(Loretta does not betray a colleague) ⊃ would(Loretta does not betray a friend)

(16b) Would(Loretta is loyal ⊃ Loretta does not betray a friend)

With the standard example (16)/(16b), the interaction of the material reading of 'if' and the contextually restricted modal reading of 'would' produces a restricted strict conditional. With the non-standard examples (13)/(13b), (14)/(14b), and (15)/(15b), the interaction simply produces a material conditional with a contextually restricted modal antecedent or consequent.

We can try making the verb in the main clause anaphoric on 'would' in the subordinate clause. For example, what stands to (14) as (16) stands to (13)? We got (16) from (13) just by changing 'is' to 'were'. Mechanically applying the same operation to (14) results in (17):

(17)# If Loretta would not betray a friend, she were loyal.

But (17) seems ill-formed, even though the modally anaphoric 'were' comes after the 'would' on which it is meant to depend, a more natural-seeming order than in (16). The intended truth-conditions are as in (17a), and so on the material interpretation (17b):

(17a)   Would(if Loretta does not betray a friend, Loretta is loyal)
(17b)   Would(Loretta does not betray a friend ⊃ Loretta is loyal)

But to express those truth-conditions properly one needs something like (18):

(18)   If Loretta were not to betray a friend, she would be loyal.

Not surprisingly, 'would' in the subordinate clause cannot normally scope over the main clause.[1]

The assignment of wide scope to 'would' in sentences like (16) and (18) may seem to run into trouble with sentences like (19):[2]

(19)   If Alice hadn't asked Jim to the ball on Thursday, then on Friday Jessica would have and Jill might have.

The trouble is that both 'would' and 'might' seem trapped in the consequent of (19), for assigning wide scope to either would result in a reading with one modal operator inside the scope of the other, contrary to what is intended. However, a similar problem arises for non-modal examples such as (20):

(20)   If put in water, most samples turned red but some turned green.

The trouble is that both 'most' and 'some' seem trapped in the consequent of (20), for assigning wide scope to either would result in a reading with one quantifier inside the scope of the other, contrary to what is intended. However, on the natural reading of (20), it is elliptical for something like (21):

(21)   Most samples turned red if put in water but some samples turned green if put in water.

Of course, the interactions of the quantifiers with the 'if' clauses in (21) pose further problems, but they are of a sort discussed in chapter 8, and pose no threat to the overall parsing of (20) as (21). Similarly, on the natural reading of (19), it is elliptical for something like (22):

(22)   Jessica would then have asked Jim to the ball on Friday if Alice hadn't asked Jim to the ball on Thursday and Jill might then have asked Jim to the ball on Friday if Alice hadn't asked Jim to the ball on Thursday.

---

[1] Note that interpreting 'if A, would C' as 'would(A ⊃ C)' is truth-conditionally equivalent to interpreting 'if A' as restricting the modal operator 'would', applied to C, since 'would' acts like a (restricted) universal quantifier over worlds.

[2] Thanks to an anonymous referee for this example.

The intended overall structure of (22) is this:

(22a)   would(if Alice did not ask Jim to the ball on Thursday, then Jessica asked Jim to the ball on Friday) and might(if Alice did not ask Jim to the ball on Thursday, then Jill asked Jim to the ball on Friday).

The dangling occurrences of 'would have' and 'might have' in (19) already indicate some ellipsis.

The interaction of 'would' with 'if' in (22)/(22a) fits the pattern already discussed. At first sight, the interaction of 'would' with 'might' in (22)/(22a) looks more worrying. Although 'might' is often used as an epistemic modal, its use in (19) in parallel with the non-epistemic 'would' suggests reading it as the equally non-epistemic dual of 'would': 'might(A)' is true just in case A is true at some relevant world, where the relevant worlds are the same as for 'would'. But there is the further worry that the material reading of 'if' makes the truth-conditions of sentences of the form 'might(if A, C)' too weak, since any relevant world at which A is false is a world at which 'if A, C' is true, and so suffices to verify 'might(if A, C)'. However, we can recover a stronger reading of 'might if' by following the pattern of section 8.5, where the explicit antecedent of a conditional under an adverb of quantification can be reused as an implicit restriction on the quantifier domain. Thus the worlds relevant for 'might' in 'might(if A, C)' may be further restricted to those at which A is true (although the restriction is unlikely to be obligatory). Under that further restriction, the truth-conditions for 'might(if A, C)' collapse into those for 'might(A and C)'. That surprising-looking prediction is confirmed. In particular, one can hear (23) and (24) as equivalent:

(23)   If Alice hadn't asked Jim to the ball on Thursday, then on Friday Jill might have.

(24)   Jill might have asked Jim to the ball on Friday while Alice didn't on Thursday.

Imposing the same restriction on the worlds relevant for 'would' in 'would(if A, C)' is harmless, because it makes no difference to its truth-conditions. In brief, the interactions of the modal operators with the 'if' clauses in (22)/(22a) pose no threat to the overall parsing of (19) as (22)/(22a). Thus examples like (19) do not undermine the treatment of 'would' as scoping over 'if' in sentences such as (16), (18), and (19).

The variations amongst (13)–(18) manifest the inadequacy of a binary distinction between 'indicative' and 'subjunctive' conditionals, and especially of attempts to base that distinction on anything like a putative ambiguity in 'if'. There is a distinction between sentences, such as (13)–(15), whose analysis takes the overall form of a bare conditional, and those, such as (16) and (18), whose analysis takes the overall form of a modally qualified conditional: call the former *unmodalized conditionals* and the latter *modalized conditionals*. However, no such binary distinction does justice to the complex play of variations to be expected from the interaction of the material conditional and a modal operator.

The literature on 'subjunctive' conditionals has mainly focussed on modalized conditionals of the 'would if' form. On the semantics just sketched, they are contextually restricted strict conditionals.

This non-epistemic modal generality of 'would if' conditionals gives hope for explaining their distinctive role in abductive reasoning, in contrast with plain conditionals (chapter 9). Usually, some worlds relevant for 'would' are non-actual, so the known truth of a material conditional does not imply the truth of the corresponding 'would if' conditional. Something more counterfactually robust is needed. A good explanatory connection between the hypothesis H and some evidence E is just the kind of thing needed for the truth of 'Would(if H, E)'.

One general methodological moral is salient. The traditional and still widespread practice of theorizing about 'would if' conditionals, in the absence of a theory about 'would' in general, is quite unsatisfactory. 'Would' has a life of its own; without a proper attempt to understand that independent life, we are poorly placed to understand what 'would' is doing in 'would if'.

Section 10.2 explains the consequences of the present semantic proposal for 'would if' conditionals. Chapter 11 discusses the application of a suppositional procedure to 'would', and its interaction with the corresponding procedure for 'if'.

10.3 The logic of 'would if'

We treat validity in standard model-theoretic terms, familiar from possible world semantics for modal logic, and more specifically from Kripke models.

We sketch only the relevant features of models. Each is equipped with a set of worlds, one of which is the designated actual world of the model, @. Sentences are evaluated as true or false at each world (given values for whatever other semantic parameters are in play). A sentence is true in the model if and only if it is true at @. The evaluation of sentences at worlds is classical: at any given world, A is false just in case A is not true; 'not(A)' is true just in case A is false; 'A and B' is true just in case A is true and B is true. In particular, conditionals are evaluated materially: at any given world, 'if A, C' is true just in case either A is false or C is true. 'Would' receives the usual possible world treatment: there is a dyadic accessibility relation R between worlds; 'would(A)' is true at a world w just in case A is true at every world to which w has R. The set of worlds accessible from w is R(w).

The relation R represents the contextual restriction on worlds. It is fixed by the model itself; validity holds context fixed. The model theory imposes no special restrictions on R: the contextual restriction on 'would' is a matter of pragmatics, not semantics. In some extreme contexts, R may be the universal relation, holding between any worlds. Then the restriction is redundant: all worlds are relevant. Towards the other extreme, R may be the identity relation, holding only between a world and itself. Then 'would' is redundant, for 'would(A)' becomes equivalent to A. But R need not be a reflexive relation; as noted with example (12), 'would(A)' does not always imply A. That can happen because, wittingly or unwittingly, speakers may impose constraints on relevance false in the actual world. In an even more extreme context, R may be the empty relation, holding between no worlds. Then 'would(A)' is vacuously true, irrespective of A. That can happen because speakers unwittingly impose mutually inconsistent constraints on relevance. Although such extreme contexts may be rare, there is no special *semantic* reason to exclude them.

The restriction R takes the form of a dyadic relation rather than a monadic property to allow for contingency in which worlds are ranged over. For example, consider a context where 'would' expresses nomic necessity. Thus 'would(A)' is true at a world w just in case A follows from the laws at w. With w as the world of evaluation, 'would' ranges over just the worlds where the laws at w hold. Similarly, with another world of evaluation x, 'would' ranges over just the worlds where the laws at x hold. But the laws at x may differ from the laws at w. Thus 'would' ranges over different sets of worlds depending on the world of evaluation. Such a reading of 'would' makes good sense. Equating the contextual restriction with a single monadic property of worlds does not allow for such variation in the domain. That is why we use a dyadic relation between worlds instead.

However, the flexibility of the model theory does *not* make it neutral between most formal semantic theories of counterfactual conditionals. For it automatically validates many highly controversial principles of counterfactual logic, as explained below, and no restriction on the class of models will invalidate them (though it may validate some principles invalidated by the unrestricted model theory). Thus the model theory is by no means anodyne.

The standard model-theoretic version of validity is truth-preservation in every model, sometimes called *real-world validity*, because truth in a model is identified with truth at @, the actual world of the model. More formally, the argument from a set of premises AA to a conclusion C is real-world valid ($AA \vDash_{RW} C$) just in case for every model M, if every member of AA is true in M then C is true in M. A formula C is real-world valid just in case the argument from the empty set of premises to C is real-world valid, in other words, just in case C is true in every model.[3]

For most purposes, real-world validity does fine. However, it has one relevant limitation: it does not automatically validate the unrestricted rule of necessitation for 'would'. In other words, this principle can fail:

Necessitation$_{RW/RW}$ # Whenever $\vDash_{RW} A$, $\vDash_{RW}$ would(A)

The reason is that the real-world validity of A means only that A is true at the actual world of each model, whereas the real-world validity of 'would(A)' requires A to be true throughout R(@), the set of worlds accessible from the actual world, for each model, which typically requires A to be true at some non-actual worlds too. This does not matter when the choice of actual world makes no difference to the truth-value of any sentence at any world, for then the falsity of A at a non-actual world w in a model M implies the falsity of A at w in a model M* just like M except that w is the actual world of M*, in which case A is not real-world valid in the first place, and so is no counterexample to Necessitation$_{RW/RW}$.

However, not all languages work like that. For example, a language may contain a rigidifying operator 'actually' with this semantics: 'actually(A)' is true at any world in a model just in case A is true at the actual world of the model. Whether the word 'actually' in English works like that does not matter (though it probably can); what

---

[3] Davies and Humberstone 1980 is the seminal treatment of the distinction between real-world and general validity and its consequences for logic.

matters is that an operator with that semantics, whatever word we use for it, is perfectly intelligible. Then sentence (19) is real-world valid:

(25)   If it will rain, actually(it will rain)

For (25) trivially comes out true at the actual world of every model, since evaluating both antecedent and consequent reduces to evaluating 'it will rain' twice over at the actual world. But the simple modalized conditional (26) is *not* real-world valid:

(26)   Would(if it will rain, actually(it will rain))

For consider a model where $R(@)$ includes a non-actual world w at which 'it will rain' is true, although it is false at @ itself, a counterexample to Murphy's Law, 'Whatever can go wrong, will'. Thus, in the model, 'actually(it will rain)' is false at w, so the indicative conditional 'if it will rain, actually(it will rain)' has a true antecedent and false consequent at w, and so is itself false at w; since w is in $R(@)$, (26) is false at @, and so is not real-world valid.

A neat way to handle this complexity is by defining a second version of model-theoretic validity, *general validity*. It requires truth-preservation at every world in every model. More formally, the argument from a set of premises AA to a conclusion C is generally valid ($AA \vDash_G C$) just in case for every model M and world w in M, if every member of AA is true at w in M then C is true at w in M. A formula A is generally valid just in case the argument from the empty set of premises to A is generally valid, in other words, just in case A is true at every world in every model. General validity entails real-world validity, because truth in a model is just truth at the actual world of that model. But real-world validity does not entail general validity; for example, (25) is real-world valid but not generally valid.

Although the rule of Necessitation fails for real-world validity, it holds for general validity:

Necessitation$_{G/G}$ Whenever $\vDash_G A$, $\vDash_G$ would(A)

Since general validity implies real-world validity, a mixed variant also holds:

Necessitation$_{G/RW}$ Whenever $\vDash_G A$, $\vDash_{RW}$ would(A)

We can sometimes still obtain the real-world validity of 'would(A)', but from the stronger assumption of the general validity of A. For Necessitation, general validity is better behaved than real-world validity. But we should not neglect the latter: for many purposes, truth at the actual world is what needs to be preserved. Moreover, the real-world validity of simple indicative conditionals like (25) compared to the real-world invalidity of the corresponding simple modalized conditionals like (26) provides a useful structural contrast between the two types of sentence. We will keep track of both versions of validity.

In non-modal ways, both versions of validity are well behaved, obeying all the standard structural rules for logical consequence (see section 3.1) and the principles of classical logic. In what follows, these features will usually be taken for granted.

We are now ready to discuss the validity or invalidity of various logical principles, and to compare the results with the pre-theoretic plausibility or implausibility of those principles. Where there is a mismatch, we can expect the pragmatics to do some explanatory heavy lifting. For instance, when we consider a simple modalized conditional sentence 'would(if A, C)', we naturally expect the contextually relevant worlds to include some in which A is true, if there are any, while otherwise not deviating too much from the actual world. But such constraints depend on which sentences are conversationally salient, which is a matter of pragmatics, not semantics; it has nothing to do with validity. David Lewis (1973) incorporated such constraints in his semantics, but that will turn out to have been a methodological error.

Many of the considerations to come in this section are closely related to points made by Kai von Fintel (2001) in defence of an account of counterfactual conditionals as dynamic strict conditionals, inspired by a suggestion from Irene Heim, though without the decomposition into the modal operator 'would' and the material 'if', and without the role of the suppositional heuristic in explaining contextual shifts. Earlier, Ken Warmbrod (1981, 1982) had defended an account of counterfactual conditionals as context-sensitive strict conditionals, though without the dynamic mechanism.

All principles displayed below are correct on the proposed semantics, except those marked '#'.

We start off with a generalization of Necessitation to arguments with premises:

Normality$_{G/G}$ Whenever $AA \vDash_G C$, {would(A): $A \in AA$} $\vDash_G$ would(C)

Roughly: arguments which *do* generally preserve truth *would* generally preserve truth. As always, the correctness of the global to global (G/G) version of a principle implies the correctness of the global to real-world (G/RW) version. The argument for Normality$_{G/G}$ is a straightforward generalization of the argument for Necessitation$_{G/G}$, which is just the special case of Normality$_{G/G}$ when AA is empty. Conversely, Normality$_{RW/RW}$ fails because its special case Necessitation$_{RW/RW}$ fails.

From Normality$_{G/G}$ we can easily draw consequences such as this:

K$_G$ would(if A, C), would(A) $\vDash_G$ would(C)

For *modus ponens* is generally valid for 'if': 'if A, C', $A \vDash_G C$. We can also derive principles where all the 'would's govern conditionals:

Normality$_{G/G}$ in the Consequent   Whenever $BB \vDash_G C$,
                                        {would(if A, B): $B \in BB$} $\vDash_G$ would(if A, C)

Roughly: what would be given A is closed under general validity. This follows from Normality$_{G/G}$, for whenever $BB \vDash_G C$, {if A, B: $B \in BB$} $\vDash_G$ (if A, C) by truth-functional logic (on the material interpretation of 'if'). By contrast, the real-world analogue, Normality$_{RW/RW}$ in the Consequent, fails, as the counterexample to Necessitation$_{RW/RW}$ shows, since we can take BB to be empty and A to be a tautology.

Normality in the Consequent is very plausible: what *would* be on a given supposition should be closed under logical consequence. Most semantic theories of counterfactual conditionals validate their own version of the principle. However, although Lewis's preferred theory validates it when BB is finite, it fails for some infinite sets of premises. Here is a version of Lewis's example (1973: 20). Let L be a line which happens to be exactly c long (all lengths in millimetres), but could have been any length between $c - 1$ and $c + 1$. For each natural number $n > 0$, let B_n be 'L is at most

$c + 1/n$ long' (with suitable designators for the numbers), $BB = \{B_n: n > 0\}$, and C be 'L is at most c long'. By the Archimedean principle for lengths, if L is more than c long, L is more than $c + 1/n$ long for some natural number $n > 0$. Contrapositively, $BB \vDash_G C$. Hence the analogue of Normality$_{G/G}$ in the Consequent for Lewis's counterfactual conditional $\square\!\!\rightarrow$ implies, for $A = \text{not}(C)$:

$$\{\text{not}(C) \ \square\!\!\rightarrow B_n: n > 0\} \vDash \text{not}(C) \ \square\!\!\rightarrow C$$

Each premise is true, on a reasonable application of Lewis's semantics: in the closest worlds in which L is more than c long (as it indeed could be), L differs from its actual length c by at most $1/n$. But the conclusion is false, for 'not(C)' is possible, and on Lewis's semantics a counterfactual conditional whose consequent is incompossible with its antecedent is true only if it is vacuous, because its antecedent is impossible. Lewis contemplated excluding such infinitary failures of deduction in the consequent by imposing the ad hoc Limit Assumption, but doing so merely swaps one awkwardness for another. It is a sign that he was trying to do in the semantics what really ought to be done in the pragmatics.

The special case of Normality$_{G/G}$ in the Consequent when BB has just one member is a monotonicity principle:

Monotonicity$_{G/G}$ in the Consequent   Whenever $B \vDash_G C$

                                                 would(if A, B) $\vDash_G$ would(if A, C)

There is a parallel principle of anti-monotonicity in the antecedent, sometimes known as 'strengthening the antecedent':

Anti-Monotonicity$_{G/G}$ in the Antecedent   Whenever $A \vDash_G B$,

                                                 would(if B, C) $\vDash_G$ would(if A, C)

This also follows from Normality, for whenever $A \vDash_G B$, 'if B, C' $\vDash$ 'if A, C', by truth-functional logic (on the material interpretation of 'if').

Many accounts of counterfactual conditionals reject Anti-Monotonicity in the Antecedent. Alleged counterexamples are legion. For instance, 'It rains hard' entails 'It rains', but (27) seems not to entail (28):

(27)   If it were to rain, it would not rain hard.
(28)   If it were to rain hard, it would not rain hard.

Again, 'I dropped the pen and you caught it' entails 'I dropped the pen', but (29) seems not to entail (30):

(29)   If I dropped the pen, it would land on the floor.
(30)   If I dropped the pen and you caught it, it would land on the floor.

The apparent failure of Anti-Monotonicity is then rationalized by the point that when A entails B, so all A-worlds are B-worlds, the converse may still fail. But when B-worlds need not be A-worlds, the closest B-worlds to actuality may be C-worlds but not A-worlds, while the closest A-worlds to actuality are more distant B-worlds but not C-worlds.

However, the alleged counterexamples to strengthening of the antecedent have never been totally convincing, for it is not clear that once the second counterfactual conditional has been rejected, the first can still be maintained (see also von Fintel 2001). Once I concede that if I dropped the pen and you caught it, it would not land on the floor, it feels obtuse simply to insist that, nevertheless, if I dropped the pen, it would land on the floor. In the familiar conversational way, raising the possibility of your catching the pen lets the cat out of the bag; getting it back in again is much harder. When I consider just (29), the relevant worlds must include some in which I drop the pen, but need not include any in which I drop the pen and you catch it. But when I go on to consider (30) too, doing so makes some worlds in which I drop the pen and you catch it relevant, and they undermine (29).

A similar argument applies to (27) and (28). When I consider just (27), the relevant worlds must include some in which it rains, but need not include any in which it rains hard. But when I go on to consider (28) too, doing so makes some worlds in which it rains hard relevant, and they undermine (27).

The sense of change in context may be less marked for (27)–(28) than for (29)–(30), and so the appearance of a counterexample more persistent for (27)–(28). For the Suppositional Procedure may create the illusion that (28) can be rejected on purely logical grounds, with no need for a world in which it rains hard to be relevant. Chapter 11 discusses the role of the Suppositional Procedure for 'would if' conditionals in more detail. But none of the standard semantic explanations for the alleged failures of strengthening the antecedent confine the phenomenon to cases where the conclusion can be rejected on purely logical grounds. They predict counterexamples like (29)–(30) too. More specifically, the 'closest worlds' explanations impose no constraints incompatible with some A-worlds being C-worlds. But apparent counterexamples like (29)–(30) are exactly those which give the most plausibility to the charge that the apparent counterexamples to strengthening the antecedent illicitly depend on a change of context. That is reason to doubt those semantic explanations.

Edgington discusses a similar defence of strengthening the antecedent, with reference to this example (1995: 252–3):

> a piece of masonry falls from the cornice of a building, narrowly missing a worker. The foreman says: 'If you had been standing a foot to the left, you would have been killed; but if you had (also) been wearing your hard hat, you would have been all right'

Using Lewis's symbolism (but not his semantics), she formalizes the foreman's discourse as '$S \,\square\!\!\rightarrow K$; but $(S \,\&\, H) \,\square\!\!\rightarrow \neg K$'; she takes the latter counterfactual to be incompatible with $(S \,\&\, H) \,\square\!\!\rightarrow K$. Against the objection, from Crispin Wright (1983) and E. J. Lowe (1990), that 'the same possible worlds should be in play throughout a single piece of reasoning or discourse', she responds: 'the building foreman's remarks above, violating transitivity as they do, constitute a single, pointful piece of discourse.' She is right that there is no intrinsic defect in the foreman's discourse. But that does not mean that it requires no change of context. Such changes are normal in ordinary coherent discourse, as the speaker continually provides new information to guide hearers' understanding of what comes next. Instead, the key issue concerns the application of logic: one cannot expect deductive rules to preserve truth from a premise in one context to a conclusion in another. Even 'A, therefore A'

becomes problematic by that standard. Edgington's comments do not address the concern that worlds in which the worker is wearing a hard hat become relevant only halfway through the foreman's discourse.

But *why* is there pressure to make the contextually relevant worlds include some where the antecedent is true? After all, the semantics makes 'would(if A, C)' truth-conditionally equivalent to 'would(not(A) or C)', and considering the latter does *not* put much pressure to make the contextually relevant worlds include some where A is true. The natural conjecture is that the difference lies in the heuristics for assessing the embedded conditional and disjunction. The former requires one to suppose A; the latter does not. When one is supposing A in the course of assessing a 'would' claim, it makes sense for the worlds contextually relevant to 'would' to include some where A is true. The difference between (27) and (28), or between (29) and (30), in *what* one is supposing, the antecedent itself, may be more directly felt, but accommodating the supposition with some relevant worlds is a natural effect.

That is not to say that such accommodation is indefeasible. Instead of (28), I could say:

(28!)   If it were to rain hard, the weather forecast would be wrong. But the weather forecast would never be wrong, so it would not rain hard.

I implicitly insist that both the content and the correctness of the past forecast be held fixed across worlds relevant to 'would' in this context, even though holding them fixed deprives the 'would if' conditional in (28!) of worlds where its antecedent is true. If you go along with (28!), and there are no contextual pressures pushing the other way, my insistence is successful. The value of a contextually set parameter is often the resultant of many forces. In such a context, (28) itself is true. But that is not the sort of context we tend to think of first in casually assessing (28).

How does this postulation of context-sensitivity for modalized conditionals square with its denial in earlier chapters for plain conditionals? After all, the difference between assessing (27) and assessing (28) feels very similar to the difference between assessing (27a) and (28a), plain unmodalized conditional analogues (the speaker is in a tunnel, ignorant of the weather outside):

(27a)   If it is raining, it is not raining hard.
(28a)   If it is raining hard, it is not raining hard.

If context-sensitivity is needed to explain our resistance to inferring (28) from (27), why is it not also needed to explain our resistance to inferring (28a) from (27a)?

The objection is powerful once we treat contextual restrictions as the explanatory bottom line. But that is not the best approach. Whether a world is relevant in a context is not a brute fact; it depends on the context's less abstract cognitive features. By supposing A, you may *make* some A-worlds relevant. They may remain relevant even after you have discharged that supposition, because the psychological salience of A as a possibility may affect how you develop other suppositions too. Thus applying a suppositional procedure can help determine contextual restrictions. But to figure in the compositional semantics, a contextual feature still needs a locus in the semantic structure of a sentence. As already argued, 'would' supplies such a locus for contextual variation, while 'if' does not.

Here is an example. In the present context, I say:

(31) Plato would not have gone to Rome.

A natural puzzled response is: in what circumstances? The present context is too bare for (31) to be clear enough for conversational purposes. By contrast, (32) is clear enough:

(32) If Plato went to Rome, so did Aristotle.

You may wonder what evidence I have for (32), but that is not unclarity about what I said, only about why I said it. 'Would' alerts hearers to context in a way 'if' does not.

At a more concrete level, resistance both to inferring (28) from (27) and to inferring (28a) from (27a) comes from the suppositional nature of the heuristics we apply to both 'if' and 'would' (see chapter 11). In both cases, the switch of antecedent is crucial in generating the illusion that 'strengthening the antecedent' is a fallacy. Under appropriate conditions, the heuristics guide us towards accepting both (27) and (27a) while rejecting both (28) and (28a). The difference is that the extra degree of freedom in the context-sensitivity of 'would' enables us to *accommodate* both acceptance of (27) and rejection of (28) as semantically correct, though in different contexts, while for (27a) and (28a) such flexibility is not available. The diagnosis of resistance to inferring (28) from (27) in terms of a contextual shift has the benefits of a familiar form of explanation, but really just goes proxy for a deeper explanation in terms of structurally similar heuristics for 'would' and 'if'. At that deeper level, resistance to the 'would if' and plain 'if' inferences has similar sources. The same applies to other alleged counterfactual fallacies discussed in this chapter. Nevertheless, on this understanding, we may continue to describe them in terms of contextual shifts, leaving the analysis of the heuristic for 'would' to chapter 11.

The contrast between the close similarities in cognitive processing for 'would if' and plain 'if' and their sharp difference in semantic structure may seem surprising. However, it is far from unprecedented. Here is another case. Pretend that all you know about Altrincham is that it is a town. Here are two questions:

(33) How likely is Altrincham to have a museum?
(34) How likely is a town to have a museum?

For you, Altrincham is in effect 'an arbitrary town', so the two questions may prompt very similar cognitive processes. Yet semantically the two questions are of quite different kinds: (33) is about a particular town, whereas (34) is about towns in general, and what it takes for an answer to be correct varies accordingly. This in turn gives (34) an extra locus of context-sensitivity, since contextual standards for the application of the word 'town' may vary, and can be adjusted to accommodate your answer to (34) charitably. By contrast, we may assume, there is no relevant context-sensitivity in the name 'Altrincham'. Thus the close similarities between (33) and (34) in cognitive processing coexist with sharp differences between them in semantic structure.

Similar issues arise for alleged counterexamples to two more principles for modalized conditionals, Contraposition and Transitivity, here analysed thus:

Contraposition$_G$ would(if A, C) $\vDash_G$ would(if not(C), not(A))

Transitivity$_G$ would(if A, B), would(if B, C) $\vDash_G$ would(if A, C)

We can derive both principles from Normality$_{G/G}$, Contraposition because 'if A, C' $\vDash_G$ 'if not(C), not(A) and Transitivity because 'if A, B', 'if B, C' $\vDash_G$ 'if A, C' (given the material interpretation of 'if').

An apparent counterexample to Contraposition$_G$ is that (27) ('If it were to rain, it would not rain hard') seems not to entail (35):

(35)   If it were to not not rain hard, it would not rain.

For the double negation cancels out, reducing (35) to 'If it were to rain hard, it would not rain'. Both Lewis's semantics and Stalnaker's rationalize failures of Contraposition: if the closest A-worlds are C-worlds, the closest 'not(C)'-worlds may be more distant A-worlds.

But again, it is unclear that once the second counterfactual conditional has been rejected, the first can still be maintained. When I consider just (27), the relevant worlds must include some in which it rains, but need not include any in which it rains hard. But once I reject (35), I take the possibility of its raining hard more seriously: the relevant worlds must include some in which it rains hard, and (27) becomes much more difficult to accept.

Someone might object that (35) can be rejected on broadly logical grounds, without need of any change in context, since its consequent is inconsistent with its antecedent. The heuristic for 'would if' conditionals casts light on that objection too (chapter 11).

For now, it helps to have an apparent counterexample to Contraposition free of that logical issue. Here is one (with double negations already cancelled):

(36)   If it were to rain, the match would not be cancelled.
(37)   If the match were to be cancelled, it would not rain.

Background: mere rain is insufficient for cancellation; a very bad storm would be required, which would typically involve rain, but such a storm is a distant possibility. Then one accepts (36) but rejects (37). However, the sense of a change in context is palpable. Once one rejects (37), one is no longer in a position to accept (36). When one originally accepted (36), some worlds in which it rains were relevant, but none in which the weather is bad enough for the match to be cancelled. Once one considers (37), some of the latter become relevant. Semantic theories predicting counterexamples to Contraposition$_G$ like (21) and (35) also predict counterexamples like (36) and (37) too. More specifically, the 'closest worlds' explanation above does not constrain how A and C pattern over more distant worlds, and so is compatible with A and C being logically independent of each other. But such apparent counterexamples are those which give the most plausibility to the charge that the apparent counterexamples to Contraposition depend illicitly on a change of context. That is reason to doubt those semantic explanations too.

Both Lewis's semantics and Stalnaker's also predict counterexamples to Transitivity. If the closest A-worlds are B-worlds and the closest B-worlds are C-worlds, it does not follow that the closest A-worlds are C-worlds, for the closest B-worlds may be closer than the closest A-worlds.

An alleged counterexample to Transitivity$_G$ is that (38) and (27) ('If it were to rain, it would not rain hard') seem not to entail (28) ('If it were to rain hard, it would not rain hard'):

(38)   If it were to rain hard, it would rain.

Here the position of someone who accepts both premises, (38) and (27), is already of doubtful stability, whether or not the conclusion (28) is rejected. Accepting (27) involves treating worlds in which it rains hard as irrelevant. Accepting (38) seems to involve treating some worlds in which it rains hard as relevant. Even the order in which the premises are presented makes a difference: accepting (38) makes it harder to accept (27) (see von Fintel 2001).

Once again, there is a potential objection: perhaps one can accept (38) on purely logical grounds, without considering any worlds, so no change of context is required. It therefore helps to consider an apparent counterexample to Transitivity$_G$ where neither premise can be accepted on purely logical grounds. Edgington provides one (1995: 253):

(39)   If Brown had been appointed, Jones would have resigned immediately afterwards.

(40)   If Jones had died before the appointment was made, Brown would have been appointed.

(41)   If Jones had died before the appointment, Jones would have resigned immediately after the appointment.

Obviously, (41) is to be rejected. But can (39) and (40) be accepted together? As Edgington herself notes, it helps to read (39) before (40); once you have read (40), it is natural to object to (39) 'But not if he was dead!' Accepting (39) involves treating worlds in which Jones died before the appointment as irrelevant; accepting (40) involves treating some of those worlds as relevant. Edgington claims that one can believe both (39) and (40). But if one can believe them together, one should also be able to believe their conjunction (perhaps, on Edgington's probabilistic view of belief, with slightly less confidence). Yet the natural response to the conjunction of (39) and (40) is disbelief. One may be both stably disposed to assent to (39) and stably disposed to assent to (40) in isolation, but not stably disposed to assent to their conjunction. The best explanation of such patterns is context-sensitivity.

The 'closest worlds' explanation above for alleged counterexamples to Transitivity does not constrain how A, B, and C pattern over more distant worlds, and so is compatible with there being no logical interdependencies between them. But such apparent counterexamples to Transitivity are exactly those which give the most plausibility to the charge that the problem illicitly depends on a change of context. That is reason to doubt those semantic explanations too.

In brief, alleged counterexamples to Anti-Monotonicity$_{G/G}$ in the Antecedent, Contraposition$_G$, and Transitivity$_G$ are unconvincing. They give strong signs of depending on illicit context-shifting. On the balance of the evidence, the attempt to accommodate them in the semantics of 'would if' conditionals is another case of trying to do in the semantics what really ought to be done in the pragmatics.

We now turn to cases where the present semantics invalidates a principle usually considered valid. Indeed, it implies that they are not real-world valid, let alone generally valid.

Here is one:

Strong Centring$_{RW}$ # A and $B \vDash_{RW}$ would(if A, B)

Strong Centring$_{RW}$ fails on the present semantics because a world w other than the actual world of the model may belong to the set R(@) of contextually relevant worlds; the accessibility relation R does not imply identity. Thus A and B may both be true at the actual world @, while A is true and B false at w, so 'If A, B' is false at w, so 'would(if A, B)' is false at @.

By contrast, many possible world semantic theories of the counterfactual conditional validate their version of Strong Centring, for since @ is closer or more similar to @ than any other world is, if A and B are true at @, the closest or most similar A-world to @ is a B-world.

On the present semantics, although Strong Centring$_{RW}$ is incorrect, we can still discuss models in which it *preserves truth*, in the sense that for any instance of the principle, if the premise (a sentence of the form 'A and B') is true in the model, so is the conclusion (the corresponding sentence of the form 'would(if A, B)'). This talk of truth-preservation generalizes in the natural way to any other principle whose instances are all of the form $AA \vDash_{RW} B$, where AA is a set of sentences of the relevant language and B is a single sentence of that language. We will also speak of one such principle *entailing* another when the latter preserves truth in any model in which the former does. All these distinctions are drawn in terms of the present semantics.

The failure of Strong Centring$_{RW}$ is quite plausible once one looks at the right examples. For instance, Peter the pessimist is betting on tails. The coin comes up heads. Self-pityingly, he says:

(42)   It would come up heads.

He is suggesting that it is no matter of chance that he lost; the world is against him. But, presumably, he is wrong: in some relevant worlds, he bets on tails and the coin comes up tails. In the first instance, this is a counterexample to the principle that 'is' implies 'would be':

T^c_{RW} # $B \vDash_{RW}$ would(B)

Strong Centring$_{RW}$ entails T^c_{RW}, for when A is a tautology, both 'A and B' and 'If A, B' are logically equivalent to B (on the material interpretation of 'if'), and 'would(if A, B)' is equivalent to 'would(B)' by Normality$_{G/G}$; thus every counterexample to T^c_{RW} generates a counterexample to Strong Centering$_{RW}$. Conversely, T^c_{RW} entails Strong Centring$_{RW}$, for by truth-functional logic 'A and B' yields 'If A, B' (on the material interpretation of 'if'), which by T^c_{RW} yields 'would(if A, B)'.

We should expect T^c_{RW} to fail in most contexts, for it holds for arbitrary B only when the actual world of the model has access to no world other than itself. If so, the set R(@) of worlds accessible from the actual world is just its singleton, if it is accessible from itself, and the empty set otherwise. In both cases, 'would' is trivial: in the former case, 'would(B)' is simply equivalent to B itself; in the latter case, 'would(B)' is vacuous, and so equivalent to a tautology. Either way, 'would' does no modal work. Such an operator would not earn its keep in the language.

For a direct counterexample to Strong Centring$_{RW}$, consider Peter's fatalistic claim:

(43) If I were to bet on tails, it would come up heads.

In the envisaged context, (43) is false, because the coin is fair, even though Peter does bet on tails and the coin does come up heads. The relevant worlds are not always restricted to the very most similar worlds to the actual world.

Although Strong Centring$_{RW}$ was a popular view, some philosophers rejected it in practice. For example, in explaining his fourth condition on knowledge, Robert Nozick wrote: 'Not only is p true and S believes it, but if it were true he would believe it', insisting that the counterfactual is stronger than the mere conjunction, suggesting that the former requires that in 'close' worlds where p is true, S believes p (Nozick 1981: 176). Similarly, Ernest Sosa glossed the *safety* of a belief in p by modalized conditionals such as 'it would have been held only if (most likely) p', whose truth he takes not to follow from the true belief conjunction (Sosa 2007: 25).

Of course, although Strong Centring$_{RW}$ is not logically valid, it may preserve truth in many contexts, where the antecedent holds in the actual world and we therefore need look no further, given our conversational purposes. Strong Centring$_{RW}$ will then preserve truth in the intended models for those contexts. But, in the intended models for many other contexts, Strong Centring$_{RW}$ will not preserve truth. Once again, we should not try to do in the semantics what is better done in the pragmatics.

The present semantics also invalidates the Weak Centring principle, which it makes equivalent to *modus ponens* for the counterfactual conditional:

Weak Centring$_{RW}$ #   would(if A, B) $\vDash_{RW}$ if A, B

WMP$_{RW}$ #   would(if A, B), $A \vDash_{RW} B$

These principles fail because the set R(@) of contextually relevant worlds need not contain @, the actual world of the model. R was not required to be reflexive, so @ may not be accessible from itself. Thus all the A-worlds in R(@) may be B-worlds, so 'would(if A, B)' is true at @, even though A is true and B false at @.

By contrast, most possible world semantic theories of the counterfactual conditional validate their own version of Weak Centring, for since no world is closer or more similar to @ than @ is, if the closest or most similar A-worlds to @ are B-worlds, and @ is an A-world, it is also a B-world.

But the failure of Weak Centring$_{RW}$ is quite plausible too, once one looks at the right examples. We noted above that, in a context where we are developing an elaborate scenario in which a tiger enters the room, (12) ('Francis would be frightened') may be true, even though 'Francis is frightened' is false. In the first instance, this is a counterexample to the principle that 'would be' implies 'is', to which T$^c_{RW}$ is the converse:

T$_{RW}$ #   would(A) $\vDash_{RW} A$

Weak Centring$_{RW}$ entails T$_{RW}$, for when A is a tautology, 'if A, B' is logically equivalent to B, by Normality$_{G/G}$ (at least, on the material interpretation of 'if'); thus every counterexample to T$_{RW}$ generates a counterexample to Weak Centering$_{RW}$. Conversely, T$_{RW}$ entails Weak Centring$_{RW}$, for the latter is just a special case of the former.

For a direct counterexample to both WMPs$_{RW}$ and Weak Centring$_{RW}$, consider (44):

(44)   If Jessica looked behind her, she would jump.

In the context where we are developing the tiger scenario, (44) may be true, even though in the actual world Jessica looks behind her and does not jump. The relevant worlds do not always include the actual world.

Of course, although Weak Centring$_{RW}$ is not logically valid, it may preserve truth in most contexts, since the actual world is presumably relevant for most conversational purposes. Weak Centring$_{RW}$ will then preserve truth in the intended models for those contexts. But, in the intended models for many other contexts, Weak Centring$_{RW}$ will not preserve truth. Yet again, that is a matter for the pragmatics, not the semantics.

An even weaker consequence of T$_{RW}$ is D$_{RW}$, equivalent to a 'would'-qualified principle of non-contradiction:

D$_{RW}$ #   would(not(A)) ⊨$_{RW}$ not(would(A))

WNC$_{RW}$ #   ⊨$_{RW}$ not(would(A)) and would(not(A)))

In other words, 'would(A)' and 'would(not(A))' are mutually incompatible. This principle preserves truth whenever the set R(@) of contextually relevant worlds is non-empty. However, we have not required R to be *serial*, in the sense that every world has R to some world; R(@) may be empty. When so, both 'would(A)' and 'would(not(A))' are vacuously true. Normally, there are some contextually relevant worlds. However, that is a pragmatic rather than semantic matter. In special cases, the relevant worlds may be only those which satisfy various assumptions, and we may have inadvertently imposed inconsistent conditions, just as there are inconsistent fictions. Even a single constraint may have a hidden inconsistency. We have no good reason to rule out such possibilities semantically. They occur naturally in some conversational contexts.

A weaker consequence of Strong Centring$_{RW}$ is this converse of D$_{RW}$, equivalent to a 'would'-qualified principle of excluded middle:

D$^c_{RW}$ #   not(would(A)) ⊨$_{RW}$ would(not(A))

WEM$_{RW}$ #   ⊨$_{RW}$ would(A) or would(not(A))

For 'A or not(A)' is valid, so Strong Centring$_{RW}$ entails WEM$_{RW}$, by classical reasoning. But WEM$_{RW}$ does not entail Strong Centring$_{RW}$, for the former but not the latter holds in models where the set R(@) contains only one world but it is not @, the actual world of the model.

WEM$_{RW}$ has the controversial 'would'-qualified principle of conditional excluded middle as a special case:

WCEM$_{RW}$ ⊨$_{RW}$ would(if A, C) or would(if A, not(C))

Conversely, WCEM$_{RW}$ entails WEM$_{RW}$, for when A is a tautology, 'would(if A, C)' and 'would(if A, not(C))' are equivalent to 'would(C)' and 'would(not(C))' respectively, by Normality$_{G/G}$ (at least, on the material interpretation of 'if'). Thus

$WCEM_{RW}$ and WEM_{RW} are equivalent, on the present semantics. They are liable to fail in models where $R(@)$ contains more than one world, since it may contain both an A-world which is also a C-world and an A-world which is not a C-world.

Stalnaker's semantics notoriously validates his version of WCEM, on a counterfactual reading of his conditional $>$, for the same reason as explained in section 3.4. By contrast, Lewis's semantics invalidates his analogue of WCEM, since he allows many closest worlds at which the antecedent is true.

From the present perspective, there is no reason to expect at most one world to be contextually relevant. For example, the point of uttering (4) ('Loretta would not betray a friend') may well be to emphasize the modal robustness of Loretta's loyal behaviour: she does not betray a friend in any nearby world. That requires 'would' to range over many worlds. That goes for 'would if' conditionals too. One can express a false fatalism by saying:

(45)   Either, if the coin were tossed, it would come up heads, or, if the coin were tossed, it would come up tails.

However, some evidence points in the opposite direction. For example, consider the two disjuncts of (45):

(46)   If the coin were tossed, it would come up heads.
(47)   If the coin were tossed, it would come up tails.

Imagine knowing the coin to be fair, and being asked for the probabilities of (46) and (47). A natural answer is ½ for each, as though they cannot both be false. Similarly, imagine being asked:

(48)   If the coin were tossed, would it come up heads?

A natural answer is 'I don't know.' It would be strange to answer 'No, it's a fair coin.'

Taking such evidence and more at face value in the present setting, one might try requiring at most one world to be accessible from any given world, thereby validating WEM. But, as already explained, doing so undermines many of the most distinctive and functional uses of the modal 'would'. In particular, the effects on 'would if' conditionals are drastic. Whenever the actual world is contextually relevant, $R(@) = \{@\}$, and the 'would if' conditional collapses into the unmodalized one. Even when the actual world is not contextually relevant, the uniqueness constraint validates a crude strengthening of standard Conditional Excluded Middle for 'would if' conditionals:

$WSCEM_{RW}$ # $\vDash_{RW}$ would(if A, C) or would(if B, not(C))

For, on the material reading of 'if', either C is true at the unique member of $R(@)$, in which case 'would(if A, C) is true at @, for any A, or C is false at the unique member of $R(@)$, in which case 'would(if B, not C)' is true at @, for any B. But $WSCEM_{RW}$ has instances such as (49) (with reference to a single counterfactual toss):

(49)   Either, if it had come up heads, it would have come up tails, or, if it had come up tails, it would not have come up tails.

Each disjunct seems obviously false. Such results would severely undermine the cognitive value of 'would if' conditionals. In the present setting, the uniqueness constraint on accessibility is hopeless.

Of course, the *un*modalized analogue of (49) also *looks* obviously false (with reference to a single past toss of unknown outcome):

(50)   Either, if it came up heads, it came up tails, or, if it came up tails, it did not come up tails.

But (50) is a tautology, on the material reading of 'if'. Our reliance on the Suppositional Procedure explains the apparent badness of (50). For each disjunct, the conditional probability of the consequent on the antecedent is zero; more generally, the consequent can be conclusively rejected on the supposition of the antecedent. Applying the Procedure, we conclusively reject each disjunct, and so their disjunction. But the Procedure is unreliable in such cases, as seen in chapter 3. These issues will be reconsidered in section 11.2.

To understand the effect of our heuristics on our assessments of 'would if' conditionals, the natural strategy is first to work out what heuristic stands to such conditionals as the Suppositional Procedure stands to unmodalized conditionals. If all goes well, we can then invoke that heuristic to explain away the apparent evidence for the uniqueness constraint, such as our pre-theoretic probability judgements about (46) and (47), without collapsing the difference between the modalized and unmodalized conditionals, for that would land us back with bad results such as (49). Chapter 11 pursues that strategy.

11

The Interaction of 'If' and 'Would'

Heuristics

11.1 An analogy between 'would' and 'if'

On the view developed so far, 'would if' conditionals are, to a first approximation, composed semantically by applying the modal operator 'would' to a sentence composed by applying the material conditional operator 'if' to a pair of sentences. However, in Part I, we saw that one cannot properly evaluate the data bearing on the semantics of conditionals without understanding the role of fallible heuristics in generating those data: specifically, the Suppositional Rule as the primary heuristic and cross-contextual reliance on memory and testimony as a secondary heuristic. That moral extends to sentences with 'if' in the scope of 'would'. Although the structure of cognition does not always mirror the compositional semantic structure of the sentences being cognized, a natural first hypothesis is that our primary cognitive assessments of such 'would if' sentences combine heuristics for 'would' with heuristics for 'if'. We start work on that basis.

What heuristic is associated with 'would'? How do we assess statements of the form 'Would(A)'? A significant clue is in the evidence for two tempting principles about the logic of 'would' provisionally rejected in section 10.3: the 'would'-qualified versions of excluded middle (WEM) and non-contradiction (WNC). Together, WEM and WNC amount to the principle that 'would' commutes with negation, in the sense that 'not(would(A))' is always equivalent to '(would(not(A)))'.[1] The evidence for WEM and WNC was strongly reminiscent of the evidence in chapter 3 for the principles of conditional excluded middle (CEM) and conditional non-contradiction (CNC), which together amount to the principle that 'if' commutes with negation in the consequent, in the sense that 'not(if A, C)' is always equivalent to 'if A, not(C)'.

A much less contentious principle for 'would' is that it commutes with conjunction, in the sense that 'would(A) and would(B)' is always equivalent to 'would(A and B)'; that is an easy consequence of the Normality principle for 'would' endorsed in section 10.3 (for 'A and B' follows from the separate premises A and B, and vice

---

[1] The exact strength of the equivalence depends on whether real-world or general validity is at issue. For purposes of this chapter, we can ignore that distinction, since the difference it makes in WEM and WNC mainly concerns sentences which apply another layer of modal operators to the 'would' sentences in WEM and WNC. This chapter does not focus on the subtleties involved in iterating 'would'.

Suppose and Tell: The Semantics and Heuristics of Conditionals. Timothy Williamson, Oxford University Press (2020).
© Timothy Williamson.
DOI: 10.1093/oso/9780198860662.001.0001

versa). Since every truth-function is definable in terms of negation and conjunction, the principles that 'would' commutes with conjunction and negation together entail that 'would' commutes with every truth-function, at least as defined in terms of conjunction and negation. Similarly, a much less contentious principle for 'if' is that it commutes with conjunction in the consequent, in the sense that '(if *A*, *C*) and (if *A*, *D*)' is always equivalent to 'if *A*, (*C* and *D*)', which the evidence in Part I strongly favoured. The principles that 'if' commutes with conjunction and negation in the consequent together entail that 'if' commutes with every truth-function in the consequent.

By themselves, the commutativity principles for 'if' are formally consistent, since they all hold on the unintended interpretation of 'if *A*, *C*' as simply equivalent to *C*. However, they generate a contradiction as soon as they are combined with the trivial principle 'If *A*, *A*' (section 3.5). Combined with other evidence, that led to the conclusion in Part I that the commutativity principles for 'if' are not all valid: appearances to the contrary result from our implicit reliance on a fallible heuristic for 'if', the Suppositional Rule. Similarly, the commutativity principles for 'would' are formally consistent, since they all hold on an interpretation where R(*w*), the set of worlds accessible from a world *w*, always contains exactly one member. However, the threat is that they generate some sort of collapse once they are combined with what should be harmless additional assumptions. If so, the appropriate moral might well be that the commutativity principles for 'would' are not all valid, with appearances to the contrary resulting from our implicit reliance on a fallible heuristic for 'would'. Indeed, the strong structural analogy between the effects of the two heuristics suggests that the heuristics themselves may be structurally analogous.

To test that line of thought properly, we need to examine the evidence in more detail. That is the work of the next section.

11.2 Does 'would' single out a unique world?

Here are two questions, asked in the same context:

(1)   How probable is it that Diana would obey the rules?
(2)   How probable is it that Diana would break the rules?

In context, breaking the rules is equivalent to not obeying them. Imagine someone answering '20 per cent' to (1) and '10 per cent' to (2). Something seems to be wrong with that pattern of answers. Something also seems to be wrong with answering '80 per cent' to (1) and '90 per cent' to (2). Apparently, one's answers to (1) and (2) should add up to 100 per cent. The natural explanation for that impression is that we are implicitly treating (3) and (4) as contradictories:

(3)   Diana would obey the rules.
(4)   Diana would break the rules.

But they are equivalent (given Normality$_{RW}$) to (4a) and (5a) respectively:

(3a)   would(Diana obeys the rules)
(4a)   would(not(Diana obeys the rules))

Of course, any contradictory of (3a) amounts to (5):

(5)   not(would(Diana obeys the rules))

Thus we are implicitly treating (4a) and (5) as equivalent, and 'would' as commuting with 'not'. That seems to support the conjunction of WEM and WNC.

This is no mere trick with probability. Other considerations point the same way. For example, the following two sentences seem to be equivalent, in a context where we can treat talking as equivalent to not being silent (ignoring non-verbal noises), and the quantifiers scope over 'would':

(6)   No student would be silent.
(7)   Every student would talk.

Since 'No student' is equivalent to 'Every student not' and 'x talks' to 'not(x is silent)' here, (6) and (7) should be equivalent to (6a) and (7a) respectively:

(6a)   Every student$_x$ not(would(x is silent)).
(7a)   Every student$_x$ would(not (x is silent)).

But (6a) and (7a) are generally equivalent only if 'not(would(x is silent))' is equivalent to 'would(not(x is silent))', so the moral is again that 'not' and 'would' are treated as commuting, adding further support to both WEM and WNC. In terms of possible world semantics, we seem to treat 'would' as singling out a uniquely accessible world.

Grammatically, 'would' is the past tense of the future auxiliary verb 'will', and we sometimes use it as such. An example is (8), written in 2019 by a historian describing Napoleon sailing into exile in 1815:

(8)   He would never see France again.

That just corresponds to (9), said in 1815:

(9)   He will never see France again.

But 'will' is treated as singling out a unique future history, even if it is somehow not yet causally or metaphysically determined which one, and so as commuting with negation (Cariani and Santorio 2018). Thus future-qualified versions of both excluded middle (FEM) and non-contradiction (FNC) look plausible:

FEM   Either will(A) or will(not(A))
FNC   Not(will(A) and will(not(A)))

Grammatically, WEM and WNC are the past tenses of FEM and FNC respectively. Thus someone might try to argue for the validity of WEM and WNC as derivative from the validity of FEM and FNC.

Of course, some philosophers reject FEM, on the grounds that the future is metaphysically open. Given that it is contingent whether there will be a sea battle tomorrow, they deny (10):

(10)   will(there is a sea battle tomorrow) or will(there is no sea battle tomorrow)

On their view, the first disjunct is false because in some genuinely possible futures there is no sea battle tomorrow, and the second disjunct is false because in some genuinely possible futures there is a sea battle tomorrow. They may still accept (11), because in every genuinely possible future either there is a sea battle tomorrow or there is not:

(11)   will(there is a sea battle tomorrow or there is no sea battle tomorrow).

But (11) is not the same thing as (10); a disjunction may hold in every possible future even though neither disjunct holds in every possible future. Such metaphysicians give the future 'will' a genuinely modal reading, and thereby bring it much closer to a fully modal reading of 'would', but their denial of FEM leaves little hope for WEM.[2]

A compromise view applies the apparatus of supervaluations (Klinedinst 2011, Cariani and Santorio 2018). Each admissible sharpening selects a unique open future, and uses it to evaluate 'will', although the intentions of language-users are neutral between different admissible sharpenings. A sentence is *supertrue* just in case it is true on every admissible sharpening, and *superfalse* just in case it is false on every admissible sharpening. On this view, (10) and not just (11) is supertrue, because on every admissible sharpening one or other disjunct is true, even though neither disjunct is supertrue (or superfalse). FEM is thus validated, as is FNC (except at the last moment of time, if there is one). This approach is analogous to Stalnaker's supervaluationist treatment of the selection function in his semantics for conditionals: for a given world of evaluation and antecedent, each admissible sharpening selects a unique world where the antecedent is true, and uses it to evaluate 'if', although the intentions of language-users are more or less neutral between different admissible sharpenings. Stalnaker's semantics validates CEM, and even CNC except when the antecedent is impossible.

In reconciling the validity of FEM with a genuinely modal metaphysics of the future, the supervaluationist approach holds out hope of validating WEM on a genuinely modal reading of 'would'. However, in section 9.2 we saw evidence that 'would' has a fully modal reading which is a fake past, not the real past of future 'will', even on a genuinely modal metaphysics of the future. The strictly temporal past future reading of 'would' still falls short of its fully modal reading. For example, in a conversation about what it *would* be like to live in the best of all possible worlds, it makes good sense to discuss possible worlds which share no common past with this world at all. Nevertheless, while distinguishing a fully modal reading of 'would' from its temporal reading, even on a modal metaphysics of the future, one might still try to validate WEM for the fully modal reading by a separate application of the supervaluationist approach, applied to selection functions each singling out a unique member of the set of contextually relevant worlds for 'would'.

However, such a supervaluationist approach is ill-suited to many modal uses of 'would' in ordinary conversation. Imagine these two mini-conversations:[3]

(12)   Speaker 1: Jasper told his wife.
       Speaker 2: He would tell her.

---

[2] Compare the dispute between Leslie 2009 and Klinedinst 2011.
[3] One can hear 'would' in speaker 2's contributions to both (12) and (13) as attributing wilfulness rather than something modal to Jasper's behaviour: roughly, he insisted on telling/not telling her. That is not the intended reading for present purposes.

(13) Speaker 1: Jasper didn't tell his wife.
      Speaker 2: He wouldn't tell her.

The two conversations have the same overall structure. In each case, Speaker 2 is adding something non-trivial to what Speaker 1 has just said, by subsuming Jasper's behaviour under a modally robust trend: in all relevant worlds in which Jasper is in those circumstances with his actual character, he behaves that way. Speaker 2 is *strengthening* what speaker 1 has said, by generalizing it over a range of contextually relevant worlds, including the actual world. Thus what speaker 2 says implies what speaker 1 says, but not conversely. Because the actual world is in the range of 'would' in this context, both (14) and (15) are ruled out:

(14) Jasper told his wife, but he would not tell her.
(15) Jasper did not tell his wife, but he would tell her.

In effect, the material conditionals (16) and (17) are common ground in this context (whether or not they are expressed with 'if'):

(16) would(not(Jasper told his wife)) ⊃ not(Jasper told his wife)
(17) would(Jasper told his wife) ⊃ Jasper told his wife

Both (16) and (17) are derivable from the T principle for 'would' rejected in section 10.3, on which 'would(A)' entails A; their status here depends not on T as a general principle of logic but on the specifics of the conversational context. Now suppose that WEM is part of the logic of 'would'. Thus, in particular, we have:

(18) would(Jasper told his wife) or would(not(Jasper told his wife))

But (18) trivializes Speaker 2's contributions in conversations (12) and (13), given what Speaker 1 has just said.

In conversation (12), Speaker 1 says 'Jasper told his wife', which together with (16) eliminates the antecedent of (16), which is the second disjunct of (18); by disjunctive syllogism, that leaves only the first disjunct of (18), 'would(Jasper told his wife)', which is just what Speaker 2 says. Thus WEM makes Speaker 2's contribution to conversation (12) redundant, adding nothing new to Speaker 1's contribution and the common ground. That reasoning is just as legitimate on a supervaluationist approach as on a classical one.

Similarly, in conversation (13), Speaker 1 says 'Jasper didn't tell his wife', which together with (17) eliminates the antecedent of (17), which is the first disjunct of (18); by disjunctive syllogism, that leaves only the second disjunct of (18), 'would(not(Jasper told his wife))', which is just what Speaker 2 says. Thus WEM also makes Speaker 2's contribution to conversation (13) redundant, adding nothing new to Speaker 1's contribution and the common ground. Again, the reasoning is just as legitimate on a supervaluationist approach as on a classical one.

But Speaker 2's contributions to conversations (12) and (13) are not redundant. They add the new information that Jasper's behaviour was not an uncharacteristic blip. The culprit is WEM, and in particular its instance (18). Assumptions (16) and (17) are harmless; they just reflect the specific context. To put the point more generally, WEM requires the set of contextually relevant worlds to have at most one member;

thus when the actual world is contextually relevant, the set of contextually relevant worlds is just the singleton of the actual world, which makes 'would' a redundant operator. But, as such examples show, the contextual relevance of the actual world does *not* always make 'would' redundant. Thus WEM is invalid, and supervaluationism does not rescue it.

Here is another way of making the point. We can understand Speaker 2 in both conversations, (12) and (13), as offering a simple proto-explanation of the behaviour reported by Speaker 1, by bringing it under a modal generalization: Jasper behaved that way *because* that is the way he would behave. For the explanation works properly only if the actual world is within the scope of the modal generalization, in other words it is in the domain of worlds for 'would', otherwise the explanandum does not follow from the explanans in the intended way. We may assume that Speaker 2 is not making the strange suggestion 'Jasper behaved that way in the actual world *because* he behaved that way in the selected non-actual world'; Speaker 2 has a more straightforward explanatory connection in mind. Equally, the explanation works properly only if the actual world is not the only world within the scope of the modal generalization, in other words it is not the only world in the domain of worlds for 'would', otherwise the explanans collapses into the explanandum. Although that is not the only possible filling-out of what is going on in the conversations, it is one perfectly possible filling-out, which is enough for the argument. The explanatory role for the 'would' claims in conversations (12) and (13) is reminiscent of the explanatory role of 'would if' claims in arguments by inference to the best explanation (chapter 9).

A real-life example comes from an incident during the Profumo Affair, a complex British political scandal, one turning-point of which was a trial at the Old Bailey in June 1963. The defence counsel James Burge was cross-examining a young woman, Mandy Rice-Davies, who claimed to have had an affair with Lord Astor. This exchange occurred:[4]

James Burge: Do you know Lord Astor has made a statement to the police saying that these allegations of yours are absolutely untrue?

Mandy Rice-Davies: Well [giggle] he would, wouldn't he? [laughter in court]

Mandy Rice-Davies' point was that Lord Astor's making the denial was modally robust given his interests, irrespective of the truth-value of her allegations, and so was evidentially worthless. She could not have made that point just by saying 'Well he did, didn't he?' She used the modal sense of 'would' to full effect.

The arguments about (12) and (13) in no way depends on using 'would' to take the evaluation back to a time before the relevant possible worlds started to diverge from each other. The same logical structure applies to possible histories differing from each other at all times. For example, imagine that the ultimate physical theory entails that there will be eternal recurrence (both backwards and forwards in time), but leaves it open exactly which types of event will eternally recur. Two scientists are having a conversation, in which all and only physically possible worlds are relevant to

⁴ See   https://www.theguardian.com/politics/2014/dec/19/mandy-rice-davies-fabled-player-british-scandal-profumo.

'would'. The first scientist has worked out that, given various known but physically contingent facts about the actual world, there will be eternal recurrence. The second scientist but not the first knows that eternal recurrence is a quite general consequence of the theory, independently of those facts. Then Scientist 2's remark is not redundant, given Scientist 1's:

(19) Scientist 1: There will be eternal recurrence.
     Scientist 2: There would be.

Scientist 2's point is that there will be eternal recurrence *because* there would be eternal recurrence. That is a non-trivial appeal to the physical theory.

Exchange (19) contrasts with exchange (20), where Scientist 3 accepts Scientist 2's deduction but mistakenly believes that only the actual history of the universe is compatible with the theory:

(20) Scientist 1: A conversation like this will happen infinitely often.
     Scientist 3: It would.

In the given context, Scientist 3's contribution is false. The physical theory leaves open histories on which nothing like that conversation ever occurs. Thus Scientist 3's contribution is not redundant, for in this context its redundancy would imply its truth, for both Scientist 1's contribution and the common ground are true.

In both exchanges (19) and (20), the actual world is contextually relevant, but that does not make the second speaker's contribution redundant. Once again, the culprit is WEM. It is invalid, and the evidence for it needs to be explained away. A natural hypothesis is that it is generated by a fallible heuristic, perhaps one structurally similar to the Suppositional Rule.

As noted in section 10.3, on the material reading of 'if', WEM is equivalent to its analogue WCEM for 'would if', the 'would' version of conditional excluded middle:

WCEM   would(if A, C) or would(if A, not(C))

WCEM entails WEM because A can be a tautology; WEM entails WCEM because 'not(if A, C)' truth-functionally entails 'if A, not(C)'. Given that we should reject WEM, we should also reject WCEM, with Lewis and against Stalnaker.

11.3 Negating 'would'

Another similarity between 'would' and 'if' is that both form sentences which lack natural wide-scope negations. In the case of 'if', that turned out in chapter 5 to be a clue to the relation between its heuristics and its semantics. We may hope for something similar in the case of 'would'. Moreover, in both cases, the lack of such negations makes the semantics harder to determine, by obscuring falsity-conditions. This section examines the case of 'would' in more detail.

Both 'will' and 'would' sentences lack natural wide-scope negations. How are we to negate (21) and (22)?

(21) She will write.
(22) She would write.

Obviously, (23) and (24) are ill-formed:

(23)   She not will write.
(24)   She not would write.

Grammar forces us to put the 'not' inside 'will' and 'would', giving (25) and (26), but for that very reason they seem not to be the wide-scope negations of (23) and (24) respectively:

(25)   She will not write.
(26)   She would not write.

The failure of WEM means that we cannot in general treat the narrow-scope negation for 'would' as equivalent to the wide-scope negation. In particular, Speaker 2's contribution to conversation (13) ('He wouldn't tell her') seems to stand in the same logical relation to Speaker 1's contribution ('Jasper didn't tell his wife') as Speaker 2's contribution to conversation (12) ('He would tell her') stands to Speaker 1's contribution ('Jasper told his wife'). In each case, the second utterance modalizes the first with 'would'. That requires formalizing the negation in Speaker 2's contribution as scoping under 'would', which is just how they were treated in the previous section.

Of course, we can always use (27) and (28) to negate (21) and (22) respectively:

(27)   It is false (/not the case) that she will write.
(28)   It is false (/not the case) that she would write.

However, (27) and (28) are clunky, cumbersome, and pedantic-sounding. Although they are not meaningless, they are too artificial to provide very telling data.

Another approach to finding a natural negation for 'would' assumes that 'might' is the dual of 'would', standing to it as 'in some world' stands to 'in every world'. Since 'might' is grammatically the past tense of 'may', perhaps 'may' (read non-deontically) is also the dual of 'will'. On that view, (30) is equivalent to the wide-scope negation of (22), and perhaps (29) to the wide-scope negation of (21):

(29)   She may not write.
(30)   She might not write.

Clearly, (29) makes trouble for (21), and (30) for (22). But the trouble is more naturally understood as epistemic doubt than as broadly logical inconsistency. To illustrate, imagine that we know a given coin to be either double-headed or double-tailed, but we do not know which. Then we can definitely assert (31) about a future toss, and (32) about a merely hypothetical one:

(31)   It will come up heads or it will come up tails.
(32)   It would come up heads or it would come up tails.

The assertions are not based on FEM for (31) or WEM for (32). Rather, their justifications are more specific to the case at hand. For either the coin is double-headed or it is double-tailed. With a future toss, if the coin is double-headed then it will come up heads, and if it is double-tailed then it will come up tails, so (31) follows by standard propositional logic. With a merely hypothetical toss, if the coin is double-headed then it would come up heads, and if it is double-tailed then it would come up tails, so (32) follows in the same way. But, without inconsistency, in our state of ignorance we can add (33) to (31) and (34) to (32), on salient readings of 'may' and 'might':

(33)   It may come up heads and it may come up tails.
(34)   It might come up heads and it might come up tails.

Clearly, if the first conjunct of (33) contradicts the second disjunct of (31) and the second conjunct of (33) contradicts the first disjunct of (31), then (33) is inconsistent with (31). But, on the salient reading of 'may', there is no such inconsistency. Thus, although 'may not' casts doubt on 'will', it does not exclude the truth of 'will'. Similarly, if the first conjunct of (34) contradicts the second disjunct of (32), and the second conjunct of (34) contradicts the first disjunct of (32), then (34) is inconsistent with (32). But, on the salient reading of 'might', there is no such inconsistency. Thus, although 'might not' casts doubt on 'would', it does not exclude the truth of 'would'.

In articulating such epistemic readings of 'may' and 'might', we must be careful about what exactly is meant to be in doubt. This is clearest for 'might' when the contextual restriction R excludes the actual world. For example, suppose that only counterfactual worlds in which a tiger enters the room are contextually relevant. Then (35) casts doubt on (36) but not on (37):

(35)   Federico might be scared.
(36)   Federico would not be scared.
(37)   Federico is not scared.

In the actual world, Federico is sitting next to us, quite obviously not scared; there is quite obviously no tiger in the room. At least in such cases, 'might' seems to be the dual neither of 'known' nor of 'would', but of their combination, in that order. We can roughly paraphrase (35) as (38):

(38)   It is not known that Federico would not be scared.

One advantage of (38) is that it avoids applying 'not' directly to 'would'; instead, it applies 'not' once to 'known', which can be negated straightforwardly, and once inside 'would'. For present purposes, we can be non-committal on how exactly the epistemic part of (38) works: for instance, whose knowledge is in question, and whether to count undrawn consequences as part of what is known. By contrast with (38), pulling 'would' to the outside gets (35) wrong:

(39)   Federico would not be known not to be scared.

For (35) is appropriate when we have no idea how Federico would react to the tiger, whereas (39) excludes the possibility that before our eyes he would confront it with manifestly relaxed courage, and so *would* be known not to be scared.

The future 'may' and 'will' behave analogously. Franz is a swaggering young soldier in the jingoistic atmosphere of August 1914, about to go off to war for the first time.

More experienced soldiers are discussing how he will perform in his first battle. In this context, (40) casts doubt on (41) but not on (42):

(40)   Franz may be scared.
(41)   Franz will not be scared.
(42)   Franz is not scared.

At present, Franz is cheerfully boasting about his future exploits, quite obviously not scared. At least in such cases, 'may' is the dual neither of 'known' nor of 'will', but of their combination, in that order. We can roughly paraphrase (40) as (43):

(43)   It is not known that Franz will not be scared.

One advantage of (43) is that it avoids applying 'not' directly to 'will'; instead, it applies 'not' once to 'known', which can be negated straightforwardly, and once inside 'will'. By contrast with (43), pulling 'will' to the outside gets (40) wrong:

(44)   Franz will not be known not to be scared.

For (40) is appropriate when it is quite unclear how Franz will react to his first battle, whereas (44) excludes the possibility that he will act with conspicuous fearlessness.

Perhaps 'may' and 'might' have other readings, non-deontic and non-epistemic, on which they are the duals of 'will' and 'would' respectively. However, those readings are not easy to recognize in practice. If they are too elusive, 'may not' and 'might not' will not give robust negations for 'will' and 'would'.[5]

Another candidate for the dual of 'would' is 'could', perhaps with 'can' as a candidate for the dual of 'will'. But that proposal too fares badly, for each of (45)–(48) is quite consistent:

(45)   She can write, but she won't.
(46)   She can talk, but she will stay silent.
(47)   She could write, but she wouldn't.
(48)   She could talk, but she would stay silent.

An ability is one thing, the disposition to exercise it quite another.

A further problem for the attempt to use 'can' and 'could' to form equivalents of 'will' and 'would' is that the former interact with negation in surprising ways. For, despite the position of 'not' after 'can', normally (49) should be paraphrased as (51), not as (50):

(49)   He cannot breathe.
(50)   It is possible for him not to breathe.
(51)   It is not possible for him to breathe.

Similarly, despite the position of 'not' after 'could', normally (52) should be paraphrased as (55) (past) or (56) (modal), not as (53) (past) or (54) (modal):

---

[5] Given the present semantics and a reading of 'might' on which it is dual to 'would', the argument that counterfactuals 'if A, would C' are mostly false because the corresponding 'if A, might not(C)' statements are mostly true (Hájek 2009) is still unsound. For we may tend to accommodate the latter statements by enlarging the domain of relevant worlds.

(52) He could not breathe.
(53) It was possible for him not to breathe.
(54) It would be possible for him not to breathe.
(55) It was not possible for him to breathe.
(56) It would be not possible for him to breathe.

Although 'not' is placed inside 'can' and 'could', semantically it scopes over them. Although one can exceptionally give 'not' in (52) narrow scope by pronouncing 'not breathe' as a single verb 'not-breathe', doing so sounds quite artificial. For multiple reasons, therefore, 'can' and 'could' are ill-suited to making natural negations for 'will' and 'would'.

The odd behaviour of 'not' after 'can' and 'could' may give one pause to wonder whether the negation in 'will not' and 'won't', and in 'would not' and 'wouldn't', is to be understood similarly, as scoping over 'will' and 'would', despite its surface position. If that analogy worked, 'will' and 'would' would have natural negations after all, and what look like instances of the controversial principle WEM would turn out to be harmless tautologies, mere instances of the simple law of Excluded Middle. But, as observed above, the structural parallel between the conversations (12) and (13) shows that the analogy with 'cannot' and 'could not' does not work, for the parallel depends on negation taking narrow scope in (13). In any case, examples such as (1)–(4), (6)–(7), (10), and (31)–(32) were set up to achieve the effect of negation in ways which did not permit any such deflationary explanation.

Here is another way to see that the analogy with 'cannot' and 'could not' fails for 'will not' and 'won't'. Someone who made the prediction (57) in advance is vindicated by Speaker 1's contribution to conversation (13) in exactly the same way as someone who made the prediction (56) in advance is vindicated by Speaker 1's contribution to conversation (12):

(56) Jasper will tell his wife.
(57) Jasper will not tell his wife.

But if (57) is semantically the negation of (56), with 'not' scoping over 'will', and not equivalent to the result of inserting negation genuinely within the scope of 'will' in (56), then the parallel between (56) and (57) breaks down, because (57) is no longer a prediction of what the future will be; it is merely the denial of such a prediction. Since the parallel is good, (57) is not the direct negation of (56); 'not' in (57) is genuinely within the scope of 'will'.

We can still negate 'will' and 'would' by brute force, as in (27) and (28), but the lack of more natural negations hints at something significant.

The awkward interaction of 'will' and 'would' with 'not' is analogous to the awkward interaction of 'if' with 'not' even in non-modal contexts. Conditionals lack natural external negations; there is a strong tendency to treat their internal negations as though they were equivalent to external negations, even though doing so creates logical paradoxes. In both cases, the problems are not merely syntactic; they are robust to paraphrasing 'not' away. For conditionals, the problems turned out to be best explained as byproducts of our reliance on a fallible heuristic, the Suppositional Procedure. By analogy, a natural hypothesis for 'will' and 'would' is that the problems are best explained as byproducts of our reliance on a similar fallible heuristic. For example, when we judge how confident to be that the coin would come up heads, we suppose that the contextual restrictions obtain—for instance, that the coin will be tossed—and, on that supposition, take an attitude to the proposition that it comes up heads—say, the attitude of 50 per cent credence; we then take that attitude unconditionally to the proposition that the coin would come up heads. The task of the next section is to develop such a hypothesis in detail.

11.4 A suppositional heuristic for 'would'

For purposes of comparison, here is the Suppositional Rule SR for plain 'if' from chapter 3:

> SR   For any attitude ||, sentences A, C, and set of sentences BB:
> BB || 'if A, C' just in case BB, A || C

What is to the left of || represents what is being supposed; what is to its right represents that to which the attitude || is to be taken on those suppositions. SR provides a general test for taking any cognitive attitude || to any conditional 'if A, C' on any set of suppositions BB. It is whether one takes the same attitude to C on the same set of suppositions plus A.

Following that precedent, we seek a modal suppositional rule, a general test for taking any attitude || to any restricted necessity claim 'would(A)' on any set of suppositions BB. Thus the left-hand side of the desired biconditional will be 'BB || would(A)'.

Just as with the Suppositional Rule, the informal version of the modal test involves an imaginative exercise. However, the process is now more complex, since in imagining a state of affairs, we are no longer imagining that it obtains in the actual world, but only that it obtains in a world w, not assumed to be actual. The difference is crucial to the proper use of modalized conditionals in abductive inference, as chapter 9 noted. Thus, in applying the modal heuristic, we must somehow distance ourselves from our background knowledge. But we do not want to lose sight of it altogether, since we often want to compare actuality with the imagined world w (though of course, what is called 'actual' in w is w rather than the actual world itself). For example, one's background knowledge of politics properly informs one's answer to questions such as 'Who would have won the last election if the health service had been the main issue?' Some suppositions must be distanced in the same way. Speaking to someone who has doubts as to Kennedy's assassin, I might say: 'Suppose that Oswald shot Kennedy. If he had not done so, would someone else have shot Kennedy?' The hearer is intended to suppose that Oswald shot Kennedy in the actual world, and on that supposition imagine what would have happened if things had gone otherwise, in another possible world. The supposition is still intended to inform the imaginative exercise, but not as holding in the scenario imagined as counterfactual, the modally distanced scenario.

To mark that difference, the formulas which are envisaged as holding in the modally distanced scenario will be underlined. The underlining does not modify their content; it merely signals the role they are playing. The original Suppositional Rule can be applied to underlined formulas in the usual way. More elaborate notation would be needed to represent modally distanced suppositions within modally distanced suppositions, but for present purposes it is best to keep things simple.

In the model theory (section 10.3), the restriction on contextually relevant worlds is a binary relation R between worlds. However, here we are not worrying about iterated modalities, and so can harmlessly simplify by treating the restriction as just a supposition R holding in just those worlds to which the actual world has the relation R. The thinker need not be able to articulate R in words. Since R is implicitly supposed to hold in the modally distanced scenario, 'R' is underlined.

Putting the pieces together yields this formalization of the Modal Suppositional Rule:

MSR   For any attitude $\|$, sentence A, restriction R, and set of sentences BB:
        $BB \parallel$ 'would(A)' just in case $BB, \underline{R} \parallel \underline{A}$

Of course, MSR only shifts the issue to the determination of the right-hand side, at which point we may need to apply far more content-specific heuristics. Nevertheless, MSR characterizes the outermost structure of the procedure, which for present purposes is what we need.

Combining the two heuristics, we can derive a heuristic for the modalized conditional, the Modal Conditional Suppositional Rule:

MCSR   For any attitude $\|$, sentences A, C, and set of sentences BB:
        $BB \parallel$ 'would(if A, C)'   just in case $BB, \underline{R} \parallel \underline{\text{'if } A, C\text{'}}$
                              just in case $BB, \underline{R}, \underline{A} \parallel \underline{C}$

This resembles the ordinary Suppositional Rule in making the final object of the putative attitude the consequent, and the foreground supposition the antecedent; it resembles the Modal Suppositional Rule in the shift of background suppositions; it resembles both in holding the attitude at issue fixed. That fits the generic similarities and differences between the primary ways of assessing plain 'if' and 'would if' conditionals.

The distinction between underlined and non-underlined suppositions is needed to handle the rigidifying 'actually' operator, discussed in section 9.1. We must block the disastrous jump from accepting instances of the trivially real-world valid schema 'if A, actually A' to accepting instances of the real-world invalid schema 'would(if A, actually A), which induces modal collapse. For instance, we must not treat the triviality of "if it is raining, it is raining in this world" as justifying the claim 'if it were raining, it would be raining in this world'.

The point can be made more formally. Let $\|^a$ be the attitude of acceptance. Then (58) is trivial:

(58) $A \parallel^a$ actually A

By SR, (58) yields (59):

(59) $\parallel^a$ if A, actually A

Similarly, by MCSR, (60) yields (61):

(60) $\underline{R, A} \parallel^a \underline{\text{actually } A}$
(61) $\parallel^a$ would(if A, actually A)

But (61) (which has no underlining) is the disastrous modal collapse schema, and so must be avoided. Therefore we must avoid (60) too. The underlining enables us to do so, for supposing in the modally distanced way that it is raining does not commit us to supposing in the modally distanced way that it is raining in this world.

The next task is to use MSR to explain the evidence which seems to show that 'would' picks out a single modally distanced world, and likewise to use MCSR to explain the corresponding evidence for 'would' conditionals. For that purpose, we can use the framework already developed in section 3.4 for analysing complex attitudes and their interaction with the original Suppositional Rule.

We start with negative attitudes. For any cognitive attitude $||$, the corresponding negative attitude $||^{\text{not}}$ was defined in terms of $||$ thus:

$||^{\text{not}}$ $BB \, ||^{\text{not}} A$ just in case $BB \, || \, \text{not}(A)$

Taking the negative attitude to something is just taking the corresponding positive attitude to its negation.

We can now show how MSR interacts with negative attitudes to treat 'would' as commuting with negation, in the sense that it mandates taking exactly the same attitudes to 'would(not(A))' as to 'not(would(A))'. Here is the proof, for any attitude $||$, background suppositions BB, and contextual restriction R:

$BB \, || \, \text{would}(\text{not}(A))$	just in case	$BB, \underline{R} \, || \, \underline{\text{not}(A)}$	(by MSR)
	just in case	$BB, \underline{R} \, ||^{\text{not}} \underline{A}$	(by definition of $||^{\text{not}}$)
	just in case	$BB \, ||^{\text{not}} \text{would}(A)$	(by MSR)
	just in case	$BB \, || \, \text{not}(\text{would}(A))$	(by definition of $||^{\text{not}}$)

In particular, consider the attitude $||^a$ of acceptance. First, let BB be {'would(not(A))'}. Trivially, {'would(not(A))'} $||^a$ 'would(not(A))'. Therefore, by the proof, {'would(not(A))'} $||^a$ 'not(would(A))'. For the converse, let BB be {'not(would(A))'}. Trivially, {'not(would(A))'} $||^a$ 'not(would(A))'. Therefore, by the proof, {'not(would (A))'} $||^a$ 'would(not(A))'. In other words, MSR mandates accepting each of 'would (not(A))' and 'would(not(A))' on the supposition of the other.

That MSR mandates treating 'would(not(A))' and 'would(not(A))' as equivalent already explains the appeal of WEM and WNC. For the equivalence reduces each modal principle to a special case for modal sentences of the corresponding classically valid principle of non-modal propositional logic: WEM ('would(A) or would(not(A))') reduces to plain excluded middle ('would(A) or not(would(A))'); WNC ('not(would(A) and would(not(A)))') reduces to plain non-contradiction ('not(would(A) and not(would(A)))').

Analogous arguments show that the suppositional heuristic mandates treating 'would' as commuting with any other truth-function. We can take conjunction as representative. Section 3.4 defined plural conjunctive attitudes. For any cognitive attitude $||$, the corresponding conjunctive attitude $||^{\text{and}}$ was defined in terms of $||$ thus:

$||^{\text{and}}$ $CC \, ||^{\text{and}} A, B$ just in case $CC \, || \, A$ and B

Taking the conjunctive attitude to some things is just taking the original attitude to their conjunction.

As in section 3.4, we need a generalization of the suppositional rule to handle plural attitudes. Here it is:

MSR+ For any attitude $\|$, sentences $A_1, \ldots, A_n$, restriction R, and set of sentences BB: $BB \parallel$ 'would(A_1)', ..., 'would(A_n)' just in case $BB, \underline{R} \parallel \underline{A_1}, \ldots, \underline{A_n}$

This is within the spirit of MSR. The operator 'would' on each of $A_1, \ldots, A_n$ indicates that we are shifting their evaluation to the modally distanced scenario, corresponding to the restriction R.

We can now show how MSR+ interacts with conjunctive attitudes to treat 'would' as commuting with conjunction, in the sense that it mandates taking exactly the same attitudes to 'would(A and B)' as to 'would(A) and would(B)'. Here is the proof, for any attitude $\|$, background suppositions CC, and contextual restriction R:

$CC \parallel$ would(A and B)	just in case	$CC, \underline{R} \parallel \underline{A \text{ and } B}$	(by MSR+)
	just in case	$CC, \underline{R} \parallel^{\text{and}} \underline{A}, \underline{B}$	(by $\parallel^{\text{and}}$)
	just in case	$CC \parallel^{\text{and}}$ would(A), would(B)	(by MSR+)
	just in case	$CC \parallel$ would(A) and would(B)	(by $\parallel^{\text{and}}$)

In particular, by an argument like that for negation, we can draw the corollary that {'would(A and B)'} $\parallel^a$ 'would(A) and would(B)' and {'would(A) and would(B)'} $\parallel^a$ 'would(A and B)', where $\parallel^a$ is the attitude of acceptance. In other words, MSR+ mandates accepting each of 'would(A and B)' and 'would(A) and would(B)' on the supposition of the other.

Since any truth-function is definable in terms of conjunction and negation, we can use these results to show that MSR+ mandates treating 'would' as commuting with any truth-function, so defined. Alternatively, we can give analogous arguments directly in terms of those other truth-functions, using suitably defined complex attitudes.

These results about our implicit suppositional heuristic for 'would' explain the evidence which seems to show that 'would' is restricted to a single modally distanced world. They also indicate why the heuristic is likely to be fallible, for reasons already suggested. Consider any context in which the actual world satisfies the relevant restriction R. Thus all material conditionals of the form 'would(A) ⊃ A' are true. Hence, as a special case, all those of the form 'would(not(A)) ⊃ not(A)' are true. But if MSR is infallible, 'would(not(A))' is equivalent to 'not(would(A))'. In that case, all conditionals of the form 'not(would(A)) ⊃ not(A)' are true; hence, by contraposition, so are all those of the form 'A ⊃ would(A)'. Therefore 'would(A)' has the same truth-value as A, for all A. In effect, the relevance of the actual world implies the irrelevance of all non-actual worlds. But in many contexts we treat both the actual world and some range of non-actual worlds as relevant, because we have reason to care about whether the actual world is exceptional. That tendency counterbalances our implicit reliance on MSR, and gives us access to its fallibility.

Corresponding to these results for plain 'would', we can establish analogous results for 'would if', the modalized conditional, using the Modal Conditional Suppositional Rule MCSR.

First, we show how MCSR interacts with negative attitudes to treat 'would' as commuting with negation in the consequent, in the sense that it mandates taking exactly the same attitudes to 'would(if A, not(C))' as to 'not(would(if A, C)'. Of course, we know from section 11.3 that explicitly negating 'would' is awkward. Nevertheless, the result shows how MCSR implicitly constrains us to view the falsity-conditions of 'would(if A, C).

Here is the proof, for any attitude $\|$, background suppositions BB, and contextual restriction R:

$BB \parallel$ would(if A, not(C))	just in case	$BB, \underline{R, A} \parallel \underline{\text{not}(C)}$	(by MCSR)
	just in case	$BB, \underline{R, A} \parallel^{\text{not}} \underline{C}$	(by $\parallel^{\text{not}}$)
	just in case	$BB \parallel^{\text{not}}$ would(if A, C)	(by MCSR)
	just in case	$BB \parallel$ not(would(if A, C))	(by $\parallel^{\text{not}}$)

In particular, we can draw the corollary in the usual way that 'would(if A, not(C))' $\parallel^a$ 'not(would(if A, C))' and 'not(would(if A, C))' $\parallel^a$ 'would(if A, not(C))': in other words, MCSR mandates accepting each of 'would(if A, not(C))' and 'not(would(if A, C))' on the supposition of the other.

That MCSR mandates treating 'would(if A, not(C))' and 'not(would(if A, C))' as equivalent explains the appeal of the 'would' versions of Conditional Excluded Middle and Conditional Non-Contradiction:

WCEM   would(if A, C) or would(if A, not(C))
WCNC   not(would(if A, C)) and would(if A, not(C)))

For the equivalence reduces each modal conditional principle to a special case for modal conditional sentences of the corresponding classically valid principle of non-modal propositional logic: WEM reduces to plain excluded middle ('would(if A, C) or not(would(if A, C))'); WNC reduces to plain non-contradiction ('not(would(if A, C)) and not(would(if A, C)))').

Analogous arguments show that the suppositional heuristic mandates treating 'would if' as commuting with any other truth-function in the consequent. As usual, we need a generalization of the suppositional rule to handle plural attitudes. We obtain it simply by combining MSR+, the generalized rule for 'would', with SR+, the generalized rule for 'if' from section 3.4:

SR+ $BB \parallel$ (if A, C_1), . . . , (if A, C_n) just in case BB, $A \parallel C_1$, . . . , C_n

Together, MSR+ and SR+ yield the generalized rule MCSR for 'would if':

MCSR+   For any attitude $\|$, sentences A, C_1, . . . , C_n, restriction R, and set of sentences BB:

$$BB \parallel \text{would(if } A, C_1), \ldots, \text{would(if } A, C_n)$$
just in case $BB, \underline{R} \parallel \underline{\text{(if } A, C_1), \ldots, \text{(if } A, C_n)}$
just in case $BB, \underline{R, A} \parallel \underline{C_1, \ldots, C_n}$

We can now show how MCSR+ interacts with conjunctive attitudes to treat 'would if' as commuting with conjunction in the consequent, in the sense that it mandates taking exactly the same attitudes to 'would(if A, (C and D))' as to 'would(if A, C) and would(if A, D)'. Here is the general proof:

$BB \parallel$ would(if A, C and D)

just in case $BB, \underline{R, A} \parallel \underline{C \text{ and } D}$	(by MCSR+)
just in case $BB, \underline{\overline{R}, A} \parallel^{\text{and}} \underline{C, D}$	(by $\parallel^{\text{and}}$)
just in case $BB \parallel^{\overline{\text{and}}} \text{would(if } \underline{A}, C), \text{would(if } A, D)$	(by MCSR+)
just in case $BB \parallel$ would(if A, C) and would(if A, D)	(by $\parallel^{\text{and}}$)

.n particular, we can draw the corollary in the usual way that 'would(if A, C and D)' $\parallel^a$ 'would(if A, C) and would(if A, D)' and 'would(if A, C) and would(if A, D)' $\parallel^a$ would(if A, C and D)': in other words, MCSR+ mandates accepting each of would(if A, C and D)' and 'would(if A, C) and would(if A, D)' on the supposition of the other.

We can use these results to show that MCSR+ mandates treating 'would if' as commuting in the consequent with any truth-function, defined in terms of conjunction and negation. Alternatively, we can give analogous arguments directly in terms of those other truth-functions, using suitably defined complex attitudes.

These results about our implicit suppositional heuristic for 'would if' explain the evidence which seems to show that 'would if' is restricted to a single modally distanced world where the antecedent holds, as in Stalnaker's semantics for the conditional. However, we have also seen evidence that the underlying suppositional heuristic for 'would' is fallible, especially for contexts in which the actual world is relevant for 'would'. With the suppositional heuristic for 'would if', we can go further, by showing that in some cases MCSR+ generates contradictions, just as the suppositional rule SR+ for plain 'if' did in section 3.5.

Here is the argument, for any sentence A and the attitude $\parallel^a$ of acceptance. Since anything is accepted on suppositions which include itself, we have:

[I] $\underline{R, A \text{ and not}(A)} \parallel^a \underline{A \text{ and not}(A)}$

Therefore, by MCSR+:

[II] $\parallel^a$ would(if A and not(A), A not(A))

Of course, [II] is compelling anyway. Since MCSR+ implies that 'would if' commutes with conjunction in the consequent, [II] implies [III]:

[III] $\parallel^a$ would(if A and not(A), A) and would(if A and not(A), not(A))

Similarly, since MCSR+ also implies that 'would if' commutes with negation in the consequent, [III] implies [IV]:

[IV] $\parallel^a$ would(if A and not(A), A) and not(would(if A and not(A), A))

But [IV] says that a contradiction is to be accepted: the conjunction of 'would(if A and not(A))' with its own negation, however mundane the sentence A.

Thus the suppositional rule for 'would if', MCSR+, generates a paradox. When one considers sentences of a suitable form, it tells one to accept a contradiction. Of course, the derivation of MCSR+ depended on SR+, the suppositional rule for plain 'if', which by itself generates a very similar paradox. One might therefore interpret the present result as showing merely that the inconsistency latent in the original rule for plain 'if' can manifest itself even in a modal setting: the prefix 'would' is not enough to fence off 'if' and prevent it from causing trouble. But that interpretation is inadequate, because one can also derive a similar result from MSR+, the suppositional rule for 'would', without using 'if' or any rule for a conditional.

The paradox for MSR+ involves the special case where the contextual restriction R is itself inconsistent. At first sight, a paradox for that case may seem trivial: if you make inconsistent assumptions, no wonder when you get inconsistent results. But that is not quite what is going on. For the relevant application of MSR+ *discharges* the supposition R. At the final stage, R is used, implicitly, only to delimit the set of worlds relevant to one's use of 'would' in that context. In the statement of MSR+, R is underlined because it is being ascribed to each of those worlds. When R is inconsistent, that just means that the set of relevant worlds is empty. Valid reasoning about the empty set does not generate a contradiction. In particular, many Kripke models for modal logic have 'dead end' worlds, from which no world is accessible. At any such world, every formula of the form $\Box A$ is vacuously true, so both $\Box(A \wedge \neg A)$ and $\Box A \wedge \Box \neg A$ are true, but no contradiction is true, not even $\Box A \wedge \neg\Box A$. The argument from MSR+ is paradoxical because it imports the contradiction from what *would* be the case to what *is* the case.

Here is the argument. The starting point is as before, except that R is assumed to be 'A and not(A)'. Thus [I] reduces to:

[I*] $\underline{R} \parallel^a A$ and not(A)

By MSR+, [I*] yields [II*]:

[II*] $\parallel^a$ would(A and not(A))

But, as already seen, MSR+ makes 'would' commute with conjunction, so [II*] yields [III*]:

[III*] $\parallel^a$ would(A) and would(not(A))

As already seen, MSR+ also makes 'would' commute with negation, so [III*] yields [IV*]:

[IV*] $\parallel^a$ would(A) and not(would(A))

In effect, the argument [I*]–[IV*] replays [I]–[IV] but with the contradiction implicit in the condition on contextually relevant worlds rather than explicit in the antecedent of a conditional. The new argument does not involve 'if' or any heuristic for it at any point.

In such a context, therefore, MSR+ commits us to accepting a contradiction. It is not modally distanced; nothing in [IV*] is underlined. In effect, once we think about the empty set of worlds, MSR+ drags us into contradicting ourselves. The problem goes deeper than the conditional; it concerns our heuristics for handling suppositions for 'would' as well as for 'if'.

Of course, when the contextual constraints on relevance exclude all worlds, we are rarely thinking of the set of relevant worlds *as* the empty set. We may be thinking of it as the set of possible worlds in which Jack the Ripper is Edward VII, or as the set of possible worlds in which there is a largest prime number, or as the set of possible worlds in which someone kicks all and only those people who do not kick themselves. We may have good reason to consider such a set of worlds.

Only on a very naïve view of human cognition do these paradoxes look like good arguments for dialetheism, or for accepting contradictions. A more reasonable moral is instead that MSR+ and MCSR+ are fallible heuristics, not universal laws of logic, just as for the non-modal SR+.

Naturally, a question arises for MSR+ and MCSR+: since these principles generate contradictions, how come our reliance on them does not lead us into disaster?

A generic response is to recall other structural paradoxes in our linguistic and cognitive habits, such as the Liar and other semantic paradoxes, and the Heap and other paradoxes of vagueness. In those cases, we are disposed to reason along lines which in principle generate contradictions, yet in practice our reliance on them does not lead us into disaster. We usually manage to skirt the danger zones. MSR+ and MCSR+ may be like that.

Still, it would be nice to say something informative about *how* we skirt the danger zones. More positively, an explanation is needed of how MSR+ and MCSR+ come to be reliable, where they are. That question is closely related to the central problems of modal epistemology, which are only just beginning to be understood. At this stage, the best we can hope for are partial and provisional hypotheses, which may nevertheless lead in the right direction.

11.5 Heuristics for 'would' in practice

Once we consider the modal heuristics MSR+ and MCSR+ in practice, the lack of a natural outer negation for 'would' starts to look like a feature rather than a bug. It acts as a bias on what verbalized conclusions we draw from the imaginative exercises at issue, arguably a bias in the direction of reliability. In particular, the formal paradoxes for MSR+ and MCSR+ depend on accepting sentences of the form 'not(would(C))': conclusions of exactly the sort which grammar makes it unnatural for us to draw. If we reject C on the contextually relevant restriction R, the natural conclusion for us to draw is instead 'would(not(C))', or one with 'not' buried even deeper inside.

The position for plain 'if' is similar, as discussed in chapter 5, where the lack of a natural outer negation for 'if' was seen to contribute to the reliability of the non-modal heuristic SR+ for 'if': it creates a grammatical bias in favour of concluding 'if A, not(C)' rather than 'not(if A, C)' when one rejects C on the supposition A, and SR+ is more reliable in the latter case than in the former.

When no world satisfies the contextual restriction R, every conclusion of the form 'would(C)' is vacuously true. Of course, when we reach both 'would(C)' and 'would(not(C))', MSR+ itself warns us to pull back, since accepting both modal claims amounts to accepting a contradiction under the operative supposition R. But we were not really accepting false modal claims outside the scope of the supposition in the first place.

Similarly, when no contextually relevant world satisfies the antecedent A of a modalized conditional 'would(if A, C)', the inner conditional 'if A, C' is true at every such world on the material interpretation, so 'would(if A, C)' is actually true. Of course, when we reach both 'would(if A, C)' and 'would(if A, not(C))', MCSR+ itself warns us to pull back, since accepting both modalized conditionals amounts to accepting a contradiction under the embedded supposition A. But we were not really accepting false modalized conditionals outside the scope of the supposition in the first place.

When C is asserted on the modally distanced supposition of the contextual restriction, the natural conclusion to draw outside the supposition is 'would(C)'. When instead C is denied on that supposition, the natural conclusion is 'would(not(C))'. Those principles are informal generalizations of the rule of Necessitation$_{G/G}$ in modal logic, which tells us that if $\vDash_G C$ then $\vDash_G \Box C$ and so, as a special case, if $\vDash_G \neg C$ then $\vDash_G \Box \neg C$.

Similarly, when C is asserted on the modally distanced supposition of the contextual restriction and A, the natural conclusion to draw outside the suppositions is 'would(if A, C)'. When instead C is denied on those suppositions, the natural conclusion is 'would(if A, not(C))'. Those principles are informal generalizations of the one-premise case of the rule of Normality$_{G/G}$ in modal logic, which tells us that if $A \vDash_G C$ then $\vDash_G \Box(A \supset C)$ and so, as a special case, if $A \vDash_G \neg C$ then $\vDash_G \Box(A \supset \neg C)$.

What happens when C is considered but neither asserted nor denied on the modally distanced suppositions at issue? Putting modalized conditionals temporarily aside, a natural conclusion to draw is of the form (62):

(62)   might(C) and might(not(C))

Section 11.3 tentatively proposed reading 'might(C)' as 'not(known (would(not(C))))', which makes (62) equivalent to (63):

(63)   not(known(would(not(C)))) and not(known(would(not(not(C)))))

For convenience we can cancel the double negation and simplify (63) to (64):

(64)   not(known(would(not(C)))) and not(known(would(C)))

In effect, (64) admits that the imaginative exercise failed to decide the question. It is epistemically appropriate in the same way as the response 'I don't know' when your attempt to determine the answer to a question has failed.

For example, let C be 'Federico is scared'. The imaginative exercise has failed to verify either of (65) and (66):

(65)   Federico would be scared.
(66)   Federico would not be scared.

The natural conclusion to draw is (67):

(67)   Federico might be scared and he might not be scared.

Here, that is to be analysed as something like (68):

(68)   It is not known that Federico would not be scared and it is not known that he would be scared.

Epistemically, that is a far more appropriate conclusion to draw from the failure of the imaginative exercise to verify one of (65) and (66) than (69):

(69) It is not the case that Federico would not be scared and it is not the case that he would be scared.

Not merely does (69) involve linguistically awkward negations of 'would'; it is epistemically less appropriate than (68). For (69) requires it to be both objectively possible for Federico not to be scared and objectively possible for him to be scared, on some contextually relevant reading of 'objectively possible'. But the failure of the imaginative exercise does not show that. It could easily result from one's ignorance of Federico's character and his dispositions, or of how the tiger will behave. Of course, to know (68) we must be aware of the limits on our knowledge: we often take ourselves to know far more than we really do. But we are also often aware of limits on our modal knowledge, as our frequent assent to such 'might' statements shows.

As a check on the preceding line of thought, we can work through the analogous considerations for 'will' and 'may'.

The question becomes: what happens when C is considered but neither asserted nor denied about some future issue? The natural conclusion to draw is of the form (70):

(70) may(C) and may(not(C))

The proposed reading of 'may(C)' as 'not(known(will(not(C))))' makes (70) equivalent to (71):

(71) not(known(will(not(C)))) and not(known(will(not(not(C)))))

For convenience we can cancel the double negation and simplify (71) to (72):

(72) not(known(will(not(C)))) and not(known(will(C)))

In effect, (72) simply admits that the predictive exercise failed to decide the question.

For example, let C be 'Franz is scared'. The predictive exercise has failed to verify either of (73) and (74):

(73) Franz will be scared.
(74) Franz will not be scared.

The natural conclusion to draw is (75):

(75) Franz may be scared and he may not be scared.

Here, that is to be analysed as something like (76):

(76) It is not known that Franz will not be scared and it is not known that he will be scared.

That is a far more appropriate conclusion to draw from the failure of the predictive exercise to verify one of (73) and (74) than (77):

(77) It is not the case that Franz will not be scared and it is not the case that he will be scared.

Not merely does (77) involve linguistically awkward negations of 'will'; it is epistemically less appropriate than (76). For (77) requires both an objectively open future history in which Franz is not scared and an objectively open future history in which he is scared, on some contextually relevant reading of 'objectively open'. But the failure of the predictive exercise does not show that. It could easily result from one's ignorance of Franz's character and his dispositions, or of what the war will be like. Of course, to know (76) we must be aware of the limits on our knowledge. But we often are aware of limits on our knowledge of the future, as our frequent assent to such future 'may' statements shows.

The analogy thus works in a natural way. Although the modal 'would' is only the fake past of a fake future, the offline heuristics underlying the modal 'would' judgement—the imaginative exercise—are structurally analogous to the online heuristics underlying the genuinely temporal 'will' judgement—the predictive exercise. In practice, both typically involve projecting a scenario into later times. This strong cognitive similarity may help to explain the naturalness of the projection from temporal to modal constructions, a feature common to many languages (see Iatridou 2000 and section 11.6).

The next task is to apply this provisional account of 'might' to 'would if' conditionals. Here too, we are interested in the case where C has been considered but neither asserted nor denied on the modally distanced suppositions at issue, but now the latter include the potential antecedent A as well as the contextual restriction R.

The two direct answers such a procedure is set up to deliver are 'would(if A, C)' and 'would(if A, not(C))', obtained by reaching one of 'if A, C' or 'if A, not(C)' respectively under the supposition R, in turn obtained by reaching one of C or 'not(C)' under the further supposition A. Hence, when the procedure yields no decision, one has in effect failed to eliminate either of the plain conditionals under the supposition R. In that case, (78) applies:

(78)   not(known(would(not(if A, C)))) and not(known(would(not(if A, not(C)))))

Under the proposed reading of 'might(.)' as 'not(known(would(not(.))))', (78) is equivalent to (79):

(79)   might(if A, C) and might(if A, not(C))

For example:

(80)   might(if a tiger enters the room, Federico is scared) and might(if a tiger enters the room, not(Federico is scared))

Slightly elliptically, we can put (80) into ordinary English thus:

(81)   If a tiger entered the room, Federico might be scared and he might not be scared.

In the circumstances, (81) is a natural, epistemically appropriate conclusion to draw.

Of course, the material reading of 'if' makes negated conditionals very strong, and denials that they are known correspondingly weak. Substituting truth-functional equivalents in (78), admittedly in epistemic contexts, results in (82):

(82)   not(known(would(A and not(C)))) and not(known(would(A and C)))

The worry is that (82) will often be too weak. For example, it may simply not be known that a tiger would enter the room, implausibly making (81) true for a trivial reason.

One could just respond by pointing out that the equivalence of 'not(if A, C)' to 'A and not(C)' is not pre-theoretically available. Our ordinary semantic ignorance and error may enable (81) to serve its conversational purpose, because speakers all tend to treat it in the same mistaken way.

However, a strategy sketched in section 10.2 gives a much more conciliatory response. For, as with restrictions on adverbs of quantification, an explicit antecedent of a conditional may be reused as an implicit restriction on a modal operator. If the antecedent A in (78) and (79) implicitly restricts the modal 'would' and 'might', they are equivalent to (83) and (84) (again modulo the substitution of truth-functional equivalents in epistemic contexts):

(83)   not(known(would(if A, not(C)))) and not(known(would(if A, C)))
(84)   might(A and C) and might(A and not(C))

These impute much more substantial ignorance, as desired.

Similar considerations apply to 'may' and 'will' as to 'might' and 'would' in relation to 'if'. The reader is spared details.

Although the relation of 'might' to 'would' is epistemically modulated, in some circumstances the modulation can be eliminated. For consider a case where (85) holds:

(85)   if would(not(A)), known(would(not(A)))

Given (85), 'might(A)', expanded as (86), *is* equivalent to the dual of 'would', articulated as (87):

(86)   not(known(would(not(A))))
(87)   not(would(not(A))

For the converse of (85) is automatic, since 'known' is factive, so (85) yields the material biconditional:

(88)   would(not(A)) if and only if known(would(not(A)))

Consequently, the negations of the two sides of (88) are also materially equivalent:

(89)   not(would(not(A))) if and only if not(known(would(not(A))))

But (89) is simply the material biconditional of (86), which paraphrases 'might(A)', and (87), the dual of 'would(A)'. Thus, when we can assume that the Suppositional Rule was properly applied on the basis of adequate background information, so that (85) holds, we can also treat 'might' as materially the dual of 'would', for the given sentence A.

For example, I may know enough about this window and its materials to take for granted this informal instance of (85):

(90)   If this window would not crack, I know that this window would not crack.

Then, in this context, I can treat (91) as materially equivalent to the awkward (92):

(91)   This window might crack.

(92)   It is not the case that this window would not crack.

In such circumstances, 'might' is in effect objectified.

For those who accept 'Would' Excluded Middle (WEM), it constitutes a special obstacle to assuming many instances of (85), such as (93), with a contextual restriction to counterfactual circumstances in which this coin is tossed:

(93)   If this coin would not come up tails, I know that this coin would not come up tails.

For even though I know that the coin is fair, by WEM either it would come up heads or it would come up tails, yet I do not know which. However, since the attraction of WEM has already been seen to be specious, this objection need not concern us.

Thus the 'Would' Suppositional Rule MSR+ indirectly provides a way of falsifying 'would' statements, as well as directly providing a way of verifying them. The way depends on some sensitivity to one's own epistemic position, in order to track conditionals like (90) and (93), but it may involve nothing like conscious reasoning; reliable correlations suffice.

Of course, for many practical purposes, there is no need to objectify 'might'. A mere epistemic possibility of danger is enough to motivate caution. But once we seek a deeper understanding of how the world works, we must discriminate between gaps in our knowledge and ways things are really capable of being. Then it matters whether a 'might' can be objectified.

11.6 Prediction and imagination

Although the modal 'would' is only the fake past of a fake future 'will', even a fake should have much in common with the original article. As already observed, there is a significant cognitive similarity between the processes of assessing future-tense statements and modal statements. Very roughly, prediction involves projecting forward from our knowledge of the past and present, using our knowledge of how the world tends to work. In real time, that is an online process. The offline analogue involves projecting forward from a partly imagined, partly known past and present, using partly imagined, partly known tendencies. To a first approximation, the structure of the process from inputs to outputs is the same: the difference between online and offline lies in the source of the inputs.

Some aspects of the process may be hardwired into us by evolution. But we should not regard the process as always primitive in nature. Its overall structure has much in common with the process of projecting forward from observed or hypothesized *initial conditions*, using established or hypothesized *laws* or dynamic equations, typical of scientific explanation.

When we think of a conditional, the salient input or initial condition is the antecedent. Then there is a neat analogy between the offline process of developing that supposition and the online process of updating on the same content, perhaps input on the testimony of a trusted source. However, we should not restrict the offline process to conditionals. Even with no antecedent, 'would' supplies its own context-dependent restrictions, which the thinker may be incapable of verbalizing in a sentence or potential antecedent. The process of prediction is not always prompted by new evidence; sometimes it is just a response to a new question that happens to occur to one: 'How long will this last?' Similarly, one can just project forward an imagined scenario, which did not originate in a sentence. Like 'will', 'would' has a life independent of 'if'.

We must be careful not to project contextual restrictions often marked by 'would' onto the semantics of 'if'. An example is our aversion to backtracking conditionals, such as (94), as contrasted with (95):

(94)   If you had robbed a bank yesterday, your whole previous upbringing would have been radically different.

(95)   If you had robbed a bank yesterday, it would have been a complete aberration.

Usually, we prefer (95) to (94) because we hold history before yesterday fixed. We prefer (97) to (96) because we do *not* hold history after yesterday fixed; rather, we hold normal causal processes fixed thereafter:

(96)   If you had robbed a bank yesterday, all trace of your doing so would soon have disappeared.

(97)   If you had robbed a bank yesterday, your subsequent life would have been ruined.

The temptation is to think that (94) and (96) sound bad because the meaning of 'if' requires us to assign much more weight to perfect similarity between possible worlds before the time of the antecedent than thereafter (Lewis 1979). One might then worry that prejudices about time, in particular an asymmetry between past and future, are somehow built into the very meaning of the 'counterfactual conditional'. Instead, the analogy with prediction may simply bias us in favour of contextual restrictions which hold the past fixed, because they require a process of imaginative development more similar to the familiar process of prediction, and so easier for us, than do alternative restrictions.

We can still work with alternative restrictions, in a suitable context. For example, suppose that the question is raised: how easily could circumstances have led us to commit crimes? A colleague says: 'If I had needed the money enough, I could easily have robbed a bank.' In that context, you might appropriately and truly reply with (94). Similarly, I may appropriately and truly assert (96) in the context of a conversation about the advantages of having a guardian genie to look after one's interests. The point is *not* that (96) is true in the fiction of the genie. In story-telling mode, the thing to say is simpler, an unmodalized conditional:

(98)   If you robbed a bank yesterday, all trace of your doing so soon disappeared.

I use the modalized conditional (96) because what I am saying, though with reference to the fiction, is not part of it.

Proposed sets of detailed rules for evaluating 'would' conditionals inevitably go wrong, because they try to codify the inherently unruly business of contextual restriction as a set of semantic laws. We should not expect semantics to do so much.

12

Is 'Would If' Hyperintensional?

12.1 Intensional and hyperintensional semantics

Chapter 10 characterized the meaning of 'would' and 'would if' in a standard framework for *intensional semantics*. Such a semantics is compositional. It assigns each semantically significant expression of the object-language a semantic value, determining the semantic value of a complex expression functionally from the semantic values of its simpler constituents and how they are put together. These semantic values may vary with the context of utterance, but the compositional determination of meaning proceeds in terms of the semantic values themselves (with respect to a given context). To a first approximation, the specifically *intensional* nature of the semantics is this. Expressions of some core types, including (declarative) sentences and predicates, have *extensions* at worlds. For example, the extension of a sentence at a world is its truth-value at that world. The *intension* of such an expression is the function taking each world to the extension of the expression at that world. For example, the intension of a sentence is the function taking each world to the truth-value of that sentence at that world. The intensional semantics identifies the semantic value of such an expression with its intension. To make the account fully adequate, we may need to include other parameters, such as time, in the input to the intension. Thus, on a more abstract view, the intension of a sentence is the function taking each *circumstance of evaluation* to the truth-value of the sentence at that circumstance, where the world of evaluation is just one component of the circumstance. For present purposes, we can ignore that issue, and speak simply of worlds rather than circumstances.

For any expression e, let $[e]$ be the semantic value of e (for a given context), on the intensional semantics. In order not to prejudge the issue against alternative analyses of counterfactuals, in this chapter we write the counterfactual conditional with antecedent A and consequent C as $A > C$. Thus an intensional semantics for $>$ yields a general equation of the form (1):

(1)　$[A > C] = [>]([A], [C])$

Here $[A > C]$, $[A]$, and $[C]$ are all functions from worlds to truth-values, while $[>]$ is a function from pairs of such intensions to such intensions (an intension only in an extended sense). A straightforward consequence of (1) is (2), for all declarative sentences A, A', C, and C':

(2)　If $[A] = [A']$ and $[C] = [C']$, then $[A > C] = [A' > C']$

Suppose and Tell: The Semantics and Heuristics of Conditionals. Timothy Williamson, Oxford University Press (2020).
© Timothy Williamson.
DOI: 10.1093/oso/9780198860662.001.0001

For $[A > C] = [>]([A], [C] = [>]([A'], [C']) = [A' > C']$. In other words, counterfactual conditionals have the same intension whenever their antecedents have the same intension and so do their consequents. The operator $>$ is *intensional* just in case operator it satisfies principle (2) quite generally (in every model). It is *hyperintensional* just in case it violates (2). The account of $>$ in chapter 10 is of the form (1) (for any given context), and so entails (2). If it is correct, $>$ is intensional, not hyperintensional.

Intensionality is sometimes confused with a subtly distinct phenomenon, *congruentiality*. To say that $>$ is congruential is to say that whenever A is logically equivalent to A^* and C logically equivalent to C^*, then $A > C$ is logically equivalent to $A^* > C^*$. We should not take for granted that $>$ is congruential if and only if it is intensional in the sense of principle (2). For sameness of intension is insufficient for logical equivalence. For example, the sentences 'Plato is not Socrates' and '$2 + 2 = 4$' presumably have the same intension, both being necessary, but they are not logically equivalent.

To see how $>$ might be intensional without being congruential, assume for the sake of argument that the standard for logical consequence is real-world validity as defined in section 10.3: an argument is real-world valid just in case whenever all its premises are true at the designated actual world of a model, so is its conclusion. Thus two sentences are logically equivalent just in case whenever one of them is true at the actual world of a model, so is the other. Consider an object-language with the rigidifying model operator 'actually', for which the rule is that 'actually(A)' is true at a world in a model just in case A is true at the actual world of that model. Trivially, 'actually(A)' is always logically equivalent to A by the standard of real-world validity. Now suppose that $>$ is congruential. Then, by substitution in the consequent, the trivial logical truth $A > A$ is logically equivalent to '$A >$ actually(A)', so the latter is a logical truth too. We can equally well argue for the same conclusion by substitution in the antecedent, starting with the trivial logical truth 'actually(A) > actually(A)'. But whenever A is a contingent falsehood, '$A >$ actually(A)' is false, and a fortiori not a logical truth. For example, (3) is false on its intended reading, since Napoleon won at Waterloo in some possible worlds, but not in the actual world:

(3)   If Napoleon had won at Waterloo, he would have actually won at Waterloo.

Thus $>$ is non-congruential, with respect to both its antecedent and its consequent, for languages of the appropriate expressive power.

Such examples do nothing to show that $>$ is not intensional in the sense of (2), even granted real-world validity as the standard of logical equivalence. For although 'Napoleon won at Waterloo' and 'actually(Napoleon won at Waterloo)' are logically equivalent in that sense, they differ in intension. The intension of 'Napoleon won at Waterloo' maps every possible world in which Napoleon won at Waterloo to truth, whereas the intension of 'actually(Napoleon won at Waterloo)' maps every world to falsity, because its truth-value at any world is the truth-value of 'Napoleon won at Waterloo' at the actual world. In particular, if the strict conditional semantics in chapter 10 is adopted for 'would if', $>$ is intensional because in each context it satisfies an equation of the form (1) for all sentences. It does so for every model of the relevant kind.

Intensionality and congruentiality are very different *kinds* of semantic feature. Intensionality concerns the behaviour of an operator within a *given* model. By contrast, when logical consequence is defined model-theoretically, congruentiality concerns the operator's behaviour across all models. Expanding the range of models may turn a non-congruential operator into a congruential one, but it cannot turn a hyperintensional operator into an intensional one.

Nevertheless, despite the differences between intensionality and congruentiality, some kinds of behaviour are incompatible with both. In particular, let A be logically equivalent to A^* in classical non-modal propositional logic: they are Boolean equivalents. Then in every (classical) model A and A^* have the same intension. Now suppose that $A > C$ is true and $A^* > C$ false. Then they are not logically equivalent, even by the standard of real-world validity, since they differ in actual truth-value. Since A is logically equivalent to A^* and C is of course logically equivalent to itself, $>$ is not congruential. Furthermore, $A > C$ and $A^* > C$ differ in intension, since they differ in extension at the actual world. Since A has the same intension as A^* and C has of course the same intension as itself, $>$ is also not intensional; it is hyperintensional. It also violates the principle of Anti-Monotonicity in the Antecedent, defended in section 10.3.

Similarly, let C be logically equivalent to C^* in classical non-modal propositional logic: they are Boolean equivalents. Then in every (classical) model C and C^* have the same intension. Now suppose that $A > C$ is true and $A > C^*$ false. Then they are not logically equivalent, even by the standard of real-world validity, since they differ in actual truth-value. Since A is logically equivalent to itself and C to C^*, $>$ is again not congruential. Furthermore, $A > C$ and $A > C^*$ differ in intension, since they differ in extension at the actual world. Since A has the same intension as itself and C as C^*, $>$ is again hyperintensional too. It also violates the principle of Monotonicity in the Consequent, also defended in section 10.3.

Kit Fine (2012) has argued that $>$ displays the former kind of behaviour: the truth-value of a counterfactual is sensitive to the substitution of Boolean equivalents in the antecedent. If so, $>$ is neither congruential nor intensional, and the account in chapter 10 is wrong. On Fine's account, what would be if it were wet may differ from what would be if it were either wet and cold or wet and not cold, even though the worlds in which it is wet are exactly the worlds in which it is either wet and cold or wet and not cold. Thus, in a given context, the compositional semantics of the 'would if' construction requires more information about the input antecedent sentence than just its intension.

To achieve the desired hyperintensional effect, Fine extends the intensional semantic framework to a more complicated one with possible partial states as well as possible worlds. He does not attempt to derive his semantic account of 'would if' sentences compositionally from independently motivated semantic accounts of 'would' and 'if', nor does his framework lend itself to any such generalization. For reasons seen in chapter 10, the failure to separate the contributions of 'would' and 'if' already casts serious doubt on his account, though of course the same applies to most of its rivals. However, the present interest is not in Fine's positive theory, but in his argument for it, and any threat it may hold to the intensional account in this book. He needs a strong argument to justify the high cost in extra complexity of his hyperintensional semantics.

12.2 Fine's examples

Fine's main argument is based on examples. He describes a possible case, and claims that it satisfies various non-logical constraints in counterfactuals terms. He then shows that the constraints jointly entail a contradiction, given plausible principles of counterfactual logic. Since all those principles preserve general validity on the semantics above, I will not challenge them.[1] However, one of them is in effect the substitution principle that if A is logically equivalent to A^* then $A > C$ is logically equivalent to $A^* > C$, where $>$ is Fine's symbol for the counterfactual conditional. He argues that, of the assumptions used to derive the contradiction, only the substitution principle can plausibly be denied. Since he takes the non-logical constraints to describe a possible case, he regards the derivation of the contradiction as unsound, and therefore rejects the substitution principle. Moreover, he rejects it for cases where the antecedents are truth-functionally equivalent while the counterfactuals allegedly differ in truth-value, so if he is right $>$ is neither intensional nor congruential in the senses explained above. In what follows, the focus is on whether Fine's examples really do satisfy his non-logical constraints.

Fine's first example involves an infinite sequence of rocks, rock 1, rock 2, rock 3, ... None of them actually falls, but each of them, if it were to fall, would always fall on the next rock in the series, in a domino effect. Fine uses R_n for the sentence 'Rock n falls' and R for the infinite disjunction $R_1 \lor R_2 \lor R_3 \lor \ldots$ (in other words, at least one rock falls). Fine's first non-logical constraint is that 'if a given rock were to fall then the next rock would fall' (2012: 222):

Positive Effect $R_n > R_{n+1}$ (for all $n \geq 1$)

His second non-logical constraint is that 'if a given rock were to fall then the previous rock would still stand':

Negative Effect $R_{n+1} > \neg R_n$ (for all $n \geq 1$)

Fine's final non-logical constraint is that 'it is a counterfactual possibility that one of the rocks falls, that is, a contradiction does not follow counterfactually from the supposition that a rock falls':

Counterfactual Possibility $\neg(R > \neg R)$

---

[1] We can represent the principles used in Fine's proof thus, where $\Rightarrow$ is logical consequence

Entailment	If $C \Rightarrow C^*$ then $\Rightarrow C > C^*$
Substitution	If $A \Leftrightarrow A^*$ then $A > C \Rightarrow A^* > C$
Transitivity	$A > B, A \land B > C \Rightarrow A > C$
Disjunction	If $A \Rightarrow \neg B$ then $A > C, B > C \Rightarrow (A \lor B) > C$
Conjunction	$A > C_1, A > C_2, A > C_3, \ldots \Rightarrow A > (C_1 \land C_2 \land C_3 \land \ldots)$

On the semantics of chapter 10, all these principles are generally valid. Transitivity and Conjunction are also real-world valid. Entailment, Substitution, and Disjunction are not real-world valid for a language with an 'actually' operator; however, they become real-world valid when the logical connections before 'then' in the statement of the rules are required to be purely Boolean, and Fine's proof requires only that special case.

The details of Fine's proof of a contradiction from the three non-logical constraints and standard principles of counterfactual logic are intricate, but its correctness is not in doubt.

To see what is puzzling about Fine's non-logical constraints, consider the conjunction of any given instance of Positive Effect with the corresponding instance of Negative Effect. It can be paraphrased as (4):

(4)   Rock $n + 1$ would fall if rock n were to fall, and rock n would not fall if rock $n + 1$ were to fall.

Pre-theoretically, how does (4) sound? Forget semantic theories of counterfactuals, those of Lewis, Stalnaker, and others, and equally the theory in chapter 10, and how they might model (4). Just hear (4) as an ordinary speaker of English hears it. The two conjuncts of (4) sound to be in tension with each other, to put it mildly.

Now (4) is not quite a contradiction. Even on the semantics in chapter 10, if rock n just would not fall, both conjuncts are true, the first vacuously, the second redundantly. But we can articulate the tension between its two conjuncts thus. Suppose that rock n falls in some relevant situation or possible world w (Fine's account uses worlds as well as states, so it is fair to mention them). Does rock $n + 1$ also fall in w? If it does, that seems to undermine the second conjunct of (4), since its antecedent is true and its consequent false at some relevant possible world. But if rock $n + 1$ does *not* fall in w, that seems to undermine the first conjunct of (4), since then its antecedent is true and its consequent false at some relevant possible world.

Of course, Fine's own semantics does not work in terms of relevant possible worlds, but that is not the issue. The point is that when we contemplate (4) pre-theoretically, as ordinary speakers of English, we cannot see a stable way to accept both conjuncts in the same context without excluding the relevant possibility of rock n falling. That Fine's semantics, or Stalnaker's, or Lewis's, predicts the existence of such a way is a *prima facie* problem for them, not for our pre-theoretic attitude to (4). Of course, Fine could try arguing that we misjudge (4), perhaps by relying on a fallible heuristic, but that is far from his actual strategy. Instead, he treats Positive Effect and Negative Effect as natural, plausible descriptions of an evidently possible case.

On the natural understanding of (4), it implies (5):

(5)   Rock n would not fall.

Analyses of the counterfactual conditional as a strict conditional, including the one in chapter 10, confirm the implication from (4) to (5). For (4) is formalized as (4*) (with a restricted reading of the necessity operator $\Box$):

(4*) $\Box(R_n \supset R_{n+1}) \wedge \Box(R_{n+1} \supset \neg R_n)$

Since $R_n \supset R_{n+1}$ and $R_{n+1} \supset \neg R_n$ are together truth-functionally equivalent to $\neg R_n$, (4*) is equivalent to (5*), the formalization of (5), by normal modal logic (corresponding to the principle Normality$_{G/G}$ for 'would' in section 10.3):

(5*) $\Box \neg R_n$

But (5)/(5*) follows from (4)/(4*), and so from Positive Effect and Negative Effect together, for all $n \geq 1$, and R is just the disjunction of R_n for all $n \geq 1$, we can conclude (6) (with the quantification over rocks restricted to those in the series), formalized as (6*), for even an infinite disjunction is impossible if each disjunct is impossible:

(6)   No rock would fall.

(6*) $\Box \neg R$

From (6)/(6*), we can conclude (7), formalized as (7*), again by normal modal logic:

(7)   If a rock were to fall, no rock would fall.

(7*) $\Box(R \supset \neg R)$

But Fine's constraint Counterfactual Possibility is just the negation of (4).

The objection to Fine's account of the example does not depend on the theoretical analysis of the counterfactual conditional as a strict conditional. Pre-theoretically, to accept Positive Effect and Negative Effect in the same context, we must exclude rock n falling as a relevant possibility, for each $n \geq 1$, which means that we must exclude a rock falling as a relevant possibility in the same context, which is to reject Counterfactual Possibility. In other words, pre-theoretically, Fine's three non-logical constraints are not cotenable in the same context. Fine shows some awareness of the problem, since he mentions the qualm (2012: 223):

> that it is not altogether clear that the counterfactual 'if the second rock were to fall then the first would not fall' is true, for in considering the counterfactual possibility that the second rock falls, one should perhaps not keep fixed the fact that the first rock does not fall but allow that the second might fall by way of the first falling.

Fine reports that he does not find the objection convincing, but he does not try to rebut it directly. Instead, he constructs a more complicated example for which, he claims, the problem does not arise.

Fine's new example involves an infinite series of matches, each causally isolated from all the others. In the initial world, no match is struck, each match is dry, and more generally there are no obstacles to its lighting if struck. For each $n \geq 1$, in place of R_n, Fine uses a sentence M_n, which says that every match is struck and every match from n on is wet and does not light. M is the disjunction of M_n for all $n \geq 1$. This possible case is supposed to obey three non-logical constraints analogous to the previous ones:

Positive Effect	$M_n > M_{n+1}$	(for all $n \geq 1$)
Negative Effect	$M_{n+1} > \neg M_n$	(for all $n \geq 1$)
Counterfactual Possibility	$\neg(M > \neg M)$	

A key feature of the new example is that whenever $m \leq n$, M_m entails M_n, so Positive Effect comes for free given any reasonable logic of counterfactual conditionals; thus it provokes no tension with the other constraints. But Negative Effect is trickier. The conjunction $M_{n+1} \wedge \neg M_n$ is equivalent to (8):

(8)   Every match is struck, every match from $n + 1$ on is wet and does not light, but match n either lights or is not wet.

Thus Negative Effect is equivalent in effect to (9):

(9)   If every match were struck, those from $n + 1$ on being wet and not lighting, match n would either light or not be wet.

When we consider (9), do we treat M_n as a relevant possibility? In other words, do we exclude as irrelevant the possibility of every match being struck, those from n on being wet and not lighting? If not, it seems to undermine (9), for it is a relevant possibility in which the antecedent of (9) is true while its consequent is false. But if we do treat M_n as an irrelevant possibility when we consider (9), for all $n \geq 1$, then we are so treating every disjunct of M, so we are in effect treating M as an irrelevant possibility too. But that undermines Fine's constraint of Counterfactual Possibility.

The underlying problem is fundamentally the same as for Fine's first example. We still have an analogue of (4), schematized to show its overall structure:

(10) M_{n+1} would hold if M_n held, and M_n would not hold if M_{n+1} held.

As before, the problem is that, pre-theoretically, we can accept (10) only by excluding the possibility of M_n as irrelevant. Since that holds for all n, we end up not accepting all Fine's constraints in the same context. Since Fine provides no story to explain why such pre-theoretic assessments might be in error, his examples do not serve his case.

Strikingly, Fine himself appeals to similar context-shifting effects when he defends the principle of Simplification, that $(A \vee B) > C$ entails both $A > C$ and $B > C$, against apparent counterexamples (Simplification is of course valid on the semantics proposed in chapter 10).[2] He invokes 'a principle of "Suppositional Accommodation", according to which we always attempt to interpret a counterfactual in such a way that its antecedent A represents a genuine counterfactual possibility' (2012: 232). We can so accommodate $(A \vee B) > C$ by treating B as a genuine counterfactual possibility, without treating A likewise, but once we consider $A > C$ we may accommodate it by treating A as a genuine counterfactual possibility too. He concludes: 'Thus what accounts for the appearance of invalidity is a shift in the relevant "space of possibilities" as we move from premise to conclusion' (ibid.). Fine's defence of Simplification is cogent, but he does not observe that the same approach undermines his descriptions of his own key examples.

No good reason has emerged to doubt the natural thought behind the substitution principle, that when A is necessarily equivalent to B, what would be if A *just is* what would be if B. If 'would' were hyperintensional rather than merely intensional, its hyperintensionality might be expected to show up in more natural and less convoluted ways than in Fine's infinitary examples (chapter 15 considers and rejects some more candidates). After all, from an evolutionary perspective, the cognitive system is

---

[2]   In the notation of footnote 1, Simplification corresponds to $(A \vee B) > C \Rightarrow A > C$ and $(A \vee B) > C \Rightarrow B > C$. On the semantics of chapter 10, it is both generally valid and real-world valid.

not very likely to rely on a conditional whose working involves the extra complexity of Fine's semantics, if a simpler one works equally well in all finitary cases.[3]

Similar problems arise for other alleged examples of the hyperintensionality which Fine has invoked elsewhere. Here is one (Fine 2017: 571–2):

(11)   If Sue were to take the pill then she would live.

(12)   If Sue were to take the pill or to take the pill and the cyanide then she would live.

The antecedents of (11) and (12) are truth-functionally equivalent, for $A \lor (A \land B)$ and A are Boolean equivalents. Fine claims that (11) might be true while (12) is false. For Fine, (12) entails (13), by the rule of Simplification:

(13)   If Sue were to take the pill and the cyanide then she would live.

He judges that (11) might be true while (13) is false, so that (12) is also false, making 'would if' hyperintensional.

On the semantics proposed here, (11) is equivalent to (12) and both entail (13). But there are some obvious contextual effects. Having denied (13), one cannot just go straight on and flatly reassert (11). By using the antecedent of (13), one has raised to relevance some possibilities in which Sue takes both the pill and the cyanide, which one cannot simply ignore by reasserting (11). One should instead say something like (14):

(14)   If Sue were to take just the pill then she would live.

Even if one does not explicitly consider (13), assessing (12) will raise to relevance possibilities in which Sue takes both the pill and the cyanide, which one cannot simply ignore in reasserting (11). Why would anyone bother to formulate the elaborate, logically redundant antecedent in (12) unless they wanted to draw attention to such possibilities and make them relevant to assessing (12)? The example cries out to be understood in terms of pragmatic effects, which Fine does not address.

The sensitivity of 'would if' to context-shifting undermines Fine's case for hyperintensionality. Chapter 13 considers that sensitivity further.

---

[3] Fine gives some cognitive considerations in favour of his hyperintensional semantics which arise even in finitary cases. However, the cases can equally well be handled by heuristics operating on 'would if' sentences, like the suppositional heuristics, rather than on their coarse-grained truth-conditions.

13

More on the Interaction of 'Would' with Context

13.1 Reverse Sobel sequences

Analyses of counterfactual conditionals as strict conditionals—including that in chapter 10—validate the principle of Strengthening the Antecedent, since $\Box(A \supset C)$ entails $\Box((A \wedge B) \supset C)$ in any normal modal logic. Moreover, in any such logic, $\Box(A \supset C)$ and $\Box((A \wedge B) \supset \neg C)$ together entail $\Box\neg(A \wedge B)$; in other words, $A \wedge B$ is impossible in the relevant sense and so vacuously but strictly implies anything, including both C and $\neg C$.

As David Lewis pointed out, with credit to Howard Sobel, often an antecedent counterfactually implies a consequent while the conjunction of the antecedent with something else counterfactually implies the negation of that consequent (1973: 10). In Lewis's examples, the latter implication is *non-vacuous*, because the conjunction does not also counterfactually imply the original example. He gave examples such as (1) and (2), which can be consistently endorsed by the same speaker:

(1)   If Otto had come, it would have been a lively party.
(2)   If both Otto and Anna had come, it would have been a dreary party.

Presumably, the speaker who asserts (2) will deny rather than assert (3):

(3)   If both Otto and Anna had come, it would have been a lively party.

(Lewis's Anna must have been a very different character from my wife Ana.)

On behalf of a strict conditional analysis, the standard response to such *Sobel sequences* is to postulate a change of context. In the context in which (1) was asserted, some possible worlds in which Otto but not Anna came were relevant, but none in which they both came; in all the worlds relevant to that context in which Otto came, it was a lively party (according to the speaker). The utterance of (2) then made some possible worlds in which they both came relevant; in all those worlds, it was a dreary party (according to the speaker).

Lewis conceded the feasibility of such a treatment of counterfactuals as context-sensitive strict conditionals, but thought he could do better, by incorporating the shifting standards of relevance into the semantics, since the shifts are prompted by a constituent of the sentence itself, its antecedent. Thus, roughly speaking, he equated the truth of a counterfactual at a world w with the truth of its consequent at the most similar worlds to w at which the antecedent is true. He aimed thereby to articulate the regularities in the shifts of relevance systematically in the semantics, rather than

Suppose and Tell: The Semantics and Heuristics of Conditionals. Timothy Williamson, Oxford University Press (2020).
© Timothy Williamson.
DOI: 10.1093/oso/9780198860662.001.0001

consigning them to the wastebasket of pragmatics (1973: 13). Thus (1) is true because in the most similar worlds in which Otto came, Anna did not come and it was a lively party, while (2) is true because in the most similar worlds in which Otto and Ana both came, it was a dreary party. Lewis's account makes it unproblematic for a speaker to assert (1) and (2) and deny (3) in the very same context. Proponents of Stalnaker's semantics for conditionals make closely related moves.

In achieving these effects, the Lewis and Stalnaker semantics validate much less elegant and simple logics for counterfactuals than does a straightforward context-sensitive strict conditional account, but the price was thought worth paying.

While an utterance of (1) followed by (2) is a Sobel sequence, an utterance of (2) followed by (1) is a *reverse Sobel sequence*. Lewis's account suggests no significant difference between Sobel sequences and reverse Sobel sequences. However, Kai von Fintel (2001) and Anthony Gillies (2007) pointed out that reverse Sobel sequences are conversationally much more problematic than Lewis's Sobel sequences, in a way troublesome to theories such as Lewis's and Stalnaker's. Having uttered (1), you can go on to utter (2), and each utterance sounds fine. By contrast, having uttered (2), if you go on to utter (1), it sounds bad. In uttering (2), you let the cat out of the bag, and cannot simply add (1), since you seem to have just provided a counterexample to it (unless the antecedent of (1) is treated as elliptical for 'Otto but not Anna came', contrary to Lewis's intentions). Even if you utter (1) and then (2), you cannot then reiterate (1) in the same words as before. Uttering (2) makes the possibility of both Otto and Anna coming relevant, and it is much harder to make it irrelevant again. That it is much easier to expand than to contract the domain of contextually relevant items is a familiar feature of conversational dynamics. All this suggests that some context-shifting must be taken into account, as strict conditional analyses do. Indeed, von Fintel and Gillies endorse versions of a strict conditional analysis, although with the twist that they build the contextual shifts of relevant worlds into a dynamic semantic theory, so that the crash of reverse Sobel sequences can be explained semantically. While making many of the same moves, chapter 10 derived the account of 'would if' conditionals as contextually restricted strict conditionals compositionally from separate non-dynamic semantic accounts of 'would' and 'if', while explaining the behaviour of reverse Sobel sequences in terms of general heuristics for 'would' and 'if' rather than specifically writing it into the semantics.

13.2 Epistemic and pragmatic effects

Sarah Moss (2012) has argued that the defectiveness of reverse Sobel sequences can be explained on epistemic grounds, without positing any contextually driven shifts in the truth-conditions of counterfactuals, in a way suited to Lewis's and Stalnaker's semantic approaches. Consider a standard example of a reverse Sobel sequence:

(4)   If Sophie had gone to the parade and been stuck behind a tall person, she would not have seen Pedro.

(5)   If Sophie had gone to the parade, she would have seen Pedro.

Moss proposes that what is wrong with uttering (4) followed by (5) is that uttering (4) raises to salience the possibility that Sophie might have been stuck behind a tall person, and so not seen Pedro, if she had gone to the parade, a possibility which the speaker may well be in no position to rule out epistemically and which is incompatible with (5). She postulates a general principle of conversation that it is *epistemically irresponsible* to utter a sentence in such circumstances. That principle is not specific to counterfactuals. By contrast, the speaker who utters the forward Sobel sequence of (5) followed by (4) violates no such epistemic principle, since the error possibility is not yet salient when she utters (5). Thus proponents of the unmodified Lewis and Stalnaker semantics can explain the asymmetry between forward and reverse Sobel sequences.

Of course, sometimes the speaker *is* in a position to rule out the relevant possibility which the first member of the sequence raises to salience. For example, it may be common knowledge that the stewards at the parade are exceptionally zealous and effective in organizing the spectators so that everyone has a good view. In that case, (6) is clearly true:

(6)   If Sophie had gone to the parade, she would not have been stuck behind a tall person.

In such cases, Moss suggests, the reverse Sobel sequence is acceptable. That point favours her account over von Fintel's and Gillies', she argues, because their semantic explanations are generic to reverse Sobel sequences, and so, unlike hers, do not predict such exceptions in epistemically favourable circumstances. She gives another example (2012: 574):

Suppose John and Mary are our mutual friends. John was going to ask Mary to marry him, but chickened out at the last minute. I know Mary much better than you do, and you ask me whether Mary might have said yes if John had proposed. I tell you that I swore to Mary that I would never actually tell anyone that information, which means that strictly speaking, I cannot answer your question. But I say that I will go so far as to tell you two facts:

(7)   If John had proposed to Mary and she had said yes, he would have been really happy.
(8)   But if John had proposed, he would have been really unhappy.

(My numbering.) According to Moss, it is 'okay' for her to utter (8) after (7), since she is in a position to rule out the possibility that, if John had proposed to Mary, she might have said yes. By contrast, dynamic semantic accounts of the context-shifting predict that uttering the reverse Sobel sequence of (7) followed by (8) will be infelicitous, irrespective of the epistemic circumstances.

Moss may well be right that there is such an epistemic effect. But does it fully explain the phenomena? To test that, it helps to consider conversations which minimize the scope for context-shifting between the members of the sequence. For instance, we can anchor the domain of worlds to a single occurrence of 'would' in a single question, asked by someone else. Here is such a form of the punchline of the conversation about John and Mary, in the same epistemic circumstances as before:

(9)   Q: How would John have felt?

A: Really happy, if he had proposed to her and she had said yes, but really unhappy, if he had proposed to her.

The answer sounds just as lame as more standard reverse Sobel utterances, with the two conjuncts in obvious tension with each other. Perhaps the comparative accept-ability of uttering (7) followed by (8) comes from their presentation as separate, pre-packaged nuggets of information ('two facts'), rather than successive stages in a continuous flow of discourse. That may make it easier to ignore contextual cues from (7) in processing (8).

A similar effect is observable in the Sophie case. Let the efficiency of the parade stewards be common ground in the conversation, and consider this dialogue:

(10)   Q: Would Sophie have seen Pedro?

      A: No, if she had gone to the parade and been stuck behind a tall person, but yes, if she had gone to the parade.

The answer sounds just as lame as that in (9). Even in uttering (4) followed by (5), there is some jolt between (4) and (5), although reminding oneself of the stewards' efficiency helps smooth it out, and may prompt the resetting of contextual param-eters to exclude far-out possibilities: in heuristic terms, their efficiency will figure in the development of the supposition that Sophie went to the parade.

Moss's account does not explain why (9) and (10) are infelicitous in the envisaged circumstances, despite being epistemically unproblematic on the Lewis and Stalnaker semantics. Such theories are missing something. A natural response is to postulate contextual shifts in the relevance of worlds, which may in turn make the complexities of the Lewis and Stalnaker semantics and logic redundant.

On the natural way of hearing (9) and (10), however well informed the answerer is, she is saying mutually inconsistent things about some of the relevant counter-factual possibilities. Even forward Sobel sequences start to sound bad in the same way when put in a format that discourages context-shifting, by encouraging coordination between the applications of the suppositional heuristics to the two antecedents:

(11)   Q: How would John have felt?

      A: Really unhappy, if he had proposed to her, but really happy, if he had proposed to her and she had said yes.

(12)   Q: Would Sophie have seen Pedro?

      A: Yes, if she had gone to the parade, but no, if she had gone to the parade and been stuck behind a tall person.

They sound even worse when the answer is phrased in a way to make clear that the second conjunct was premeditated, rather than just an afterthought:

(13)   Q: How would John have felt?

      A: On the one hand, really unhappy, if he had proposed to her, but on the other hand, really happy, if he had proposed to her and she had said yes.

(14)   Q: Would Sophie have seen Pedro?

      A: On the one hand, yes, if she had gone to the parade, but on the other hand, no, if she had gone to the parade and been stuck behind a tall person.

There is the same flavour of inconsistency.

'Might if' sentences occur naturally in such cases:

(15)　If Sophie had gone to the parade, she might have been stuck behind a tall person and not seen Pedro.

The use of 'had gone' suggests that the 'if' clause is within the scope of 'might', as in (15a):

(15a)　might(if Sophie went to the parade, she was stuck behind a tall person and did not see Pedro)

This raises the usual problem that the material reading of 'if' seems to make 'might if' conditionals too weak, since (15a) follows from (16)/(16a):

(16)　Sophie might not have gone to the parade.
(16a)　might(not(Sophie went to the parade))

As in sections 10.2 and 11.5, a good solution is to treat the explicit antecedent as also implicitly restricting the modal operator. Given that restriction in (15)/(15a), it is equivalent to (17)/(17a):

(17)　Sophie might have gone to the parade, been stuck behind a tall person, and not seen Pedro.

(17a)　might(Sophie went to the parade and she was stuck behind a tall person and did not see Pedro)

Oddly, the modalized conditional (15) and the modalized conjunction (17) *do* seem to have readings on which they are equivalent. That is evidence for the implicit restriction hypothesis, since it predicts the surprising fact. In heuristic terms, the suppositional procedure for a modalized conditional assigns very similar roles to the implicit restriction on the modal operator and the explicit antecedent of the embedded conditional (compare MCSR in section 11.4), so it is not unnatural for the antecedent to be assimilated into the restriction too. Both (15) and (17) can play a somewhat similar role to (4) in Sobel sequences and their reverses, though whether 'might' is to be understood as strictly the dual of 'would' or as having an epistemic ingredient too is not obvious (see section 11.5).

To summarize: the peculiarities of forward and reverse Sobel sequences involve two kinds of effect: one epistemic, diagnosed by Moss, and one contextual, diagnosed by proponents of the strict conditional analysis. When the speaker is known to be well enough informed, the epistemic effect disappears, but the contextual effect remains.

We can further explore the contrast between the two effects by considering forward and reverse Sobel sequences with plain unmodalized conditionals. Part I argued that the semantics of 'if' itself is truth-functional and so non-modal: it invokes no category of contextually relevant possible worlds, in either an epistemic or an objective sense of 'possible'. There is nothing for context to shift. Yet, as Robbie Williams has noted (2008), they exhibit a similar asymmetry between forward and reverse Sobel sequences. For example, consider the simple past tense analogues of (4) and (5):

(18)   If Sophie went to the parade and got stuck behind a tall person, she did not see Pedro.

(19)   If Sophie went to the parade, she saw Pedro.

For a speaker who does not know whether Sophie went to the parade, uttering (18) followed by (19) sounds much worse than uttering (19) followed by (18). One might try explaining that epistemically, perhaps using Moss's framework. Having uttered (18), the speaker cannot blithely add (19), for he has just provided an objection to it: what if she got stuck behind a tall person? By contrast, if he utters (19) followed by (18), both utterances sound fine at the time of utterance, although (18) casts retrospective doubt on (19).

When the speaker is known to be fully informed, the epistemic effect disappears. Consider the past tense analogue of (9), when the answerer has already admitted knowing whether John proposed to Mary and, if so, whether she said yes:

(20)   Q: How did John feel?

A: Really happy, if he proposed to her and she said yes, but really unhappy, if he proposed to her.

Of course, the answer sounds cagey and initially puzzling, as in Moss's scenario, but not inconsistent, by contrast with (9). The question is simply how John did feel, not how he felt in various epistemic possibilities; the answer gives incomplete but consistent information about the former, not inconsistent information about the latter. What it enables the questioner to work out is that if John proposed to Mary, she did not say yes.

Similarly, if the speaker of (18) and (19) has already admitted knowing whether Sophie went to the parade and, if so, whether she got stuck behind a tall person, but wants to sound clever, she can felicitously (though rather archly) utter them in either order, and likewise with the question-and-answer format:

(21)   Q: Did Sophie see Pedro?

A: No, if she went to the parade and got stuck behind a tall person, but yes, if she went to the parade.

Again, the answer sounds cagey and initially puzzling, but not inconsistent, by contrast with (14). The question is simply whether Sophie did see Pedro, not whether she saw him in various epistemic possibilities; again, the answer gives incomplete but consistent information about the former, not inconsistent information about the latter. What it enables the questioner to work out is that if Sophie went to the parade, she did not get stuck behind a tall person.

Of course, the standard problems for the material interpretation arise in the vicinity. For example, it makes (19) ('If Sophie went to the parade, she saw Pedro') entail (22), which they take to be in tension with (18) ('If Sophie went to the parade and got stuck behind a tall person, she did not see Pedro'):

(22)   If Sophie went to the parade and got stuck behind a tall person, she saw Pedro.

But, asked to estimate their probabilities, speakers may find (19) more probable than (22). This is just the usual tendency to judge the probabilities of conditionals by conditional probabilities, explained in chapter 3 as resulting from use of the Suppositional Procedure. But reverse Sobel sequences for unmodalized conditionals pose no *special* problem for the material interpretation.

'Would if' conditionals about non-contingent matters behave similarly to plain conditionals, since the modalization is redundant. Here is an example. I am interested in a particular mathematical theory T. I note that a result in one article in a journal of mathematical logic implies that if T were complete and consistent, it would not be recursively enumerable (the definitions of these terms do not matter for present purposes). Thus I assert (23):

(23)   If T were complete and consistent, it would not be recursively enumerable.

I also note that a result in an article in another journal of mathematical logic implies that if T were complete, it would be recursively enumerable. Thus I assert (24):

(24)   If T were complete, it would be recursively enumerable.

In that order, (23) and (24) constitute a reverse Sobel sequence. But uttering (23) followed by (24) does not provoke a crash, semantic or otherwise. I simply deduce (25):

(25)   If T were complete, it would be inconsistent.

Since mathematical matters are non-contingent, the antecedent of (23) is impossible, because false. It is not true at any possible world. Thus (23) is vacuously true. But that is no problem; it is common in logic and mathematics. Of course, (23)–(25) involve technical vocabulary, but that does not matter; their grammatical structure and use of 'if' and 'would' is quite standard.

At this point, one might feel tempted to invoke contextually relevant impossible worlds for the antecedent of (23) to be true at. But such a move casts no light on how a conditional such as (23) works in mathematics. The truth of such a mathematical statement does not depend on what happens in worlds which violate the laws of logic and mathematics. Of course, mathematicians must be able to argue validly from impossible suppositions like the antecedent of (23), for instance in proofs by *reductio ad absurdum*, but when the rules of proof systems in logic and mathematics are articulated, there is normally no reference to impossible worlds, and no need for any such reference. Introducing them would be a mere distraction (chapter 15 discusses counterpossible 'would if' conditionals in more detail). That is one reason to reject any form of dynamic semantics, such as those of von Fintel and Gillies, on which, in the evaluation of a counterfactual, the consequent must be evaluated in a context where the relevant possible worlds include some at which the antecedent is true.

In any case, Moss's examples already show that the crashing of reverse Sobel sequences is not purely semantic: it depends on the epistemic position of the speaker. It is better explained in the pragmatics, and ultimately by the heuristics. That is another reason to doubt theories of counterfactual conditionals such as Lewis's and Stalnaker's, which transfer so much of the work to the semantics.

14

Thought Experiments and 'Would'

14.1 Thought experiments and counterfactuals

There is a natural connection between 'would if' thinking and thought experiments. The connection does not depend on the specific semantic analysis of 'would if' sentences in chapter 10, so for immediate purposes we can maintain neutrality between different analyses by formalizing them just with '>'.

Let T be a philosophical or scientific theory which entails the necessity of something:

(1) $\Box X$

Here $\Box$ expresses whatever kind of necessity T attributes to X. In principle, a thought experiment can refute (1), and thereby T. We suppose that a scenario S obtains, and use our imagination, our capacity for offline cognition, to determine what else would obtain on that supposition. In particular, we assess X on the supposition S. When the thought experiment goes well, we reject X on that supposition, and so, by applying the suppositional procedure for 'would if' sentences (chapter 11), come to know the counterfactual conditional (2):

(2) $S > \neg X$

Of course, as a bare formula, (2) introduces no inconsistency with (1), since the result of substituting $\neg X$ for S in (2) is trivially true. However, in many cases we can verify the possibility of the scenario S (on which more below):

(3) $\Diamond S$

Here $\Diamond$ is the dual modality of $\Box$ in (1). By a widely accepted principle, the counterfactual consequences of a possibility are also possibilities; schematically:

$$\text{POSSIBILITY} \quad (A > C) \supset (\Diamond A \supset \Diamond C)$$

Instantiating A by S and C by $\neg X$ in POSSIBILITY, we can derive (4) from (2) and (3):

(4) $\Diamond \neg X$

By normal modal logic, (4) contradicts (1). Thus, in favourable epistemic circumstances, our thought experiment has refuted (1), and thereby the original theory T, by identifying a possible situation in which the allegedly necessary truth is false.

Suppose and Tell: The Semantics and Heuristics of Conditionals. Timothy Williamson, Oxford University Press (2020).
© Timothy Williamson.
DOI: 10.1093/oso/9780198860662.001.0001

Of course, deriving (4) from (3) helps only when (3) is easier to know than (4). Typically, the scenario S is given in more clearly imaginable terms than the bare negation $\neg X$, which enables us to verify the possibility of the former, (3), more easily than the possibility of the latter, (4).

The verification of premise (3) can also be understood in terms of counterfactual conditionals. That requires another principle connecting strict and counterfactual implication, to the effect that all strict consequences are counterfactual consequences too; schematically:

$$\text{NECESSITY} \quad \Box(A \supset C) \supset (A > C)$$

Suppose that we come to know (perhaps by using a suppositional procedure) that some counterfactual conditional with antecedent S does *not* hold:

(5) $\neg(S > Z)$

What Z is does not matter for present purposes, provided that (5) is known. Our capacity to assess counterfactual conditionals includes some capacity to falsify some of them. Then, by contraposing the relevant instance of NECESSITY, we can derive (6) from (5):

(6) $\neg\Box(S \supset Z)$

But (6) entails (3) by normal modal logic, since an impossibility strictly implies anything. Thus, in favourable epistemic circumstances, we have established the possibility premise (3), as required. In previous work, I developed an account of thought experiments along such lines (Williamson 2007a).

Why interpret thought experiments in terms of the counterfactual conditional (2) rather than the corresponding unrestricted strict conditional (2*)?

(2*) $\Box(S \supset \neg X)$

After all, logically, (2*) supports the move from (3) to (4) just as well as (2) does.

One reason for preferring (2) to (2*) is to demystify thought experiments by explaining how the way they use the imagination is only a special case of the general way our ordinary assessments of counterfactual conditionals, including very mundane ones, typically use the imagination. It involves no magic or pseudo-magic.

A second reason for preferring (2) to (2*) is that (2*) is more vulnerable to deviant realizations of S, perhaps in very far-out worlds which violate assumptions normally taken for granted in processing the supposition S: in some such worlds, X may be true, even though X is false at the closest worlds to actuality at which S is true. Any one such far-out deviant possibility falsifies (2*). It is usually very hard to describe the scenario for a thought experiment in perfectly watertight terms, to exclude all unwanted deviant realizations. Even if a description is in fact watertight, it may be very hard to know that it is watertight. In most thought experiments, the argument from the description of the scenario (S) to the verdict on it ($\neg X$) is not purely logical, otherwise one would need the imagination much less in reaching the verdict. In imagining the scenario, one may do so in ways which in effect prejudge how it will be realized, and so embody reasonable but not mandatory assumptions, on which the verdict may depend. Using the counterfactual conditional (2) instead of the strict conditional (2*) insulates the argument from such unnecessarily far-out deviant realizations, thereby making the methodology of thought experiments more robust.

A more specific feature of my 2007 account is that it followed David Lewis and much of the rest of the literature in treating the counterfactual conditional 'if' as a primitive two-place operator. The next section explains how the account works, and indeed benefits, once 'if' and 'would' are treated as independently moving parts, which combine compositionally to make a restricted conditional.

14.2 Interrelations between necessity and 'would'

In what follows, we continue working with formalizations, to present the logical structure more perspicuously. As before, $\Box$ expresses the kind of necessity attributed by the theory T, and $\Diamond$ the dual kind of possibility. We also have the contextually restricted necessity-like operator 'would'. Following chapter 10, we analyse $A > C$ as 'would$(A \supset C)$'.

The first task is to check the status of the resulting versions of POSSIBILITY and NECESSITY:

POSSIBILITY*   would$(A \supset C) \supset (\Diamond A \supset \Diamond C)$

NECESSITY* $\Box(A \supset C) \supset$ would$(A \supset C)$

Of those two principles, the more straightforward case is NECESSITY*. It is just a special case of the general principle that what necessarily holds would hold:

RESTRICTION $\Box A \supset$ would(A)

Conversely, the schema NECESSITY* implies the schema RESTRICTION, given that the logics of both modalities are normal. For then, when we simultaneously substitute a tautology for A and A for C, the left-hand side of NECESSITY* becomes equivalent to $\Box A$ and its right-hand side to 'would(A)'. Still given normality, RESTRICTION is of course equivalent to the dual principle that the dual of 'would' implies possibility:

RESTRICTION$_D$ $\neg$would$(\neg A) \supset \Diamond A$

We saw in sections 11.3 and 11.5 that directly negating 'would' tends to be awkward and unnatural, though it can be done, and we have indirect ways of achieving the same effect.

For the original account of thought experiments to work under the new account of 'would if' sentences, the necessity attributed by 'would' must restrict the necessity attributed by the theory T in (1), in the sense of RESTRICTION. That requirement is typically quite plausible when T attributes metaphysical necessity—'necessity in the highest degree' (Kripke 1980: 99)—and even when T attributes only nomic necessity, necessity given the laws of nature. What *could* not have been otherwise, metaphysically or nomically, *would* not have been otherwise.

One might suspect that RESTRICTION will fail in special contexts when we are talking about metaphysical or nomic impossibilities. Chapter 15 discusses that issue in detail. For present purposes, we assume that RESTRICTION holds. Consequently, NECESSITY* holds too.

Once we have NECESSITY*, we may be able to use it to verify the possibility premise (3), by applying the suppositional rule MSR for 'would' to falsify 'would(¬S)' (falsifying the counterfactual conditional 'would($S \supset Z$)' is just one way of doing that), giving (5*) in place of (5):

(5*) ¬would(¬S)

From (5*), we can derive (3) by RESTRICTION$_D$, which follows from NECESSITY*.

Unfortunately, POSSIBILITY* is much more problematic, for in the present setting it is equivalent to the converse of RESTRICTION:

DERESTRICTION would(A) $\supset \Box A$

The schema DERESTRICTION implies the schema POSSIBLITY*, for we can substitute $A \supset C$ for A in DERESTRICTION, and $\Box(A \supset C)$ entails $\Diamond A \supset \Diamond C$ by normal modal logic. Conversely, the schema POSSIBILITY* entails the schema DERESTRICTION, given that the logics of both modalities are normal, for we can simultaneously substitute $\neg A$ for A and a contradiction for C in POSSIBILITY*, making its left-hand side equivalent to 'would(A)' and its right-hand side to $\Box A$. But combining DERESTRICTION with RESTRICTION collapses the contextually restricted modality 'would' into the typically much less restricted modality $\Box$, making the contextual restriction redundant. That is a disaster. Usually, 'would' is far weaker than metaphysical or even nomic necessity; the contextual restriction is far from redundant. Thus POSSIBILITY* cannot hold in full generality.

By contrast, the original schema POSSIBILITY induces no such collapse. The key difference is that POSSIBILITY concerns a binary conditional operator, which makes more room for conditionals with inconsistent consequents to behave differently from other conditionals, whereas in normal modal logic *every* strict conditional is equivalent to a strict conditional with an inconsistent consequent, since $\Box A$ is equivalent to $\Box(\neg A \supset (A \wedge \neg A))$.

As so often, what is left out of the semantics can be put back in the pragmatics. In the thought experiment, the counterfactual conditional (2) ($S > \neg X$) is assessed by the modal conditional suppositional rule MCSR. Formally, the reparsed analogue of (2) is (2W):

(2W) would($S \supset \neg X$)

Since the focus is on the supposition S, and what it implies, we can expect some possibilities in which it holds to be contextually relevant, and so to be added to the domain of 'would' if not already there—if S is possible at all. More generally, in a context where a particular sentence 'would(if A, C)' is being assessed by something like MCSR, (7) will typically hold:

(7) would($\neg A$) $\supset \Box \neg A$

That is, some worlds where the antecedent A at issue holds will be contextually relevant unless A is impossible. In particular, when one considers the thought experiment, and so assesses the modalized conditional (2W), the antecedent is S, so in that context (7S) should hold:

(7S) would($\neg S$) $\supset \Box \neg S$

Furthermore, from (2W) we can derive (8) by the normal modal logic of 'would' in section 10.3, since X and $S \supset \neg X$ together entail $\neg S$ by standard truth-functional logic:

(8) would(X) $\supset$ would($\neg S$)

Moreover, by an instance of the comparatively uncontroversial principle RESTRICTION, we have:

(9) $\Box X \supset$ would(X)

Chaining (9), (8), and (7) together yields (10):

(10) $\Box X \supset \Box \neg S$

In many philosophical thought experiments, the possibility of the scenario, $\Diamond S$ (premise (3)), is not in doubt. Since $\Diamond S$ is inconsistent with $\Box \neg S$, we can combine it with (10) to refute the necessity claim (1) ($\Box X$), and thereby refute the theory T, which entailed (1), thereby achieving the purpose of the thought experiment. Thus an independently plausible pragmatic consideration enables us to achieve the intended effect without appeal to the disastrous schema POSSIBILITY*.

Such pragmatic considerations also explain the pre-theoretic plausibility of the principle that if C would hold if A held, and A could hold, then C could hold.

For purposes of the thought experiment, we may even be able to dispense with the pragmatic premise (7S). For we may indeed be able to establish (5*), the contextually relevant possibility of the scenario, in other words ¬would(¬S), as discussed above. Then one can use (8) (derived from (2W)), (9) (derived from RESTRICTION), and (5*) to derive a contradiction from $\Box X$, and again achieve the purpose of the thought experiment, without appeal to the pragmatic premise or anything like POSSIBILITY*.

The latter version of the argument works purely with 'would' except for the final stage, at which it uses RESTRICTION to make the connection with the necessity claim made by T. This version is in an even more anti-exceptionalist spirit than its predecessor, since it works with the kind of modality typical of ordinary, non-philosophical cognition until the point when the clash with a theoretical ascription of unrestricted necessity has to be made explicit. The key premises (2W) and (5*) are based on the verification and falsification respectively of a 'would' claim by standard ways of assessing such claims. The normality of the logic of 'would' is guaranteed by the standard form of possible world semantics for it, irrespective of the details of the contextual restriction.

As for RESTRICTION, it depends only on 'would' being a restriction of the modality ascribed by the theory T. The style of argument is not confined to theoretical ascriptions of *absolutely* unrestricted necessity. Provided that $\Box$ is at least as broad as 'would', the argument still works. For example, thought experiments in natural science presumably target claims of physical rather than metaphysical necessity, but they can still be analysed along the same lines, for a context where 'would' ranges only over (presumed) physical possibilities.

Of course, this is only a rational reconstruction. In practice, people doing thought experiments almost never consciously go through such articulated modal reasoning. They may well not even do it unconsciously. Some of the intermediate steps may serve only us, as theorists, as checks that a bigger move is truth-preserving. At this early stage of inquiry, we cannot expect to analyse the stages of the underlying psychological process in fine detail. What we can hope for is to explain in principle how the practice of testing theories by thought experiments can be reliable, given normal human cognitive capacities, in particular the capacity to assess conditionals reliably in favourable circumstances by some sort of suppositional procedure. That the thought experimenter does apply such a procedure in reaching the verdict 'would(if S, not(X))', (2)/(2W), is quite plausible. To do the thought experiment at all, you must suppose S and assess X on that supposition, although the assessment need not involve verbalized reasoning from S.

The role of the possibility premise, (3) or (5*), is often psychologically less salient. The question whether the scenario is possible may not consciously arise. However, since the relevant supposition (S) is the same as for the verdict, some unconscious monitoring of its possibility is presumably going on in the application of the suppositional procedure. After all, once the impossibility of a supposition is recognized, continuing to work out its implications is typically a waste of time and energy. Its impossibility may be recognized through the occurrence of something inconsistent with it in its modally distanced development. If serious warning signs of its impossibility arise in that process, an alarm should go off, and reach consciousness. When that does not happen, the suppositional procedure is implicitly giving the green light to the possibility premise. Since the other principles used in the argument are broadly logical, the rational reconstruction shows how our practice of testing theories by thought experiments can be reliable in circumstances favourable to our normal ways of assessing 'would if' conditionals.

The verification of the contextually restricted possibility premise (5*) need not always go via the falsification of a 'would if' conditional. Humans may have developed some capacity to recognize the restricted modal status of a supposition non-inferentially, in simple cases, either from the individual's own experience of how the world works or by an innate capacity calibrated by evolution. Pure metaphysical possibility may not matter for survival, but contextually restricted possibility does, and contextually restricted possibility entails metaphysical possibility.

The semantic analysis of counterfactual conditionals as contextually restricted strict conditionals does not simply validate the original account of thought experiments in terms of 'would if' conditionals. In several ways it is an improvement. As already seen, it allows the most challenging stages of the argument to focus purely on the restricted modalities, where our ordinary cognitive capacities are likely to be most reliable. It is also an improvement in several further ways. Two will now be discussed in more detail.

14.3 The problem of deviant realizations

For interpreting thought experiments, the use of the 'would if' conditional (2), especially when interpreted as the contextually restricted strict conditional (2W), was seen to have a significant advantage over the unrestricted strict conditional (2*): it reduces the threat of deviant realizations of the scenario. However, it does not entirely eliminate the threat. For what prevents a deviant realization from being possible, or even actual, in a contextually relevant way?

Unfortunately, the convenient term 'deviant' obscures the nature of the problem. What is meant to be deviant about deviant realizations? It cannot just be that the desired verdict $\neg X$ is false in them, for that would apply to any thought experiment proposed against a *true* theory T which entails $\Box X$, since then $\neg X$ is impossible. Some thought experiments fail. Even if $\neg X$ is possible, and so T false, the mere truth of X in a realization of a scenario S does not make the realization deviant, for S may strictly imply X. A thought experiment deployed against a falsehood may still fail. Even if $S \wedge \neg X$ is possible, the mere truth of X in a realization of S still does not make it deviant. For example, suppose that S has nothing to do with X, contrary to the hopes of the philosopher who devised the thought experiment. Actually, S is true and X false, although $S \wedge \neg X$ is possible. That does not make the actual realization of S deviant in any relevant sense. The use of S as a thought experiment against (1) simply fails.

Many philosophers have noticed that the counterfactual conditional (2) is stronger than logically needed to get from (3), the possibility of S, to (4), the required possibility of $\neg X$ (for example, Anna-Sara Malmgren 2011). They have therefore proposed weakening premise (2) in order to interpret the thought experiment as more robust. The weakest auxiliary premise to secure the move from (3) to (4) is just the material implication from (3) to (4):

(11) $\Diamond S \supset \Diamond \neg X$

A slightly stronger alternative requires S to be compossible with $\neg X$, although it violates the spirit behind (11) of 'Say the least you can make do with':

(12) $\Diamond S \supset \Diamond (S \wedge \neg X)$

Given POSSIBILITY, (2) entails (11) and (12)—in the latter case, with help from elementary logical principles about counterfactuals. The converse entailments obviously fail. For (11) and (12) may be true while (2) is false, because realizations of S in which X is true are more relevant to 'would if' in (2) in whatever sense matters than are realizations of S in which X is false.

Of course, we saw in section 14.2 that once counterfactual conditionals are analysed as contextually restricted strict conditionals, POSSIBILITY becomes POSSIBILITY*, which is unexpectedly and unacceptably strong. However, we also saw how to derive (11) from reasonable pragmatic considerations, and (12) can be derived in a similar way. Although the logical relations between (2) on one side and (11) and (12) on the other are less straightforward than they first looked, in a loose sense (11) and (12) are still less vulnerable than (2) to the threat of falsification by a deviant realization of S.

One may take (11) or (12) as a more cautious moral of the thought experiment than (2), but still enough for refuting the theory T. But what is the point of replacing (2) by (11) or (12) in the official argument if one's only basis for asserting (11) or (12) is (2)? For one gets to (11) or (12) in the first place by assessing X on the supposition S. Thus one accepted $\neg X$ with respect to the very same case in which one supposed S to hold. One was therefore committed at the very least to (12), not just (11). But not even (12) properly expresses what one came to know, if anything. For (12) corresponds to the thought 'If S is true in some possibility, then $S \wedge \neg X$ is also true in some possibility, but perhaps not the one I first thought of.' Such hedging is quite gratuitous; it reflects nothing in the execution of the thought experiment itself. In principle, someone can truly assert (11) or (12) while (2) is false, but if they got there in the usual way, by accepting $\neg X$ on the supposition S and applying the suppositional rule MCSR for 'would if' conditionals, it was a mere stroke of luck that their hedging got them something true out of something false.

Our reliance on a suppositional rule emerges in the case of opposed verdicts on a thought experiment. For example, *Gettier cases* are usually taken to show that a subject can have justified true belief without knowledge. A philosopher who judges that the subject in a Gettier case *does* know is naturally understood as disagreeing with the orthodox judgement that the subject does *not* know. In terms of the present schematic framework, someone who accepts X on the supposition S is naturally understood as contradicting the orthodox judgement $\neg X$ on the same supposition, and indeed the suppositional rule projects that disagreement onto their conditional judgements. The orthodox judgement on the thought experiment, the counterfactual conditional (2), now analysed as the contextually restricted strict conditional (2*), contrasts with the unorthodox judgement on the same thought experiment, the counterfactual conditional (2-), now analysed as the contextually restricted strict conditional (2-*):

(2-) $S > X$
(2-*)   would$(S \supset X)$

But, on a Lewis-Stalnaker semantics for counterfactual conditionals, (2) and (2-) are true together just in case their shared antecedent S is metaphysically impossible. On the present semantics, (2*) and (2a-*) are true together just in case S is false in all contextually relevant worlds (it would not hold). Both views make the conflict genuine, provided that S holds in some contextually relevant possibility. By contrast, the proposed weaker versions of the orthodox judgements, (11) and (12), correspond to these weaker versions of the unorthodox judgements:

(11-) $\Diamond S \supset \Diamond X$
(12-) $\Diamond S \supset \Diamond(S \wedge X)$

Even given that S is possible, (12) is quite compatible with (12-). Those who use S as the scenario for a thought experiment against the theory T can happily accept that there are abnormal realizations of S in which X is true as well as normal realizations of S in which X is false; the latter suffice to refute T. A fortiori, even given that S is possible, (11) is quite compatible with (11-). Thus the weaker interpretations fail to capture the way in which verdicts on a thought experiment can conflict.

Parallel considerations tell against alternative suggestions for weakening the upshot of the thought experiment to analogues of (11) or (12) but with contextually restricted rather than unrestricted possibility:

(11*) $\neg$would$(\neg S) \supset \neg$would(X)
(12*) $\neg$would$(\neg S) \supset \neg$would$(\neg(S \wedge \neg X))$

Since the structure of the arguments against these suggestions is the same as for those against (11) and (12), the details will not be spelled out. Again, they fail to capture the key cognitive role of the suppositional procedure in assessing a thought experiment, and the way in which verdicts can conflict.

The output of a thought experiment is normally a judgement made using a suppositional procedure. A full account of the thought experiment should make that dependence on the procedure plain, not hide it under the table. That requires articulations strong enough to make explicit the supposition and the verdict when the procedure was applied. As just seen, (11), (12), (11*), (12*) and other variants on the same theme are far too weak for that purpose. A conditional more like (2) or (2*) is needed.

Fortunately, the apparatus of restricted strict conditionals allows more room for manoeuvre than the 2007 account had. For the contextual restriction for 'would' may eliminate some relevantly abnormal worlds, and thereby some 'deviant realizations' of the scenario. In particular, even the actual world may be eliminated, if it counts as relevantly abnormal in the context. As observed in section 10.3, 'would(A)' does not always imply A. The development of the scenario may itself inform the contextual restriction.

Of course, it is not a case of 'Anything goes'. The tighter the contextual restriction, the more demanding is the restricted possibility premise. Moreover, thought experiments are no mere private fantasies: in both philosophy and natural science, they are normally intended as contributions to a shared inquiry or conversation. For that purpose, the relevant context is shared too. Thus idiosyncratic quirks in someone's development of a scenario are insufficient to restrict the worlds relevant to a shared conversation.

Of course, those remarks on the pragmatics of thought experiments are quite schematic. Much remains to be explored about specific ways in which contextual domain restrictions interact with the need for a shared imaginative exercise to provide reliable modal information. But the proper setting for such an inquiry is not a study of the epistemology of thought experiments, philosophical or otherwise, for they provide too narrow a range of data, and too much scope for metaphilosophical prejudices to distort the conclusions. Rather, we can achieve a more appropriate level of generality by investigating the relation between the cognitive role of the imagination and the semantics and pragmatics of modal words such as 'would' in ordinary human thought and talk. That will enable us to understand philosophical thought experiments as, for better or worse, much less exceptional than they are often made out to be—and if there *is* anything exceptional about them, it will enable us properly to identify what it is. Such an enterprise is only at the beginning.

For now, we turn to one specific issue which complicates the semantics of thought experiments, but obviously concerns a far wider range of cases.

14.4 The interaction of 'would if' with quantifiers

In a thought experiment, the scenario is usually described in quantified terms. *Someone* believes at 3 o'clock that it is 3 o'clock on the basis of a clock which happens to have stopped at 3 o'clock. Of course, the character may be given a local habitation and a name ('Tom believes at 3 o'clock...'), but normally the author and the audience have no specific person in mind. The name is there merely for purposes of vividness and anaphora, not to pass on reference to someone in particular. The vividness and the clarity of the anaphora are cognitive as well as aesthetic virtues; they make it easier for us to apply the suppositional procedure properly. But when the cognitive role of the thought experiment depends on the possibility of the scenario, the fictional name obscures just what it is that needs to be possible. That is more clearly articulated in quantificational terms ('Someone named "Tom" believes at 3 o'clock...'). Most discussions of thought experiments, including the one above, suppress their quantificational complexity.

For example, a particular Gettier case may concern an unspecified person x who stands in a specified relation S to a proposition p, which is not fully specified because it depends on x. Here 'x' and 'p' are variables, not names. We can write the scenario $S(x, p)$, and let $JTB(x, p)$ and $K(x, p)$ abbreviate 'x has a justified true belief in p' and 'x knows p' respectively. In applying the suppositional procedure, the crucial step is from supposing $S(x, p)$ to reaching verdicts on $JTB(x, p)$ and $K(x, p)$. Of course, we may start off making the existentially quantified supposition (13) that some person and some proposition stand in the specified relation:

$$(13) \quad \exists x\, \exists p\, S(x,p)$$

But properly imagining an existential state of affairs involves imagining an instance of it, in the sense in which we can imagine a cat scratching a dog without there being any cat or dog of which we imagine it. Just so, when we imagine (13), we go on to imagine $S(x, p)$ without there being any person x or proposition p of which we imagine it. $S(x, p)$ becomes a new supposition. In effect, we use the free variables like arbitrary names in many systems of proof by tableaux or natural deduction for predicate logic. Such imagined instantiation is the analogue of a standard rule of inference in such proof systems for existential formulas in premise position.

When the Gettier case works as intended, on the supposition $S(x, p)$, we assert $JTB(x,p)$ and deny $K(x, p)$, and so assert $JTB(x,p) \wedge \neg K(x, p)$. Our verdict is anaphoric on the pronouns or free variables in the supposition itself, with no clutter of quantifiers to confuse or distract us. We then apply a suppositional rule such as MCSR to assert the corresponding counterfactual conditional, still without quantificational clutter. On the Lewis-Stalnaker parsing of such conditionals, the result is (14):

$$(14) \quad S(x,p) > (JTB(x,p) \wedge \neg K(x,p))$$

The trouble with (14) is that it does not express a proposition, evaluable as true or false, for the variables 'x' and 'p' occur free in it. The obvious solution is to bind them with universal quantifiers, since (14) was reached independently of any particular values for them. The result is (15):

$$(15) \quad \forall x\, \forall y\, (S(x,p) > (JTB(x,p) \wedge \neg K(x,p)))$$

However, (15) has a different problem. The original point of using a counterfactual conditional instead of an unrestricted strict conditional was to insulate the thought experiment from deviant realizations of the scenario in unanticipated, wildly counterfactual circumstances. On the Lewis-Stalnaker approach to the semantics, the truth-value of a counterfactual conditional is determined by the truth-value of its consequent in the closest worlds (if any) to the world of evaluation at which the antecedent is true (in some appropriate sense of 'closest'); the rest make no difference. In the present case, that might be expected to mean that we need only worry about the closest worlds (if any) in which the scenario is realized, in other words, at which (13) is true. But that is not what (15) says. Even if (13) is true at the actual world, or a very close one, that only requires that *some* values of 'x' and 'p' satisfy $S(x, p)$ at such a nearby world. For some *other* values of 'x' and 'p', the closest worlds at which they satisfy $S(x, p)$ may be very distant indeed from the actual world. If they do not also satisfy '$JTB(x, p) \wedge \neg K(x, p)$' at that world, (15) is false. Our ordinary counterfactual thinking seems not to guard against such threats.

My original account gave the problem a somewhat unnatural solution, by using the counterfactual conditional:

(16) $\exists x \, \exists p \, S(x, p) > \forall x \, \forall y \, (S(x, p) \supset (JTB(x, p) \wedge \neg K(x, p)))$

As required, the truth-value of (16) depends only on the truth-value of its consequent at the closest worlds (if any) to the world of evaluation in which the scenario is realized, in other words, at which (13) is true. Moreover, (16) can easily be shown to entail (17) in a standard quantified logic for counterfactuals:

(17) $\exists x \, \exists p \, S(x, p) > \exists x \, \exists p \, (JTB(x, p) \wedge \neg K(x, p))$

Consequently, given the standard premise that its antecedent is possible, (16) is strong enough to yield a counterexample to the target theory, in this case the claim that justified true belief is necessary and sufficient for knowledge:

(18) $\Box \forall x \, \forall y \, (K(x, p) \equiv JTB(x, p))$

Why does (16) feel unnatural? The problem is that it corresponds only indirectly to the application of the suppositional procedure on which it is based. The repetition of $S(x, p)$ in (16) corresponds to nothing in that application. The quantifiers in (16) also correspond to nothing in the application of the procedure; as noted with respect to (14), the key move is just from $S(x, p)$ to $JTB(x, p) \wedge \neg K(x, p)$. There is no such step of generalization under the supposition $S(x, p)$.

The semantics of the counterfactual conditional as a contextually restricted strict conditional solves the problem. Initially, applying the suppositional procedure yields an unquantified conditional, which the proposed semantics interprets materially:

(19) $S(x, p) \supset (JTB(x, p) \wedge \neg K(x, p))$

Since the application of the procedure made no special further assumptions about the values of the free variables, one can universally quantify (19), to obtain (20):

(20) $\forall x \, \forall y \, (S(x, p) \supset (JTB(x, p) \wedge \neg K(x, p)))$

Since (20) was reached without appeal to any specific feature of the actual world (as opposed to its more generic or lawlike features), one can apply a limited step of necessitation to (20), for a suitably restricted form of necessity, to obtain (21):

(21) $\text{would}(\forall x \, \forall y \, (S(x, p) \supset (JTB(x, p) \wedge \neg K(x, p))))$

Roughly, in English: if a case did realize the scenario, it would be a case of justified true belief without knowledge. By uncontroversial normal quantified modal logic, (21) implies that if the scenario of the thought experiment is restrictedly possible, so is a counterexample to (18), the JTB analysis of knowledge:

(22) $\neg\text{would}(\neg\exists x\ \exists p\ S(x,p)) \supset \neg\text{would}(\exists x\ \exists p\ (JTB(x,p) \wedge \neg K(x,p)))$

From there, the argument goes as before. But, unlike (16), (21) preserves the direct connection between the supposition and the verdict in the application of the suppositional procedure. Unlike (15), it does not threaten to depend on what goes on in wildly counterfactual worlds, far more distant than the closest ones in which the scenario is realized. In a typical context for (21), or its equivalent in a natural language, $S(x, p)$ will be restrictedly possible for *some* values of 'x' and 'p', but not for all—not even for all those pairs for which it is metaphysically possible. That suffices to make the antecedent of (22) true, as required. Thus (21) corresponds more closely than (16) to the underlying cognitive process in a thought experiment.

We can roughly paraphrase (21) in English with a sentence like (23) (where 'that relation' indicates S):

(23)   If someone were related that way to a proposition, they would have a justified true belief in it without knowledge.

Here is a possible objection: other sentences similar in form to (23) do not entail a corresponding modalized universal generalization in (21). To vary a standard example, consider (24):

(24)   If John had a coin in his pocket, he would put it in the meter.

It is widely thought that (24) is true even when John had two coins in his pocket and only put one in the meter. However, that may be to pay too much respect to loose speech. For if (24) is true in those circumstances, so is (25), in the very same context:

(25)   If John had a coin in his pocket, he would keep it in his pocket.

But (24) and (25) are naturally understood as in conflict with each other, at least when the antecedent is possible. Moreover, we react differently to examples where more is at stake, like (26):

(26)   If John had a daughter in the burning school, he would rescue her.

To judge (26) true when John had two daughters in the burning school and only bothered to rescue one suggests callous insensibility. Loose talk is more acceptable about trivia, like (24), than about matters of life and death. In any case, as long as (23) has *a* reading with truth-conditions like (21), even if it also has other readings, (23) can still paraphrase (21) in English.

Fidelity to the underlying cognitive process matters to the epistemology of thought experiments, as section 14.3 explained. For example, one could weaken (21) to (27) or (28):

(27) $\text{would}(\exists x\ \exists p\ S(x, p) \supset \exists x\ \exists p\ (JTB(x, p) \wedge \neg K(x, p)))$

(28)   would($\exists x\ \exists p\ S(x, p) \supset \exists x\ \exists p\ (S(x, p) \wedge JTB(x, p) \wedge \neg K(x, p))$)

For (21) entails (27) and (28) by uncontroversial quantified modal logic, while they do not entail (21). Moreover, each of (27) and (28) suffices for the rest of the argument. They weaken the quantificational form of (21) in a structurally similar way to that in which (11) and (12) weaken the modal form of (2), and face structurally similar problems. They fail to reflect the role of the suppositional procedure in the underlying cognitive process. For example, (28) corresponds to the thought 'If someone were related that way to a proposition, someone would be related that way to a proposition and have justified true belief in it without knowledge, but perhaps not the person and the proposition I first thought of', which omits the anaphoric links between $S(x, p)$ and $JTB(x, p) \wedge \neg K(x, p)$ when the suppositional procedure was applied. The problem is even worse with (27), which does not even require the case of justified true belief without knowledge to realize the original scenario.

As before, the weakenings fail to capture the way in which opposite verdicts on the thought experiment conflict. Parallel to (27) and (28), the benighted philosopher who judges that it *is* a case of knowledge is represented as endorsing only (29) or (30):

(29)   would($\exists x\ \exists p\ S(x, p) \supset \exists x\ \exists p\ (JTB(x, p) \wedge K(x, p))$)
(30)   would($\exists x\ \exists p\ S(x, p) \supset \exists x\ \exists p\ (S(x, p) \wedge JTB(x, p) \wedge K(x, p))$)

But the consequent of (30) is quite consistent with the consequent of (28), and a fortiori the consequent of (29) is quite consistent with the consequent of (27).

As before, there is no legitimate gain in hiding the first part of the argument under the table. The suppositional procedure yielded (21) if it yielded anything at the level of generality of (27) or (28); the latter were merely derived from the former. Although in principle (27) and (28) can be true while (21) is false, someone who asserts (27) or (28) but not (21) in that case, having relied on the normal suppositional procedure to get there in the first place, has just had a stroke of luck. To understand the underlying epistemology, one needs to come clean about the dependence on (21).

However, just as a more legitimate dose of caution can be applied to the contextual domain restriction on the modal operators, as seen in section 14.3, so an equally legitimate dose of caution can be applied to the contextual domain restriction on the overt quantifiers, such as those ranging over people and propositions in the formulas of this section. If the corresponding English sentences involve covert quantification over situations or events, yet another legitimate dose of caution can be applied to the contextual domain restrictions for them too. In all those ways, the Gettier judgement (21) can legitimately be defended against putative counterexamples; likewise for other thought experiments.

What we should not do is slip into a confused way of thinking on which any thought experiment that can be fixed up to work worked all along. In science, as elsewhere, it is commonplace for hits to be preceded by long sequences of misses and near-misses. It would be astonishing if thought experiments were an exception.

15

Worlds and Meaning

15.1 Worlds as parameters of evaluation

'Would' has been characterized, on a non-temporal reading, as in effect a context-ually restricted universal quantifier over worlds. One might then wonder: what *are* worlds?

A first response is that it is not the business of semantics to worry about that question. After all, semanticists may characterize the word 'always', on one reading, as in effect a contextually restricted universal quantifier over times, but that does not make it the business of semantics to worry about what times are. The question can be left to physics and metaphysics, in disputed proportions. Many semantic theories treat worlds and times in parallel, as parameters of evaluation. Such theories recursively assign expressions of the object-language semantic values relative to a world w, a time t, and usually other parameters too. The aim is to describe the meanings of those expressions in terms of worlds and times, not the other way round. Of course, semanticists need some informal understanding of the words 'world' and 'time' in the metalanguage, for the semantic theory is *interpreted*; it is not a mere formal device. To swop the interpretations of the two parameters, reading 'w' as a variable for times and 't' as one for worlds, while leaving the theory orthographically as before, would yield a new theory, presumably a worse one. But that informal understanding need not involve a developed theory of the nature of worlds and times.[1]

Of course, a semanticist will normally have a pre-theoretic informal understanding of the object-language too, in order to construct and test the formal semantic theory for it. That is most blatant when the object-language is the semanticist's own native language, and the metalanguage is just a slight extension of it. To some extent, therefore, the theory can also be read in reverse, as constraining worlds and times in terms of the interpreted object-language. Consider, for example, homophonic clauses for negation, conjunction, and disjunction as operators on sentences, where A and B are variables for declarative sentences of some fragment of English, and 'true$_{w,t}$' abbreviates 'true at world w and time t':

---

[1] Whether the world and time parameters are completely independent of each other is unclear. What times there are may depend on which world one is in: time may be finite in some worlds and infinite in others. 'Necessarily' may entail 'always' (Dorr and Goodman 2019). For present purposes, these issues can be left undecided.

Suppose and Tell: The Semantics and Heuristics of Conditionals. Timothy Williamson, Oxford University Press (2020).
© Timothy Williamson.
DOI: 10.1093/oso/9780198860662.001.0001

[not]   'It is not the case that A' is true$_{w,t}$ if and only if A is not true$_{w,t}$.
[and]   'A and B' is true$_{w,t}$ if and only if A is true$_{w,t}$ and B is true$_{w,t}$.
[or]   'Either A or B' is true$_{w,t}$ if and only if either A is true$_{w,t}$ or B is true$_{w,t}$.

Using these clauses and non-modal, non-temporal classical logic in the metalanguage, one can show that classical logic holds for these operators in the object-language with respect to any world and time. For instance, since A is either true$_{w,t}$ or not true$_{w,t}$ (by the usual classical law of excluded middle), 'either A or it is not the case that A' is true$_{w,t}$ for any world w and time t (by [not] and [or]); thus excluded middle holds necessarily always.

Such conclusions are not completely uncontentious in the setting of an intensional semantic framework for conditionals. Their effect is to validate vacuously many provocative counterpossibles, counterfactual conditionals with impossible antecedents, such as (1) (for simplicity, we ignore the slight difference in the grammatical role of 'not' between (1) and [not]):

(1)   If the universe were not either finite or not finite, the universe would be either finite or not finite.

At first sight, (1) looks bad. For even some philosophers and others who accept excluded middle as necessary may reject (1), on the grounds that its antecedent directs us to consider how things would be if excluded middle failed, and so undermines the claim in the consequent, that it still holds in that very case.

The problem is *not* that homophonic clauses such as [not], [and], and [or] by themselves automatically impose classical logic on the object-language. They do not. Many deviant logicians who reject excluded middle can accept such equivalences and simply reject the assumption that A is either true$_{w,t}$ or not true$_{w,t}$. In general, the problem arises for those who *accept* a logical law, but want a non-vacuous treatment of conditionals with antecedents on which the law fails. For homophonic intensional clauses such as [not], [and], and [or] project the law from the actual world and present time to arbitrary worlds and times. They trivialize the question 'What if things were otherwise?' for logical matters.

In order to avoid such trivialization, some philosophers have invoked *impossible worlds*, at which even laws of logic may fail. That can be done without metaphysical outrage, since such a world may simply be identified with a set of sentences of the object-language, treated as the set of sentences true at the world. To take account of the time parameter, such a world may instead be identified with a function mapping each time to a set of sentences, understood as the set of sentences true at the world at the time. The truth clause for such cases will be something like this:

[imp] A is true$_{w,t}$ if and only if $A \in w(t)$.

Uncontroversially, many sets of English sentences contain the sentence 'The universe is not either finite or infinite' but not the sentence 'The universe is either finite or not finite': for example, the singleton set of the former quoted sentence. Thus, by [imp], for many pairs of an impossible world w and a time t, the antecedent of (1) is true$_{w,t}$ while its consequent is not true$_{w,t}$. That creates plenty of scope for a semantics of counterfactual conditionals to falsify (1). Other counterpossibles can be handled similarly.

In a model theory, a similar effect can be achieved by having each model fence off a subset of its worlds as 'impossible' and leaving the evaluation of simple and complex sentences unconstrained at such worlds.[2]

From a semantic perspective, however, [imp] is quite unpromising. It gives up on the central ambition of compositional semantics: to explain how the meanings of complex expressions are determined by the meanings of their constituents and the form of their combination. The clause ignores the meanings of the constituents of A. It even ignores the meaning of A itself. One consequence is that A may be true$_{w,t}$ while a synonymous sentence B is not true$_{w,t}$: [imp] deals only with the linguistic expression itself, not its meaning.[3]

One could impose by hand the constraint that if an impossible world contains a sentence, it also contains any synonymous sentence. But that is unsatisfying. It is ad hoc, merely stipulating something which standard semantic theories *explain*. It also risks undermining the rationale for invoking impossible worlds in the first place: if we want to take impossibilities seriously, why exclude the impossibility in which A but not B is true, where those sentences are in fact synonymous? The next section considers such cases in more detail, from another perspective.

Another challenge for [imp] is whether it is proposed for *all* worlds w or only for some. After all, even a *possible* world determines a set of sentences, so why not prefer a uniform approach, on which [imp] is a universal truth clause? One could still distinguish the 'possible' worlds as those which satisfy conditions such as [not], [and], and [or].

To treat [imp] as the basic, general clause is to give up on serious semantics altogether. By itself, [imp] is explanatorily impotent. The explanatory power of standard compositional semantic theories comes from clauses such as [not], [and], and [or], or more complex variations on them, which characterize the distinctive compositional ways in which the relevant expressions of the object-language work.

One might therefore propose clauses such as [not], [and], and [or] for 'primary' worlds, presumably possible ones, and [imp] for 'secondary' worlds, whether possible or impossible, used for auxiliary purposes, such as the semantics of counterfactual conditionals, at least for counterpossibles. That separation of worlds into two kinds adds a significant layer of complexity to the semantic theory; that layer would need to pay its way by adding significant explanatory power. A further concern is that the meanings of the connectives have been specified only for primary worlds. For

---

[2] For a recent example of this approach, applied to counterpossibles, see Berto, French, Priest, and Ripley, 2018: 695–9; they restrict clauses like [not], [and], and [or] to possible worlds. Their presentation has the virtue of making the ad hoc character of the addition of impossible worlds to the model theory clearly visible.

[3] Here is an illustration of the failure of compositionality in the framework of Berto, French, Priest, and Ripley 2018. Consider a model with just two worlds, the possible @ and the impossible w. Let p and q be two atomic formulas, with the following evaluations at worlds: $v_w(p) = v_w(q) = 0$; $v_@(p) = v_@(q) = 1$; $v_w(\neg p) = 1$; $v_w(\neg q) = 0$. The accessibility relations R_p and R_q for counterfactuals with antecedents p and q respectively are stipulated to hold from @ to w and not to @. Then the evaluation clause for their counterfactual conditional $\Box\!\!\rightarrow$ makes $p \Box\!\!\rightarrow \neg p$ true at @ but $q \Box\!\!\rightarrow \neg q$ false at @. Thus, although p and q have exactly the same semantic profiles in the model, the semantic profiles of their negations differ at the impossible world w, and the semantic profiles of the counterfactuals built out of the respective atoms differ even at the possible world @. This is a clear failure of compositionality.

secondary worlds, there is only [imp], which leaves them completely unspecified. After all, if we had been given [not], [and], and [or] only for some possible worlds w, not for all, the meanings of the connectives would obviously have been specified only in part. Once impossible worlds are invoked, why does the restriction of those semantic clauses to possible worlds not count as a similar failure to specify their meanings in full?

There is also a more general methodological suspicion. To invoke arbitrary impossible worlds is to grant oneself infinitely many degrees of freedom. Doing so may enable one to model almost anything, but for the same reason it typically results in a loss of explanatory power, because so little is excluded. In short, [imp] smells of bad science.

In the light of those concerns, we should be very suspicious of would-be semantic theories which invoke worlds violating conditions such as [not], [and], and [or]. Still, it would be hasty to dismiss such theories outright, without examining the phenomena supposed to motivate them. The rest of this chapter examines those phenomena, concluding that they are better explained in other ways.

15.2 Worlds and epistemic possibility

Some natural readings of some 'would if' sentences are in some sense epistemic. Here is an example.

Two pupils are discussing how to read their teacher's facial expressions. They imagine someone saying that 323 is prime.

(2)   Pupil 1: If she smiled, 323 would be prime. If she frowned, 323 would be composite.

Pupil 2: No, both claims are false. If she smiled, 323 would be composite. If she frowned, 323 would be prime.

For Pupil 2's first correction to work, there must be a world where the teacher smiles and 323 is composite. For Pupil 2's second correction to work, there must be a world where the teacher frowns and 323 is prime. Since mathematics is not contingent, no two metaphysically possible worlds differ in whether 323 is prime. Thus Pupil 2's corrections require a metaphysically impossible world to be in play. The pupils are not foolish enough to think that the teacher can really change arithmetical reality by smiling or frowning. Rather, they are holding fixed both the imaginary initial set-up and the actual epistemic significance of her facial expressions—about which they disagree—while not worrying about the actual mathematical facts. Such examples seem to beg for the postulation of epistemically but not metaphysically possible worlds.

The 'would if' sentences at issue are not even counterpossibles. Beyond dispute, it is both metaphysically possible for the teacher to smile and metaphysically possible for her to frown. One might agree, but still insist that only metaphysically possible worlds—and not all of them—are relevant to 'would'. Since both the initial set-up and the actual epistemic significance of the teacher's facial expressions are being held fixed, the upshot may be that either no world where she smiles or no world where she frowns is contextually relevant, so at least one of Pupil 1's conditionals comes out vacuously true. If so, something is wrong with Pupil 2's corrections, even if Pupil 2 is right about the epistemic significance of the teacher's facial expressions. On the other hand, there is a strong pull towards saying that, in the conversational context, there must be some relevant worlds where she smiles and some where she frowns. To achieve that, postulating epistemically but not metaphysically possible worlds may seem a small price to pay. After all, it may be argued, we need them anyway, in order to give a plausible semantics for propositional attitude ascriptions, irrespective of their role in the semantics of 'would if' sentences.

For example, consider (3):

(3) Jack does not know whether 323 is prime.

On a standard intensional semantics for 'know whether', the truth of (3) requires both an epistemically possible world (for Jack) where 323 is composite and an epistemically possible world (for Jack) where 323 is prime. The latter world is mathematically impossible (323 = 17 × 19). The same point can be made by considering epistemic modals, as in (4), uttered by Jack:

(4) 323 may be prime and it may be composite.

For this strategy to work with enough generality, the epistemically possible worlds will have to include some which are logically, not just mathematically, impossible. For example, consider (5) (with 'it is not the case that' elided for brevity):

(5) Jack does not know whether it is not not not not not not not not not not both raining and not raining.

Depending on whether the string of negations is odd or even in length, the complement of 'whether' in (5) expresses a truth-functional tautology or a truth-functional contradiction. By reasoning like that for (3), on a standard intensional semantics for 'know whether', the truth of (5) requires a world where a truth-functional contradiction holds. The same point can be made by consideration of (6) as uttered by Jack:

(6) It may be not not not not not not not not not not not both raining and not raining and it may be not not not not not not not not not not not both raining and not raining.

Thus we are back to the problems noted in the previous section with rejecting semantic clauses such as [not], [and], and [or].

A familiar strategy for handling examples like (3) and (5) is to insist that they are literally false, but are naturally reinterpreted as ascribing metalinguistic ignorance:

(3*) Jack does not know whether the sentence '323 is prime' expresses a true proposition.

(5*) Jack does not know whether the sentence 'It is not not not not not not not not not not not both raining and not raining' expresses a true proposition.

The same applies to (4) and (6) as uttered by Jack:

(4*)   The sentence '323 is prime' may express a true proposition and it may express a false proposition.

(6*)   The sentence 'It is not not not not not not not not not not both raining and not raining' may express a true proposition and it may express a false proposition.

The underlying ignorance is then a failure to know which proposition the given (syntactically individuated) sentence expresses, a contingent matter.

That metalinguistic strategy is notoriously problematic. In any reasonable sense in which humans normally know which proposition a simple sentence of their native language expresses, Jack can know which proposition the sentence '323 is prime' expresses, yet still be in the sort of state which moves us to assert (3), and him to assert (4). At first sight, the strategy may look more promising for (5) and (6), since there the syntax of the complement sentence is computationally more challenging. The tempting picture is that Jack knows the basic general rules for computing which proposition a sentence of his native language expresses, but has not yet applied them recursively to the specific complex sentence at issue, and so does not yet know which proposition it expresses. The trouble is that those basic general rules will either logically determine that the quoted sentence in (3*) and (5*) expresses a truth-functional tautology or logically determine that it expresses a truth-functional contradiction. Either way, there are not two metaphysically possible worlds compatible with everything Jack knows, including the basic semantic rules, one world where the quoted sentence expresses a true proposition and another world where it expresses a false proposition. Thus the proposed metalinguistic strategy merely postpones the problem rather than solving it.

The general problem does not depend on computational complexity. Consider a standard example of synonymy: the English words 'furze' and 'gorse', which refer to exactly the same natural kind, a type of shrub, without even differing in connotation. Nevertheless, someone can understand both terms without realizing that they refer to the same kind, in circumstances like this: in summer. Robin is taught the word 'furze' on a walk by being shown some bushes, green with yellow flowers; in winter, he is taught the word 'gorse' on a walk elsewhere by being shown other bushes, brown with no flowers. The correlation between the two words and the different appearances of the shrub is purely accidental; he could just as well have been taught the terms the other way round. By normal standards, such ostensive explanations are quite adequate for ordinary linguistic understanding. However, Robin is no expert botanist; he does not realize that the samples all belong to exactly the same natural kind. In describing Robin's state of knowledge, one is inclined to assert (7) while denying (8):

(7)   Robin knows that furze has yellow flowers.
(8)   Robin knows that gorse has yellow flowers.

The temptation is to explain the supposed difference in truth-value between (7) and (8) within an intensional semantic framework by postulating that the sentence 'Furze has yellow flowers' is true at all worlds compatible with what Robin knows, while the sentence 'Gorse has yellow flowers' is false at some worlds compatible with what Robin knows. We can use subscripts to indicate relativization to a world in the intensional semantics. Thus the idea is that for some world w compatible with what Robin knows, 'Furze has yellow flowers' is true$_w$ while 'Gorse has yellow flowers' is not true$_w$. Unless the semantics is altering the meaning of the sentences across worlds, 'Furze has yellow flowers' is true$_w$ just in case the referent$_w$ of 'furze' belongs to the extension$_w$ of 'has yellow flowers', and the referent$_w$ of 'furze' is simply furze, the natural kind it picks out, so 'Furze has yellow flowers' is true$_w$ just in case furze belongs to the extension$_w$ of 'has yellow flowers'. Similarly, 'Gorse has yellow flowers' is true$_w$ just in case gorse belongs to the extension$_w$ of 'has yellow flowers'. Thus, by hypothesis, furze belongs to the extension$_w$ of 'has yellow flowers' while gorse does not belong to the extension$_w$ of 'has yellow flowers'. Therefore, since furze belongs to a class to which gorse does not belong, furze and gorse are distinct, by the logic of identity. But that is absurd, for furze and gorse are the very same kind!

The absurd conclusion was not just that furze and gorse would be distinct in the impossible world w. It was that furze and gorse *are* distinct, in the actual world. For the reasoning was conducted in the standard extensional metalanguage for intensional semantics. It quantifies over worlds, rather than using modal operators. Merely allowing 'impossible worlds' does not explain the difference between (7) and (8). Much more violence has to be done to the structure of intensional semantics to differentiate between (7) and (8) in truth-value.

Another way to put the problem is that the sentences (7) and (8) are arguably *synonymous*, for they differ only in that (7) has the word 'furze' where (8) has its synonym 'gorse'. Since (7) and (8) are built out of constituents with exactly the same meanings put together in exactly the same way, by compositional semantics (7) and (8) have exactly the same meaning too. Thus they also have the same truth-value, in the given context. Something is wrong with taking opposite attitudes to (7) and (8).

The same problem arises for synonymous pairs with epistemic modals, such as (9) and (10):

(9)   Furze may lack yellow flowers.
(10)   Gorse may lack yellow flowers.

One is inclined to judge (9) false and (10) true, as uttered by Robin.

Of course, all this is much-disputed territory. Descriptivists will insist that 'furze' and 'gorse' have different meanings *for Robin*, even if their meaning in the public language is the same. However, the significance of that claim for (7) and (8) is unclear, since the words 'furze' and 'gorse' are used rather than mentioned in them: what matters for the meaning of (7) and (8) is primarily what their speaker or writer means by 'furze' and 'gorse', not what Robin does. After all, knowledge ascriptions like (7) and (8) are often true of people unacquainted with the language in which they are made.

It might be suggested that, at least in the relevant context, the words 'furze' and 'gorse' are implicitly mentioned as well as used in (7) and (8), and in (9) and (10), since the speaker is claiming that Robin knows the truth under the mode of presentation of the complement sentence. That suggestion faces the immediate challenge to provide evidence that the claim about modes of presentation is a genuine entailment rather than a mere conversational implicature. The implicature could be independently predicted to arise in the envisaged circumstances, and such a pragmatic alternative neatly avoids messy semantic complications.

For present purposes, we need not spend time on such descriptivist and quotational proposals. For in the doubtful event that they work at all, they do so without postulating metaphysically impossible worlds. Instead, they rework the contents of the knowledge ascribed by (7) and (8), and denied by (9) and (10), to make them differ in truth-value at some metaphysically *possible* world. On a simple version of the descriptivist proposal, the relevant metaphysical possibility is one where the natural kind with property F has yellow flowers while the distinct natural kind with property G lacks yellow flowers, where Robin actually associates the term 'furze' with property F and the term gorse with property G, and the two properties are possessed actually by the same kind but counterfactually by different kinds. On a simple version of the quotational proposal, the relevant metaphysical possibility is one where the word 'furze' (syntactically individuated) refers to a kind with yellow flowers while the word 'gorse' (syntactically individuated) refers to a kind without yellow flowers. Many more complex and sophisticated variations have been played on the same themes. The point is that such strategies are quite different from those which postulate metaphysically impossible but epistemically possible worlds. The latter strategies interpret attitude ascriptions such as (7) and (8), and (9) and (10), more or less at face value, applying an intensional semantics to them, and employing metaphysically impossible worlds, hoping to deliver the truth-values we first thought of. In the end, the impossible worlds are of no real help in resolving the main difficulty: to make pairs like (7) and (8) or (9) and (10) differ in truth-value on such an approach, one must individuate meanings so finely as to restrict synonymy to self-synonymy, and thereby render the conception of meaning theoretically useless, because it filters nothing out.

Of course, someone who asserts (7) and denies (8) in the envisaged context is responding in a natural way to a genuine difference, as is Robin in asserting (9) and denying (10). But natural responses are often opportunistic, convenient shortcuts, which play fast and loose with strict and literal meanings. A more accurate way of expressing the difference may be to assert (11) and deny (12):

(11)   Robin knows under the guise of the sentence 'Furze has yellow flowers' that furze has yellow flowers.

(12)   Robin knows under the guise of the sentence 'Gorse has yellow flowers' that gorse has yellow flowers.

These involve a three-place epistemic relation between a thinker, a guise, and a proposition (or perhaps a four-place relation, with time as the fourth parameter). Here, the first argument is Robin; the second argument is the sentence 'Furze has yellow flowers' for (11) and the sentence 'Gorse has yellow flowers' for (12); the third argument is the proposition that furze has yellow flowers, which *is* the proposition that gorse has yellow flowers. Thus (11) and (12) can differ in truth-value even though they are identical in their first and third arguments for the relation, because they differ in their second argument. A given thinker can know a given truth under one guise without knowing it under another. But (11) and (12) are cumbersome. Moreover, in many contexts, asserting (7) may conversationally implicate something like (11), and denying (8) may conversationally implicate something like the negation of (12). One sometimes gets one's point across, by saying something literally false. We can leave it to literal-minded pedants to find that surprising.

Correspondingly, Robin can deny (13) and assert (14):

(13)   Under the guise of the sentence 'Furze lacks yellow flowers', it may be that furze lacks yellow flowers.

(14)   Under the guise of the sentence 'Gorse lacks yellow flowers', it may be that gorse lacks yellow flowers.

We can state the literal truths underlying (3) and (4) in similar style:

(15)   Jack does not know under the guise of the sentence '323 is prime' whether 323 is prime.

(16)   Jack does not know under the guise of the sentence 'It is not not not not not not not not not both raining and not raining' whether it is not not not not not not not not not both raining and not raining.

Correspondingly, Jack can state (17) and (18):

(17)   That 323 is prime under the guise of the sentence '323 is prime' may be and may not be.

(18)   That it is not not not not not not not not not not both raining and not raining under the guise of the sentence 'It is not not not not not not not not not not both raining and not raining' may be and may not be.

Again, it is no wonder that we prefer the more concise but less accurate (3) and (4) to (15) and (16) respectively, and likewise (5) and (6) to (17) and (18).

Often, when we play fast and loose with the literal truth, we are aware of doing so, as when we say 'Millions of people came to the party.' But sometimes we lack such awareness. We rely on fallible heuristics, without being aware of doing so—for example, when making perceptual judgements. This book has provided evidence that conditional judgements are another such case. Similarly, once we find that someone lacks a given attitude to a proposition under the guise of a given sentence which they understand, we typically do not bother to check whether they have the attitude to the sentence under some other guise. That method of assessment is comparatively fast and frugal, and in practice fairly reliable, but still fallible. Nevertheless, its fallibility is not transparent to us. We come to recognize its limits only through theoretical reflection.

Given where the argument has led, what should we make of the conversation above between Pupil 1 and Pupil 2? Here it is again:

(2)   Pupil 1: If she smiled, 323 would be prime. If she frowned, 323 would be composite.

Pupil 2: No, both claims are false. If she smiled, 323 would be composite. If she frowned, 323 would be prime.

Suppose that all worlds are metaphysically possible, and mathematics is not contingent. Since 323 is in fact composite, Pupil 1's statement 'If she smiled, 323 would be prime' and Pupil 2's statement 'If she frowned, 323 would be prime' are both either false or vacuously true. One might be tempted to rule out the vacuous truth option, on the grounds that the teacher's facial expressions are contingent. However, that reaction ignores the contextual restrictions on 'would'. They exclude worlds which deviate from the imagined situation in various ways, such as those in which nobody said that 323 is prime. They also exclude worlds which differ from the actual world in various ways, such as in what the teacher's facial expressions mean, for the conversation is about their actual meanings. Together, the restrictions may eliminate all worlds where the teacher smiled, or all worlds where she frowned (or neither). In the first case, no worlds are left where the teacher smiled, but presumably there are some where she frowned; thus both Pupil 1's statement 'If she smiled, 323 would be prime' and Pupil 2's statement 'If she smiled, 323 would be composite' are vacuously true; Pupil 1's statement 'If she frowned, 323 would be composite' is non-vacuously true, while Pupil 2's statement 'If she frowned, 323 would be prime' is false. In the second case, no worlds are left where the teacher frowned, but presumably there are some where she smiled; thus both Pupil 1's statement 'If she frowned, 323 would be composite' and Pupil 2's statement 'If she frowned, 323 would be prime' are vacuously true; Pupil 1's statement 'If she smiled, 323 would be prime' is false, while Pupil 2's statement 'If she smiled, 323 would be composite' is non-vacuously true. In both of these cases, Pupil 2's statement 'Both claims are false' is itself false. Speakers may fail to allow for the vacuous case; that should not be surprising, when the focus of the conversation is not on the non-contingency of mathematics. The absence of meta-physically impossible worlds may have led Pupil 2 into error, but not *inexplicable* error.

So far in this chapter, we have been considering attempts within the framework of compositional intensional semantics to accommodate speakers by postulating epistemically possible but metaphysically impossible worlds. Such attempts tend to trivialize the intensional conception of meaning. They look like yet more cases where trying to load the epistemology into the semantics results in bad semantics, as well as bad epistemology. Impossible worlds provide a lazy fix, covering up the deeper cognitive nature of the problems. We do better by not trivializing the semantics; instead, we can impute explicable mistakes to speakers. One motivation for postulating metaphysically impossible worlds then falls away. The postulate will be revisited later, in discussion of claims and examples from other philosophers, but they will not revive its plausibility.

15.3 Epistemic readings of 'would'

Questions of epistemic possibility naturally arise in the context of apparently epistemic readings of 'would if' sentences. It is therefore worth considering some more examples of such readings. Dorothy Edgington has given several. Here is one (Edgington 2008: 16–17, borrowing and adapting from Grice 1989):

> There is a treasure hunt. The organizer tells me 'I'll give you a hint: it's either in the attic or the garden.' Trusting the speaker, I think 'If it's not in the attic it's in the

garden.' We are competing in pairs: I go to the attic and tip off my partner to search the garden. I discover the treasure. 'Why did you tell me to go to the garden?' she asks. 'Because if it hadn't been in the attic it would have been in the garden: that's (what I inferred from) what I was told.' That doesn't sound wrong in the context.

The relevant restriction on 'would' is to worlds compatible with what the narrator knew when she told her partner to go to the garden: that is why the reading feels epistemic. If the organizer had irresponsibly spoken at random, having forgotten where he hid the treasure, the narrator's claim 'If it hadn't been in the attic it would have been in the garden' would have been incorrect. A contrasting reading, unrelated to the narrator's epistemic state, is also available, though conversationally less relevant: if the organizer was secretly determined all along not to hide the treasure in the garden, because it would be damaged if it got wet, he could truly say to himself 'If it hadn't been in the attic it would not have been in the garden.'

Edgington gives another example, this conversation (Edgington 2008: 17):

'Why did you hold Smith for questioning?' 'Because we knew the crime was committed by either Jones or Smith—if it hadn't been Jones, it would have been Smith'

Here the restriction on 'would' is to worlds compatible with what 'we' (including the second speaker) knew when holding Smith for questioning. The second speaker's claim 'If it hadn't been Jones, it would have been Smith' may be correct even though Smith later turned out to be of exemplary law-abiding character.

Edgington also describes an example from van Fraassen (1981):

[T]he conjuror holds up a penny and claims he got it from the boy's pocket. 'That didn't come from my pocket', says the boy. 'All the coins in my pocket are silver. If that had come from my pocket, it would have been a silver coin.'

Here the restriction on 'would' is to worlds compatible with what the boy knows. On the assumption that the coin's being non-silver is essential to it, and that the boy is using 'that' as a rigid designator, in no metaphysically possible world is that coin silver. But that does not matter, for the boy is explaining why it is incompatible with what he knows for the coin to have come from his pocket. For that purpose, a vacuously true modalized conditional will do fine. The boy knows both that all coins from his pocket are silver and that this coin is not silver; thus it is incompatible with what he knows that this coin is from his pocket. He says 'If that had come from my pocket, it would have been a silver coin' to show that the antecedent is false in all worlds where everything he knows is true. For that purpose, it is irrelevant whether the consequent is false in some worlds where *not* everything he knows is true; it does not matter whether the coin is essentially non-silver.

Barbara Vetter (2016: 781) provides an example where the apparent violation of essence is in the antecedent rather than the consequent:

Imagine that, while walking in the wilderness, we enter a clearing and something runs past us from an unexpected angle. In fact, it is a harmless gazelle; but we both know that there are dangerous tigers about. Shocked, you say to me:

(19)   If that had been a tiger, we would be dead now.

(Numbering changed.) On even a mild form of essentialism, the gazelle could not have been a tiger. Still, as Vetter points out, the counterfactual may serve as a useful reminder to be warier in the future; how can it do so if it is vacuously true, and both participants in the conversation know that?

It is not clear that the demonstrative pronoun 'that' is being used as a rigid designator in (19). Some evidence for that hypothesis is that (19*) does not sound equivalent to (19) in the context:

(19*)   If that very animal had been a tiger, we would be dead now.

On an alternative non-rigid reading of (19), 'that' denotes whatever animal appeared unexpectedly as they entered the clearing, with respect to counterfactual possibilities in which one did, so the antecedent of (19) is metaphysically possible ('that' in van Fraassen's example may have a parallel non-rigid reading). By contrast, 'that very animal' in (19*) is more resistant to such a non-rigid reading. Similarly, (20) sounds much better than (20*):

(20)   That could have been a tiger.
(20*)   That very animal could have been a tiger.

The former alerts us to the same danger as (19), whereas the latter sounds like an implausible anti-essentialist claim. The danger is a straightforward physical possibility, not a metaphysically impossible epistemic possibility. It is no mere apparent danger.

In general, the semantic case for invoking worlds of a special kind to handle epistemic readings of 'would' looks flimsy. If invoked unselectively, they undermine the compositionality of the semantics. If invoked selectively, they provide only a partial solution, whereas the alternative approach used to solve the remaining cases is also applicable to the cases for which impossible worlds were invoked. An epistemically motivated restriction on worlds does not imply anything epistemic in the nature of the worlds themselves. Compare the world parameter of evaluation with the time parameter. An epistemically motivated restriction on *times* does not imply anything epistemic in the nature of the times themselves. This exchange is an example:

(21)   Q: What was he like, as far as you knew him?
       A: When he was not boasting, he was complaining.

In this context, 'when' is restricted to times when the speaker experienced the man under discussion, an epistemically motivated restriction. But 'when' still ranges over ordinary times, not special 'epistemic' times. Given the evidence so far, the role of worlds in epistemic readings of 'would' looks similar.

15.4 Objective modality

Despite the talk of metaphysically possible worlds, in this book metaphysical modality has so far been characterized quite schematically. A few constraints have been suggested: the structure of a standard intensional semantics was seen to validate the eternal necessity of classical logic in the tensed modal object-language, given classical logic in the untensed non-modal metalanguage.

Similar considerations can be taken a little further, in particular to the eternal necessity of identity and distinctness. More specifically, assume, as is natural, that the semantics evaluates the open identity formula '$x = y$' as true with respect to a world w, a time t, and an assignment $\underline{a}$ of values to variables just in case $\underline{a}('x') = \underline{a}('y')$, and false otherwise. Then, trivially, on any given assignment, '$x = y$' is either true with respect to every world and time or false with respect to every world and time: identity and distinctness are neither temporary nor contingent. Even if exceptions are made for worlds and times at which the value of a variable is supposed not to exist, some mildly qualified version of the eternal necessity of identity and distinctness will still hold. Since identity and distinctness are in general objective matters, not up to us to stipulate, necessary identity and necessary distinctness are equally objective. That indicates an objective quality of the modality itself.

However, such considerations do not take us all the way to the thick conception associated with the terms 'metaphysical necessity' and 'metaphysical possibility' in contemporary philosophy. For example, they do not suffice to vindicate rich essentialist claims about origins and kind-membership of the sort advanced by Kripke in *Naming and Necessity*. In those respects, the intensional semantics is neutral (Salmon 1982).

On some thick conceptions of metaphysical modality, not all the worlds required for the intensional semantics are metaphysically possible. For it is tempting to judge that there is some flexibility, but not total flexibility, in how a given object is in various respects, for instance in the constituents of which a given artefact was originally made. Schematically, the idea is that, necessarily, if a given object had a given origin, then it could have had a slightly different origin, but could not have had a radically different origin, and furthermore that the necessities and possibilities at issue are metaphysical. Call a world w *metaphysically accessible* from a world v if and only if in v, w is metaphysically possible. Under precise assumptions about modal flexibility of the sort just sketched, one can show that there is a sorites series of worlds $w_0, w_1, \ldots, w_n$, where w_0 is the actual world, w_{i+1} is metaphysically accessible from w_i whenever $0 \leq i < n$, and w_n is metaphysically inaccessible from w_0 (Salmon 1989). On that view, metaphysical accessibility is non-transitive. Correspondingly, that it is metaphysically possible that it is metaphysically possible that a state of affairs S obtains does not entail that it is metaphysically possible that S obtains; the S4 axiom of modal logic fails for metaphysical modality. To build the intended model within the framework of the intensional semantics, whenever one includes a world w in the model, one must also include all the worlds metaphysically accessible from w, in order to evaluate sentences with metaphysical modal operators properly at w. Just which sentences are crucial depends on the details of the assumptions about modal flexibility. Thus, once the actual world w_0 is included in the model, successive members of the sorites series must also be included, finally including the world w_n, even though it is metaphysically impossible (in the actual world). In short, the intensional semantics for metaphysical modality needs metaphysically impossible worlds—according to the argument.

The argument does not suggest that anything goes. It does not motivate postulating worlds where classical logic fails, or actually identical things are distinct, or actually distinct things identical. The worlds at issue violate the thick conception of metaphysical modality, but not the thin conception.

Of course, the modal flexibility assumptions may be challenged. For present purposes, however, that dispute is irrelevant. In this book, metaphysical modality has been understood more thinly, as something like the maximal objective modality; this is reminiscent of Kripke's phrase 'necessary in the highest degree' (1980: 99; he adds 'whatever that means'). Thus if $\Diamond$ represents any kind of objective possibility, and $\Diamond_M$ represents metaphysical possibility, then $\Diamond p$ entails $\Diamond_M p$. Moreover, if $\Diamond_1$ and $\Diamond_2$ represent any kinds of objective possibility, then $\Diamond_1 \Diamond_2$ also represents a kind of objective possibility. Consequently, since metaphysical modality is itself a kind of objective modality, $\Diamond_M \Diamond_M$ also represents a kind of objective modality, so $\Diamond_M \Diamond_M p$ entails $\Diamond_M p$. In other words, the S4 axiom automatically holds for metaphysical modality, so understood (Williamson 2016a). At best, the modal flexibility assumptions hold for some more restricted objective modality. For example, the world w_n is metaphysically possible but not possible in some more restricted sense.

Given the prominence of specific essentialist claims in the literature on 'metaphysical' modality, some may argue that the richer conception has the best right to govern the use of the term. This book follows the opposing thought that we should use our technical terms to mark the distinctions of most theoretical significance. Thus 'metaphysical' modality will be equated with the universal modality in the intensional semantics, which quantifies over all worlds in the model. Such worlds satisfy the thin logical conception of metaphysical possibility. For present purposes we can remain neutral on whether they also satisfy the thick conception.

In principle, the thin conception leaves open an alternative treatment of some apparent counterpossibles, on which their antecedents count as metaphysically possible after all. For instance, in Vetter's (19), 'If that had been a tiger, we would be dead now', we might treat 'that' as rigid and still propose that the gazelle could have been a tiger, in the metaphysical sense, perhaps constructing a long sorites series of evolutionary variations from one scenario to the other. However, even putting plausibility aside, such moves would not get us very far, because there are usually variant examples which violate the thin conception of metaphysical possibility as well as the thick one, for instance by flouting the necessity of distinctness rather than the necessity of kind-membership ('If that had been Tigger, we would be dead now'). It is better to go straight to the heart of the issue by simply granting the metaphysical impossibility of the antecedent in question, at least for the sake of argument.

One consequence of the universality of metaphysical modality is that in any context 'would' restricts metaphysical necessity: whatever is metaphysically necessary would be the case. Thus when A expresses something metaphysically impossible, so 'If A, C' expresses something metaphysically necessary (given the material reading of 'if'), 'Would(if A, C)' expresses a truth (this principle is RESTRICTION in section 14.2). In other words, counterpossibles are vacuously true. Although this result follows from most standard semantic accounts of counterfactuals, many philosophers still find it implausible, and are willing to complicate the semantics to avoid it, typically, by invoking worlds that violate even the thinnest conception of metaphysical possibility. Moreover, some constructive philosophical theories rely on supposedly non-vacuous counterpossibles. The issue is of sufficient philosophical significance to merit extended discussion. The next section considers various alleged clear counterexamples to the vacuity thesis, to test whether they really demand metaphysically impossible worlds, or can be adequately understood in terms of the cognitive dispositions already discussed.

15.5 Counterpossibles

We start with a standard pair (derived from Nolan 1997). What are the truth-values of (22) and (23)?

(22)   If Hobbes had secretly squared the circle, sick children in the mountains of South America at the time would have cared.

(23)   If Hobbes had secretly squared the circle, sick children in the mountains of South America at the time would not have cared.

Pre-theoretically, the natural snap answer is that (22) is false and (23) true. The sick children in the mountains of South America were in no position to know about Hobbes's secret geometrical reasoning thousands of miles away; even if they had known, they had far more pressing things to care about. But, as was proved in the nineteenth century, squaring the circle is mathematically impossible. Thus (22) and (23) are both counterpossibles.

Notably, the natural responses to (22) and (23) do not come from considering the modal status of the antecedent. The sick children would not have cared about any geometrical reasoning Hobbes secretly got up to. That is already worrying news for theories of counterfactual conditionals which build a bifurcation between counter-possibles and non-counterpossible counterfactuals into their semantic framework, since it indicates a failure of sensitivity to patterns in our pre-theoretic cognitive processing of (22) and (23). That goes some way towards undermining the authority of our pre-theoretic verdicts on them.

Given the analysis in chapter 10, (22) and (23) have the overall form of (22*) and (23*) respectively:

(22*)   Would(if Hobbes secretly squared the circle, sick children in the mountains of South America at the time cared)

(23*)   Would(if Hobbes secretly squared the circle, sick children in the mountains of South America at the time did not care)

The modal operator 'would' is restricted to contextually relevant worlds. Here, the salient contextual feature is just the presentation of (22) and (23), with the matters they raise. Under that constraint on relevance, one can apply the standard Suppositional Rule SR to the conditionals embedded within (22*) and (23*). On the modally distanced supposition that Hobbes secretly squared the circle, one asks oneself: did sick children in the mountains of South America at the time care? The obvious answer is 'No'. It comes immediately, with no need to consider the details of Hobbes's secret geometrical reasoning. Thus one accepts the consequent of the embedded conditional in (23*), and rejects the consequent of the embedded conditional in (22*), on the supposition of the shared antecedent. Applying the Suppositional Rule, one therefore accepts the embedded conditional in (23*), and rejects the embedded conditional in (22*). All that is still under the restriction contextually associated with 'would', but, using the closely related heuristic MCSR explained in section 11.4, one finally discharges that restriction by accepting (23)/(23*) and rejecting (22)/(22*) themselves. Thus our independently confirmed reliance on those heuristics predicts and explains our pre-theoretic verdicts on (22) and (23).

Authors who treat our pre-theoretic verdicts on such counterpossibles as obviously correct seem happy enough with the observation that we assess them by assessing the consequent on the supposition of the antecedent, even if they do not analyse such sentences as contextually restricted strict conditionals like (22*) and (23*). What they do not focus on is the need for some cognitive mechanism to get *from* the assessment of the consequent on the supposition of the antecedent *to* an assessment of the whole conditional outside the supposition of the antecedent, which is exactly what the suppositional rule MCSR supplies. As far as I know, they suggest no alternative to that rule.

Like other commonly used heuristics, the suppositional rule is not perfectly reliable. We saw in chapter 3 that it is inconsistent. Closer to the present case, we saw in section 11.4 that the modal suppositional rule MCSR gives logically inconsistent verdicts for some counterpossibles with logically inconsistent antecedents. Thus to reject the vacuous truth of counterpossibles on the basis of pre-theoretic verdicts on examples like (22) is in effect to rely on a heuristic in just the sort of cases where it can be shown to give incorrect results. That naïve attitude is not good enough. To avoid being suckered by our own heuristics, we must be willing to approach the data in a more critical spirit.

To reinforce the point, we may consider some similar examples, offered by Cian Dorr as 'manifestly false counterfactuals whose antecedents seem to be metaphysically impossible' (2008: 37):

(24)   If I were a dolphin, I would have arms and legs.

(25)   If it were necessary that there be donkeys, it would be impossible that there be cows.

(26)   If there were unicorns, none of them would have horns.[4]

We can take a first step towards analysing them in the same way as before:

(24*)   Would(if I am a dolphin, I have arms and legs).

(25*)   Would(if it is necessary that are donkeys, it is impossible that there are cows).

(26*)   Would(if there are unicorns, none of them have horns).

All three examples pattern like (22)/(22*).

---

[4] For (26), Dorr cites Kripke's influential argument for the impossibility of unicorns (1980: 156).

With (24)/(24*), the envisaged negative verdict clearly depends on imaginatively holding fixed the nature of dolphins while adjusting the nature of the speaker, and not vice versa. Under such a contextual restriction for 'would', one supposes that the (actual) speaker is a dolphin, and asks: have they arms and legs? The obvious answer is 'no'. The nature of dolphins is to have flippers, not arms and legs. From that point, everything proceeds as for (22)/(22*).

With (25)/(25*), under whatever contextual restriction is appropriate for 'would', one supposes that it is necessary that there are donkeys, and asks on that supposition: is it impossible that there are cows?[5] The answer feels less obvious than in the previous cases, but it is natural to hold imaginatively fixed the assumption that different types of farm animal have the same general modal status, so switching the existence of donkeys from contingent to necessary would if anything switch the existence of cows from contingent to necessary too, rather than switching it to impossible. With that assumption built into the contextual restriction, the natural answer to the question is indeed 'no'. From that point, everything again proceeds as for (22)/(22*).

With (26)/(26*), it is natural to hold imaginatively fixed the assumption that unicorns have horns. Then, under the appropriate contextual restriction for 'would', one supposes that there are unicorns, and asks under that supposition: do no unicorns have horns? The obvious answer is 'no'. From that point, everything yet again proceeds as for (22)/(22*).

Thus none of Dorr's examples fares substantially better than the original one. In each case, the negative verdicts are predicted by the hypothesis that we are using the fallible heuristic MCSR—in just the sorts of cases for which it is known to generate inconsistencies. Of course, even when one focusses on the impossibility of the antecedents, it does not become *obvious* that (22), (24), (25), and (26) are true. Instead, they may just feel harder to assess. One may wonder what the point is of supposing the impossible. But those are symptoms of lost confidence in the suppositional procedure. Even as a native speaker, one is at a loss to see what is going on. The semantic facts are not transparent to one. But that does not mean that they are beyond the reach of theoretical inquiry.

Berto, French, Priest, and Ripley (2018) give a couple of interestingly different examples:

(27) If it were raining and not raining, it would be Tuesday.
(28) If it were raining and not raining, it would not be Tuesday.

According to them, both (27) and (28) 'appear to be false' (2018: 707). That seems to exceed the natural pre-theoretic response to the examples. Rather, both (27) and (28) appear pre-theoretically to be unassertible, because the consequent seems unassertible on the basis of the antecedent. That is also the reaction one would get from

---

[5] The intended readings of 'necessary' and 'impossible' in (25)/(25*) are metaphysical, not epistemic; on an epistemic reading, it is not counterpossible.

applying the suppositional rule MCSR. First, we take the usual first step towards analysing the counterfactuals:

(27*)   Would(if it is raining and not raining, it is Tuesday).
(28*)   Would(if it is raining and not raining, it is not Tuesday).

Then, under an appropriate contextual restriction, one makes the modally distanced supposition that it is raining and not raining, and asks on that supposition: is it Tuesday? One possibility, which the authors do not consider, is simply to treat the supposition as irrelevant to the question and to answer 'Yes' or 'No' depending on whether it actually is or is not Tuesday, in which case one would take asymmetric attitudes to (27) and (28). Evidently, that is not what they have in mind, and it ignores the modal distancing of the supposition. A more natural response to the question may be 'Don't know', because the supposition says nothing about which day of the week it is. Applying MCSR, one would then extend one's agnosticism from the consequent on the modally distanced supposition of the antecedent to the whole modalized conditional on no supposition. That explains the pre-theoretically natural unwillingness to assert either of (27) and (28).

Of course, someone who knows a little classical logic will notice that the antecedent of the embedded conditionals in (27*) and (28*) entails their consequent. Like Nolan (1997), Berto, French, Priest, and Ripley are careful not to rest their joint discussion of counterpossibles on objections to classical logic, even those which in other settings they endorse. On theoretical reflection, one might well take those entailments as decisive reason to accept both (27)/(27*) and (28)/(28*). But that is not a natural first reaction to them. Even after one is accustomed to it, accepting them feels awkward and artificial. But that is hardly surprising. For many practical purposes, we need to be able to process conditionals fluently in real time. That is best achieved by our heuristics for processing them being hardwired into us (which does not mean that we cannot consciously inhibit our assent to their verdicts). A moment's reflection on a conditional is liable to trigger the suppositional procedure. With counterpossibles like those under consideration, the result will be to sound an alarm. The hard lesson is that instincts are not always to be trusted.

15.6 Comparison with quantifiers

'Would' is a covert restricted universal quantifier over worlds. Consequently, there is a natural comparison between conditionals with 'would' and sentences with overt universal quantifiers instead, for example, between 'If this sugar lump were put in coffee, it would dissolve' and 'All sugar lumps put in coffee dissolve.' Indeed, we often apply a kind of suppositional procedure in assessing quantified sentences as well as conditional ones: imagine a sugar lump put in coffee and ask yourself whether it will dissolve. Of course, we have other ways of assessing universally quantified sentences too, most notably by assessing particular instances of them—negatively, by a counter-instance, and positively, by complete enumeration of an easily surveyed collection ('Everyone in the room is wearing trousers'). But for universal quantification over large or scattered or open or partially or wholly inaccessible collections, we may find ourselves with no feasible alternative to some kind of suppositional procedure. To gain a further perspective on counterpossibles, we can therefore compare them with the vacuous case for universal quantification, sentences of a form such as 'Every F is a G' when there are no Fs.

For example, since there are no dolphins in Oxford, we can compare (24) ('If I were a dolphin, I would have arms and legs') with a sentence like this:

(29)   Every dolphin in Oxford has arms and legs.

At first hearing, (29) sounds wrong. 'Dolphins have flippers, not arms and legs!', one wants to protest, and then 'Anyway, there are no dolphins in Oxford'. Some people will dismiss (29) as false, others as lacking in truth-value, but each classifications is already a step towards theorizing the initial felt resistance.

The semantic status of vacuous universal generalizations was long a matter of controversy. The predominant contemporary view is that they are all true: the absence of a counter-instance is enough (Peters and Westerståhl 2006: 124–7). Amongst medieval logicians, the predominant seems to have been that they are all false: 'Every F is a G' entails 'Some F is a G', as Berto, French, Priest, and Ripley 2018 emphasize. The modern view is best for streamlining the logic. Unlike its medieval alternative, it makes 'Every F is an F' trivially true. The modern view corresponds to doing set theory with the subset relation $\subseteq$, rather than the relation which holds between X and Y if and only if X is a subset of Y *and* X is non-empty; working day-to-day with the latter instead of the former would cause a host of minor complications. In practice, one would define $\subseteq$ in terms of the other relation and then work with $\subseteq$, which is exactly the point. The third option, that 'Every F is a G' lacks a truth-value when there are no Fs (Strawson 1952:163–79), makes a far worse mess of the logic. Such alternative views are systematically violated by mathematicians writing in their native language. In ordinary conversation, one is normally expected to alert one's partners when one suspects a generalization of being vacuous, but that can be explained on general Gricean grounds, removing the motivation to postulate truth-value gaps. In short, the standard and best view of vacuous universal generalizations is the analogue of the view that counterpossibles are all false.

Not all methodological implications of the analogy with vacuous universal generalizations depend on the superiority of the modern view. The long-standing controversy over their semantic status already illustrates how the semantics of basic logical particles can be opaque to intelligent native speakers, even on extended reflection. Whoever is right, their opponents are systematically misevaluating simple sentences of their native language.

There is also a more specific point. All three main views of vacuous universal generalizations are *uniform*: they assign them all the *same* semantic status, whether it be true, false, or gappy. That is hardly surprising, since it is well established that quantifiers in natural language are *conservative*, in the sense that the truth-value of $Q(F, G)$ depends only on whether the set-theoretic relation R_Q corresponding to the quantifier Q holds between the set of Fs and the set of Fs which are also Gs (Peters and Westerståhl 2006: 138, assuming a set-sized domain); the universal quantifier is no exception. But if there are no Fs, there are no Fs which are also Gs, so both sets are the empty set $\{\}$, and the truth-value of $Q(F, G)$ depends only on whether $\{\}$ has R_Q to tself, which is independent of the meanings of 'F' and 'G'. Informally, there are no nstances for the truth-value to depend on.

By contrast, those who deny that counterpossibles are all true normally defend a *non-uniform* treatment of them. They do not claim that counterpossibles are all false, or all truth-valueless. Rather, they insist that some are false (for example, (22)), while others are true (for example, (23)), in line with their pre-theoretic reactions to cases. But pre-theoretic reactions to cases of vacuous universal generalizations also seem to support a non-uniform treatment. For example, this generalization sounds much better than (29):

(30)   Every dolphin in Oxford has flippers.

Similarly, (32) sounds much better than (31):

(31)   Every unicorn is hornless.
(32)   Every unicorn has a horn.

Mutilated or malformed unicorns aside, (32) sounds true and (31) false. But, for vacuous universal generalizations, there is widespread agreement that such apparent evidence of semantic non-uniformity should be set aside, whereas for counterpossibles very similar evidence is taken at face value.

Berto, French, Priest, and Ripley (2018: 708) have suggested that the analogy between counterpossibles and vacuous universal generalizations is question-begging, because counterpossibles are *not* vacuous in a framework with impossible worlds. But that misses the methodological point. Once one acknowledges that (29) and (31) have the same semantic status as (30) and (32), despite the difference in our pre-theoretic reactions to the former pair and the latter, one should not treat a similar difference in our pre-theoretic reactions to counterpossibles like (22) and those like (23) as compelling evidence for a difference in semantic status between the former counterpossibles and the latter. Of course, Meinongians may try to restore the analogy, by postulating non-existent dolphins in Oxford and unicorns to falsify (29) and (31) and so denying that (29)–(32) are vacuous. But most proponents of impossible worlds aim to show that their explanations have no need of extravagant metaphysics.

More soberly, we can rely on our orthodox contemporary theoretical grasp of universal generalizations, as far more secure than our theoretical grasp of counterfactuals. Thus vacuous universal generalizations are all true, despite appearances to the contrary, and we should be open to theoretical considerations which imply that all counterpossibles are true, despite similar appearances to the contrary. Both illusions are well explained by our pre-theoretic reliance on fallible heuristics of closely related kinds. To sacrifice properly compositional semantics for counterfactual conditionals in order to vindicate our naïve pre-theoretic assessments of examples would be a fool's bargain.

How analogues of the suppositional procedure apply to non-modal generalizations requires much fuller investigation. Notably, such heuristics seem especially well-suited to the assessment of *generic* generalizations, like 'Birds fly', whose truth tolerates exceptions, like penguins, by contrast with 'Every bird flies'. A natural paraphrase of 'Birds fly' is 'If it's a bird, it flies'. Although the contrast between generic generalizations and universal generalizations is more salient for falsification than for verification, it affects the latter too: asserting the universal generalization 'Every *F* is a *G*' requires a more secure epistemic link from the supposition 'It's an *F*' to the conclusion 'It's a *G*' than does asserting the generic generalization '*F*s are *G*s'. For both sorts of generalization, we should expect the link to provide knowledge only in favourable conditions. As theorists of generics have noted, tying their truth-conditions too tightly to the conditions under which they are commonly asserted would put semantics in the service of bigots, by laundering their prejudices as generic truths (Sterken 2015 makes this point). That provides another salutary warning of the dangers inherent in not properly separating heuristics from semantics.

15.7 Philosophical repercussions

In linguistic practice, counterpossibles are a comparatively minor phenomenon, which is one reason why it is implausible to complicate the semantics of modalized conditionals in natural language just to achieve a desired outcome for them: the extra cognitive load would be inefficient, a special design to let the tail wag the dog. In some philosophical debates, however, counterpossibles have played a key role. Chapter 14 noted their relevance to the treatment of thought experiments in contemporary metaphilosophy. In contemporary metaphysics, some constructive strategies have relied crucially on the assumption that some, but not all, counterpossibles are false. It is instructive to work through an example.

Nominalists crave the scientific advantages which Platonists gain from quantifying over numbers, sets, and other abstract objects. How to emulate them? A common strategy, in this and similar cases, is *fictionalist*. One treats the envied rival metaphysical theory as a useful fiction. The proposal deserves to be taken seriously only if accompanied by a properly worked-out account of how intermediate reasoning on the basis of a fiction can nevertheless constitute a reliably truth-preserving way from non-fictional premises to a non-fictional conclusion. For instance, if one reasons validly from true premises purely about concrete reality combined with a false (by nominalist lights) auxiliary mathematical theory about abstract objects to a conclusion purely about concrete reality, the conclusion needs to be true too (Field 1980). But why should the nominalist expect it to be true?

One way to implement the fictionalist strategy is by using counterfactual conditionals. In effect, the nominalist reasons about *how things would be if the mathematical theory obtained and concrete reality was just as it actually is*. Let *M* be the Platonist mathematical theory, *A* a complete description of concrete reality in all relevant respects (in effect, the statement that concrete reality is just as it actually is), and *C* the conclusion purely about concrete reality. Thus the Platonist reasoning delivers this conditional:

(33) $(M \wedge A) > C$

The idea underlying the fictionalist strategy is that the truth of the counterfactual (33) guarantees the truth of its consequent *C*, because the relevant concrete worlds coincide with the actual world over concrete reality, which *C* is purely about—even though, by nominalist lights, the antecedent of (33) is false because its conjunct *M* is.

The trouble is that nominalists typically regard Platonism as not just *false* but *metaphysically impossible*: for instance, there seems to be no contingency in the hierarchy of pure sets or the sequence of natural numbers, if there are such. For such nominalists, M is impossible, so the counterfactual (33) is a counterpossible. But then, if all counterpossibles are true, (33) is true irrespective of the specific content of C. In particular, the opposite counterpossible (34) is also true:

(34) $(M \land A) > \neg C$

In that case, the truth of (33) cannot guarantee the truth of C, since if it did the truth of (34) would guarantee the truth of $\neg C$, by parity of reasoning. Since C is purely about concrete reality, so is $\neg C$. Thus fictionalists who implement their view by the counterfactual strategy and regard the rival metaphysical theory as a useful but impossible fiction have been compelled to assume that some counterpossibles, such as (34), are false, while others, such as (33), are true (for instance, Dorr 2008).

The logical situation can be articulated more exactly. The natural route from (33) to C is this. Suppose that C is true. Since A is a complete description of concrete reality in all relevant respects, and C is a truth purely about concrete reality, C is a necessary condition of A and so of the conjunction $M \land A$; hence (33) is true. Thus C entails (33). Moreover, by parallel reasoning, $\neg C$ entails (34). By contraposition, therefore, $\neg$(34) entails C. Consequently, if opposite counterpossibles such as (33) and (34) are mutually incompatible, (33) entails $\neg$(34), so by transitivity (33) entails C. Consequently, (33) is equivalent to C. But, on the orthodox view that all counterpossibles are true, (33) and (34) are mutually compatible (if M is impossible).

From a semantic perspective, the fictionalist counterfactual strategy looks very unpromising: opposite counterpossibles are mutually compatible. Fictionalists will have to scratch around elsewhere for some alternative way of legitimating scientists' reliance on the alleged impossible fictions.

16

Conclusion
Semantics, Heuristics, Pragmatics

In the methodology of contemporary semantics, one may observe a widespread though far from universal tendency to overcomplicate. In well-meaning attempts to fit all the data, elaborate, rickety semantic structures are attributed to the ordinary nuts and bolts of natural language. For cognitive purposes, the whole edifice looks remarkably inefficient. This postulated complexity is often treated as an indicator of theoretical prowess, rather than a warning sign of a degenerating research programme. Of course, there is plenty of evidence from other branches of linguistics that natural languages are much more complex than they first appear. Nevertheless, semantics could benefit from greater awareness of the danger of what natural scientists call *overfitting*: the willingness to add extra parameters to an equation until its curve goes almost exactly through all the data points. That may sound like a laudably empirical attitude, but hard experience shows its typical result to be theoretical instability. As new data points come in, still more parameters have to be added to the equation, and their coefficients oscillate wildly.

With his theory of conversational implicature, Paul Grice demonstrated how many linguistic effects are better explained pragmatically, in terms of interactions between quite simple truth-conditional semantics and very general principles of sociolinguistic dynamics, rather than having to be separately written into the semantic clauses for each relevant term by hand. In particular, he defended the simple truth-functional semantics for 'if' along such lines. However, it is widely accepted that, despite the numerous explanatory successes of Grice's theory, his account of the conditional is not one of them. Although it can reconcile the material interpretation with some of the data, his account leaves other aspects of the evidence unexplained. For example, since it attributes the same truth-condition to 'If A, C' and 'Not(A) or C', it cannot explain why the two types of sentence behave so differently in terms of interactions between the truth-conditional semantics and general principles of conversational dynamics. Grice's theory is ill-suited to analysing cases where identity of truth-conditions is deeply hidden from speakers and hearers. Semantics and pragmatics together are not enough.

Pragmatic mechanisms are not the only ones which can mediate between semantics and unreflective native speaker verdicts on sentences. Unconscious reasoning may play a role at this stage, as well as later. Normally, it operates on semantically interpreted structured representations, rather than residues of semantic abstraction such as sets of possible worlds. Such reasoning often involves *heuristics*.

Suppose and Tell: The Semantics and Heuristics of Conditionals. Timothy Williamson, Oxford University Press (2020).
© Timothy Williamson.
DOI: 10.1093/oso/9780198860662.001.0001

Recognizing whether a well-understood sentence correctly describes a given hypothetical case is a far from trivial cognitive task. To accomplish such tasks efficiently, humans predictably resort to fast and frugal heuristics, reliable enough under normal conditions, but not perfectly reliable. Such heuristics may be built into our natural ways of thinking, with no warning of their limitations written on the packet. Some perceptual heuristics are like that. When our unreflective processing of natural language turns out to involve such heuristics too, it should occasion no surprise. In this book, I have identified the heuristics central to unreflective use of 'if' and 'would', which enable us to exploit our capacity for supposing and imagining to good cognitive effect, in ways which are reliable enough under normal conditions, but not perfectly reliable. They cannot be perfectly reliable, because they are in principle inconsistent, even though in practice we usually manage to steer clear of the lurking inconsistencies, without even suspecting their presence.

Just as observational data are often mediated by heuristics in our perceptual systems, so the data on which we rely in semantics are often mediated by heuristics in our linguistic systems. In neither case does that warrant total distrust of the data, but it does recommend a critical attitude. Our data points are bound to include some errors: not just random ones but systematic errors. Sometimes, the robustly shared verdict of native speakers on a sample sentence will be simply false, the predictable output of a fallible human heuristic. Complicating one's semantic theory to fit all such data points is bad science.

The appropriate response to these difficulties is not scepticism but better science: specifically, a more sophisticated treatment of the data. We must identify the heuristics in play, and allow for their role in the generation of the data. That will not make the semantics redundant, for it is needed to define a target for the heuristic to aim at. Without the semantics, the heuristics are pointless. Indeed, by taking account of the heuristics in moulding speaker verdicts, one can greatly simplify the semantics, so that it defines a target more worth aiming at. This book has explained how that can be done in the case of 'if' and 'would'. It has also sketched a more general sort of heuristic relevant to vague language quite generally. But the implications of heuristics for semantics are likely to be much more extensive than that. It is a field ripe for study.

References

Abbott, Barbara 2004: 'Some remarks on indicative conditionals', *Proceedings of SALT: Semantics and Linguistic Theory*, 14: 1–19.

Adams, Ernest 1965: 'The logic of conditionals', *Inquiry*, 8: 166–97.

Adams, Ernest 1975: *The logic of conditionals*. Dordrecht: Reidel.

Anderson, Alan Ross 1951: 'A note on subjunctive and counterfactual conditionals', *Analysis*, 12: 35–8.

Barker, Stephen 1997: 'Material implication and general indicative conditionals', *Philosophical Quarterly*, 47: 195–211.

Belnap, Nuel 1970: 'Conditional assertion and restricted quantification', *Noûs*, 4: 1–12.

Belnap, Nuel 1973: 'Restricted quantification and conditional assertion', in Hugues Leblanc (ed.), *Truth, Syntax and Modality*: 48–75. Amsterdam: North-Holland.

Bennett, Jonathan 2003: *A Philosophical Guide to Conditionals*. Oxford: Clarendon Press.

Berto, Francesco, French, Rohan, Priest, Graham, and Ripley, Dave 2018: 'Williamson on counterpossibles', *Journal of Philosophical Logic*, 47: 693–713.

Bradley, Richard. 2002. 'Indicative conditionals', *Erkenntnis*, 56: 345–78.

Brogaard, Berit, and Salerno, Joe 2013: 'Remarks on counterpossibles', *Synthese*, 190: 639–60.

Byrne, Ruth, and Johnson-Laird, Philip 2009: ' "If" and the problems of conditional reasoning', *Trends in Cognitive Science*, 13: 282–7.

Cariani, Fabrizio, and Santorio, Paolo 2018: '*Will* done better: selection semantics, future credence, and indeterminacy', *Mind*, 127: 129–65.

Carnap, Rudolf 1947: *Meaning and Necessity: A Study in Semantics and Modal Logic*. Chicago: Chicago University Press.

Davies, Martin, and Humberstone, Lloyd 1980: 'Two notions of necessity', *Philosophical Studies*, 38: 1–30.

de Finetti, Bruno 1935: 'La logique de la probabilité', in *Actes du Congrès International de Philosophie Scientifique, Sorbonne, Paris, 1935, IV: Induction et Probabilité*. Paris: Hermann and Cle: 31–9.

de Finetti, Bruno 1995: 'The logic of probability', *Philosophical Studies*, 77: 181–90. Translation by R.B. Angell of de Finetti 1935.

Dietz, Christina 2019: 'Conditional emotions', MS.

Dorr, Cian 2008: 'There are no abstract objects', in Ted Sider, John Hawthorne, and Dean Zimmerman (eds.), *Contemporary Debates in Metaphysics*. Oxford: Blackwell.

Dorr, Cian 2018: 'Conditionals under quantifiers'. Hand-out.

Dorr, Cian, and Goodman, Jeremy 2019: 'Diamonds are forever', *Noûs*.

Dorr, Cian, and Hawthorne, John 2018: *If . . . : A Theory of Conditionals*. MS.

Douven, Igor 2008: 'Kaufmann on the probabilities of conditionals', *Journal of Philosophical Logic*, 37: 259–66.

Douven, Igor 2016: *The Epistemology of Indicative Conditionals: Formal and Empirical Approaches*. Cambridge: Cambridge University Press.

Dummett, Michael 1959: 'Truth', *Proceedings of the Aristotelian Society*, 59: 141–62.

Dummett, Michael 1973: *Frege: Philosophy of Language*. London: Duckworth.

Dummett, Michael. 1982: 'Realism', *Synthese*, 52: 55–112.

Dummett, Michael 1992: *The Logical Basis of Metaphysics*. London: Duckworth.

Edgington, Dorothy 1986: 'Do conditionals have truth-conditions?', *Crítica*, 18: 1–30.

Edgington, Dorothy 1995: 'On conditionals', *Mind*, 104: 235–329.

Edgington, Dorothy 1997: 'Commentary', in Woods 1997: 93–137.

Edgington, Dorothy 2008: 'Counterfactuals', *Proceedings of the Aristotelian Society*, 108: 1–21.

Eklund, Matti 2002: 'Inconsistent languages', *Philosophy and Phenomenological Research*, 64: 251–75.

Evans, Jonathan, Handley, Simon, and Over, David 2005: 'Suppositions, extensionality, and conditionals: a critique of the mental model theory of Johnson-Laird and Byrne (2002)', *Psychological Review*, 112: 1040–52.

Evans, Jonathan, and Over, David 2004: *If.* Oxford: Oxford University Press.

Field, Hartry 1980: *Science without Numbers: A Defence of Nominalism.* Oxford: Blackwell.

Field, Hartry 1989: *Realism, Mathematics, and Modality.* Oxford: Blackwell.

Fine, Kit 1994: 'Essence and modality', *Philosophical Perspectives*, 8: 1–16.

Fine, Kit 2012: 'Counterfactuals without possible worlds', *Journal of Philosophy*, 109: 221–46.

Fine, Kit 2017: 'Truthmaker semantics', in Bob Hale, Crispin Wright, and Alexander Miller (eds.), *A Companion to the Philosophy of Language, Second Edition*, vol. II. Oxford: Blackwell: 556–77.

von Fintel, Kai 1994: *Restrictions on Quantifier Domains.* Ph.D. thesis, University of Massachusetts at Amherst.

von Fintel, Kai 1998a: 'The presupposition of subjunctive conditionals'. In U. Sauerland and O. Percus (eds.), *The Interpretive Tract* (MIT Working Papers in Linguistics 25): 29–44.

von Fintel, Kai 1998b: 'Quantifiers and "if"-clauses', *Philosophical Quarterly*, 48: 209–14.

von Fintel, Kai 2001: 'Counterfactuals in a dynamic context', in Michael Kenstowicz (ed.) *Ken Hale: A Life in Language*: 1233–152. Cambridge, MA: MIT Press.

von Fintel, Kai, and Gillies, Anthony 2010: '*Must...stay...strong!*', *Natural Language Semantics*, 18: 351–83.

von Fintel, Kai, and Iatridou, Sabine 2002: 'If and when *if*-clauses can restrict quantifiers'. http://citeseerx.ist.psu.edu/viewdoc/download?doi=10.1.1.11.7629&rep=rep1&type=pdf

van Fraassen, Bas 1980: 'Review of Brian Ellis, *Rational Belief Systems*', *Canadian Journal of Philosophy*, 10: 497–511.

van Fraassen, Bas 1981: 'Essences and laws of nature', in R. Healey (ed.), *Reduction, Time and Reality.* Cambridge: Cambridge University Press: 189–200.

Geach, Peter 1962: *Reference and Generality.* Ithaca, NY: Cornell University Press.

Geurts, Bart 2004: 'On an ambiguity in quantified conditionals'. http://ncs.ruhosting.nl/bart/papers/conditionals.pdf

Gibbard, Allan 1981: 'Two recent theories of conditionals', in W. Harper, R. Stalnaker, and C. Pearce (eds.), *Ifs.* Dordrecht: Reidel: 211–47.

Gigerenzer, Gerd, Hertwig, Ralph, and Pachur, Thorsten (eds.) 2011: *Heuristics: The Foundations of Adaptive Behavior.* New York: Oxford University Press.

Gillies, Anthony 2009: 'On truth-conditions for *if* (but not quite only *if*)', *Philosophical Review*, 118: 325–49.

Gillies, Anthony 2010: 'Iffiness', *Semantics and Pragmatics*, 3: 1–42.

Grice, Paul 1989: *Studies in the Ways of Words.* Cambridge, Mass.: Harvard University Press.

Hájek, Alan 1989: 'Probabilities of conditionals—revisited', *Journal of Philosophical Logic*, 18: 423–8.

Hájek, Alan 2009: 'Most counterfactuals are false'. https://docplayer.net/32738-Most-counterfactuals-are-false-alan-hajek.html

Hájek, Alan 2011: 'Triviality pursuit', *Topoi*, 30: 3–15.

Hájek, Alan 2012: 'The fall of "Adams' Thesis"?', *Journal of Logic, Language and Information*, 21: 145–61.

Hájek, Alan, and Hall, Ned 1994: 'The hypothesis of the conditional construal of conditional probability', in Ellery Eells and Brian Skyrms (eds.), *Probabilities and Conditionals: Belief Revision and Rational Decision*. Cambridge: Cambridge University Press: 75–110.

Higginbotham, James 1986: 'Linguistic theory and Davidson's program in semantics', in Ernest LePore (ed.), *Truth and Interpretation: Perspectives on the Philosophy of Donald Davidson*. Oxford: Blackwell: 29–48.

Higginbotham, James 2003: 'Conditionals and compositionality', *Philosophical Perspectives*, 17: 181–94.

Huitink, Janneke 2009: 'Domain restriction by conditional connectives'. https://www.semanticsarchive.net/Archive/zg2MDM4M/Huitink-domainrestriction.pdf

Huitink, Janneke 2010: 'Quantified conditionals and compositionality', *Language and Linguistics Compass*, 4: 42–53.

Iatridou, Sabine 2000: 'The grammatical ingredients of counterfactuality', *Linguistic Inquiry*, 31: 231–70.

Jackson, Frank 1979: 'On assertion and indicative conditionals', *Philosophical Review*, 88: 565–89.

Jackson, Frank 1987: *Conditionals*. Oxford: Blackwell.

Johnson-Laird, Philip, and Byrne, Ruth 2002: 'Conditionals: A theory of meaning, pragmatics, and inference', *Psychological Review*, 109: 646–78.

Kahneman, Daniel, Slovic, Paul, and Tversky, Amos 1982: *Judgment under Uncertainty: Heuristics and Biases*, Cambridge: Cambridge University Press.

Kasper, Walter 1992: 'Presuppositions, composition, and simple subjunctives', *Journal of Semantics*, 9: 307–31.

Kaufmann, Stefan 2004: 'Conditioning against the grain: abduction and indicative conditionals', *Journal of Philosophical Logic*, 33: 583–606.

Kaufman, Stefan 2009: 'Conditionals right and left: probabilities for the whole family', *Journal of Philosophical Logic*, 38: 1–53.

Kaufmann, Stefan 2015: 'Conditionals, conditional probabilities, and conditionalization', in Hans-Christian Schmitz and Henk Zeevat (eds.), *Bayesian Natural Language Semantics and Pragmatics*. Berlin: Springer: 71–94.

Khemlani, Sangeet, Byrne, Ruth, and Johnson-Laird, Philip 2018: 'Facts and possibilities: a model-based theory of sentential reasoning', *Cognitive Science*, 42: 1887–924.

Khoo, Justin 2015: 'On indicative and subjunctive conditionals', *Philosophers' Imprint*, 15.32: 1–40.

Klinedinst, Nathan 2011: 'Quantified conditionals and conditional excluded middle', *Journal of Semantics*, 28: 149–70.

Kment, Boris 2014: *Modality and Explanatory Reasoning*. Oxford: Oxford University Press.

Kneale, William, and Kneale, Martha 1962: *The Development of Logic*. Oxford: Clarendon Press.

Koralus, Philipp, forthcoming: *Reason and Inquiry: The Erotetic Theory*. Oxford: Oxford University Press.

Koralus, Philipp, and Mascarenhas, Salvador 2013: 'The erotetic theory of reasoning: bridges between formal semantics and the psychology of deductive inference', *Philosophical Perspectives*, 27: 312–65.

Kratzer, Angelika 1981: 'The notional category of modality', in Hans-Jurgen Eikmeyer and Hannes Rieser (eds.), *Words, Worlds, and Contexts: New Approaches in Word Semantics*. Berlin: de Gruyter: 38–74.

Kratzer, Angelika 1986: 'Conditionals', in Anne Farley, Peter Farley, and Karl Eric McCollough (eds.), *Papers from the Parasession on Pragmatics and Grammatical Theory*. Chicago: Chicago Linguistics Society: 115–35.

Kratzer, Angelika 2012: *Modals and Conditionals*. Oxford: Oxford University Press.

Kratzer, Angelika 2020: 'Chasing hook: quantified indicative conditionals', forthcoming in John Hawthorne and Lee Walters (eds.), *Conditionals, Probability, and Paradox: Themes from the Philosophy of Dorothy Edgington*. Oxford: Oxford University Press.

Kripke, Saul 1980: *Naming and Necessity*. Oxford: Blackwell.

Leslie, Sarah-Jane 2009: '"If", "unless", and quantification', in Robert Stainton and Christopher Viger (eds.), *Compositionality, Context and Semantic Value: Essays in Honour of Ernie Lepore*. Dordrecht: Springer: 3–30.

Lewis, David 1970: 'General semantics', *Synthese*, 22: 18–67.

Lewis, David 1973: *Counterfactuals*. Oxford: Blackwell. Page references to 2nd ed., 1986.

Lewis, David 1975: 'Adverbs of quantification', in E.L. Keenan (ed.), *Formal Semantics of Natural Languages*. Cambridge: Cambridge University Press: 3–15.

Lewis, David 1976: 'Probabilities of conditionals and conditional probabilities', *Philosophical Review*, 85: 297–315.

Lewis, David 1979: 'Counterfactual dependence and time's arrow', *Noûs*, 13: 455–76.

Lewis, David 1986a: 'Postscript to "Probabilities of conditionals and conditional probabilities"', in his *Philosophical Papers*, vol. 2. Oxford: Oxford University Press: 152–6.

Lewis, David 1986b: 'Probabilities of conditionals and conditional probabilities II', *Philosophical Review*, 95: 581–9.

Lowe, E.J. 1990: 'Conditionals, context and transitivity', *Analysis*, 50: 80–7.

Lycan, William 2001: *Real Conditionals*. Oxford: Oxford University Press.

Mackay, John 2015: 'Actuality and fake tense in conditionals', *Semantics and Pragmatics*, 8: 1–12.

Malmgren, Anna-Sara 2011: 'Rationalism and the content of intuitive judgments', *Mind*, 120: 263–327.

Mandelkern, Matthew 2018: 'Import-Export and "and"', *Philosophy and Phenomenological Research*.

Mandelkern, Matthew 2019: 'What "must" adds', *Linguistics and Philosophy*, 42: 225–66.

McCullagh, Mark, and Yli-Vakkuri, Juhani (eds.) 2017: *Williamson on Modality*, London: Routledge.

McGee, Vann 1985: 'A counterexample to modus ponens', *Journal of Philosophy*, 82: 462–71.

McGee, Vann 1994: 'Learning the impossible', in Ellery Eeells and Brian Skyrms (eds.), *Probability and Conditionals*. Cambridge: Cambridge University Press: 179–99.

McGee, Vann 2000: 'To tell the truth about conditionals', *Analysis*, 60: 107–11.

Moss, Sarah 2012: 'On the pragmatics of counterfactuals', *Noûs*, 46: 561–86.

Moss, Sarah 2018: *Probabilistic Knowledge*. Oxford: Oxford University Press.

Nolan, Daniel 1997: 'Impossible worlds: a modest approach', *Notre Dame Journal for Formal Logic*, 38: 535–72.

Nozick, Robert 1981: *Philosophical Explanations*. Oxford: Clarendon Press.

Oaksford, Mike, and Chater, Nick 2007: *Bayesian Rationality*. Oxford: Oxford University Press.

Peters, Stanley, and Westerståhl, Dag 2006: *Quantifiers in Language and Logic*. Oxford: Clarendon Press.

Pizzi, Claudio 1977: 'Boethius' thesis and conditional logic', *Journal of Philosophical Logic*, 6: 283–302.

Pizzi, Claudio, and Williamson, Timothy 1997: 'Strong Boethius' thesis and consequential implication', *Journal of Philosophical Logic*, 26: 569–88.

Pizzi, Claudio, and Williamson, Timothy 2005: 'Conditional excluded middle in systems of consequential implication', *Journal of Philosophical Logic*, 34: 333–62.

Ramsey, Frank 1929: 'General propositions and causality', reprinted in Hugh Mellor (ed.), *Foundations: Essays in Philosophy, Logic, Mathematics and Economics*: 133–51. London: Routledge & Kegan Paul, to which page numbers refer.

Rieger, Adam 2006: 'A simple theory of conditionals', *Analysis*, 66: 233–40.

Rieger, Adam 2013: 'Conditionals are material: the positive argument', *Synthese*, 190: 3161–74.

Rieger, Adam 2015: 'Defending a simple theory of conditionals', *American Philosophical Quarterly*, 52: 253–60.

Rothschild, Daniel 2013: 'Do indicative conditionals express propositions?', *Noûs*, 47: 49–68.

Rumfitt, Ian 2013: 'Old Adams buried', *Analytic Philosophy*, 54: 157–88.

Russell, Bertrand 1950: *An Inquiry into Meaning and Truth*. London: Allen and Unwin.

Salmon, Nathan 1982: *Reference and Essence*. Oxford: Blackwell.

Salmon, Nathan 1989: 'The logic of what might have been', *Philosophical Review*, 98: 3–34.

Schulz, Katrin 2014: 'Fake tense in conditional sentences: a modal approach', *Natural Language Semantics*, 22: 117–44.

Schulz, Katrin 2017: 'Fake perfect in X-marked conditionals', *Proceedings of SALT: Semantics and Linguistic Theory*, 27: 547–70.

Sennet, Adam, and Weisberg, Jonathan 2012: 'Embedding "if and only if"', *Journal of Philosophical Logic*, 41: 449–60.

Sosa, Ernest 2007: *A Virtue Epistemology: Apt Belief and Reflective Knowledge, Volume I*. Oxford: Clarendon Press.

Stalnaker, Robert 1968: 'A theory of conditionals', *American Philosophical Quarterly Monographs*, 2: 98–112.

Stalnaker, Robert 1970: 'Probability and conditionals', *Philosophy of Science*, 37: 64–80.

Stalnaker, Robert 1975: 'Indicative conditionals', *Philosophia*, 5: 269–86.

Stalnaker, Robert 1981: 'A defense of conditional excluded middle', in William Harper, Robert Stalnaker, and Glenn Pearce (eds.), *Ifs: Conditionals, Belief, Decision, Chance, and Time*. Dordrecht: Reidel: 87–104.

Stalnaker, Robert 1984: *Inquiry*. Cambridge, Mass.: MIT Press.

Starr, William 2014: 'A uniform theory of conditionals', *Journal of Philosophical Logic*, 43: 1019–24.

Sterken, Rachel 2015: 'Generics, content and cognitive bias', *Analytic Philosophy*, 56: 75–93.

Strawson, Peter 1952: *Introduction to Logical Theory*. London: Methuen.

Veltman, Frank 1986: 'Data semantics and the pragmatics of indicative conditionals', in Elizabeth Closs Traugott et al. (eds.), *On Conditionals*. Cambridge: Cambridge University Press.

Vetter, Barbara 2016: 'Williamsonian modal epistemology, possibility based', *Canadian Journal of Philosophy*, 46: 766–795, and in McCullagh and Yli-Vakkuri 2017: 314–43.

Warmbrod, Ken 1981: 'Counterfactuals and substitution of equivalent antecedents', *Journal of Philosophical Logic*, 10: 267–89.

Warmbrod, Ken 1982: 'A defense of the Limit Assumption', *Philosophical Studies*, 42: 53–66.

Weisberg, Jonathan 2019: 'Belief in psyontology', *Philosophers' Imprint*.

Weisberg, Michael 2013: *Simulation and Similarity: Using Models to Understand the World*. Oxford: Oxford University Press.

Willer, Malte 2010: 'New surprises for the Ramsey test', *Synthese*, 176: 291–309.

Williams, J. Robert G. 2008: 'Conversation and conditionals', *Philosophical Studies*, 138: 211–23.

Williams, J. Robert G. 2010: 'Defending conditional excluded middle', *Noûs*, 44: 650–68.

Williamson, Timothy 1988: 'Bivalence and subjunctive conditionals', *Synthese*, 75: 405–21.

Williamson, Timothy 1994: *Vagueness*. London: Routledge.

Williamson, Timothy 2000: *Knowledge and its Limits*. Oxford: Oxford University Press.

Williamson, Timothy 2006: 'Indicative versus subjunctive conditionals, congruential versus non-hyperintensional contexts', in E. Sosa and E. Villanueva (eds.), *Philosophical Issues, Volume 16: Philosophy of Language*. Oxford: Blackwell: 310–33.

Williamson, Timothy 2007a: *The Philosophy of Philosophy*. Oxford: Blackwell.

Williamson, Timothy 2007b: 'How probable is an infinite sequence of heads?', *Analysis*, 67: 173–80.

Williamson, Timothy 2009: 'Conditionals and actuality', *Erkenntnis*, 70: 135–50.

Williamson, Timothy 2013: *Modal Logic as Metaphysics*. Oxford: Oxford University Press.

Williamson, Timothy 2016a: 'Modal Science', *Canadian Journal of Philosophy*, 46: 453–92, and in McCullagh and Yli-Vakkuri 2017: 1–40.

Williamson, Timothy 2016b: 'Reply to Vetter', *Canadian Journal of Philosophy*, 46: 796–802, and in McCullagh and Yli-Vakkuri 2017: 344–50.

Williamson, Timothy 2016c: 'Knowing by imagining', in Amy Kind and Peter Kung (eds.), *Knowledge through Imagination*. Oxford: Oxford University Press: 113–23.

Williamson, Timothy 2017a: 'Counterpossibles in semantics and metaphysics', *Argumenta*, 4. https://www.argumenta.org/article/counterpossibles-semantics-metaphysics/

Williamson, Timothy 2017b: 'Model-building in philosophy', in Russell Blackford and Damian Broderick (eds.), *Philosophy's Future: The Problem of Philosophical Progress*. Oxford: Wiley-Blackwell: 159–73.

Williamson, Timothy 2018: 'Counterpossibles', *Topoi*, 37: 357–68.

Williamson, Timothy, forthcoming: 'The counterfactual-based approach to modal epistemology', in Otávio Bueno and Scott Shalkowski (eds.), *Routledge Handbook of Modality*, London: Routledge.

Williamson, Timothy 2019: 'Knowledge, credence, and the strength of belief'. http://media.philosophy.ox.ac.uk/docs/people/williamson/Strengthofbelief.pdf

Woods, Michael 1997: *Conditionals*, ed. David Wiggins. Oxford: Clarendon Press.

Wright, Crispin 1976: 'Language-mastery and the sorites paradox', in Gareth Evans and John McDowell (eds.), *Truth and Meaning: Essays in Semantics*. Oxford: Clarendon Press: 223–47.

Wright, Crispin 1983: 'Keeping track of Nozick', *Analysis*, 43: 134–40.

Yalcin, Seth 2007: 'Epistemic modals', *Mind*, 116: 983–1026.

Yalcin, Seth 2012: 'A counterexample to Modus Tollens', *Journal of Philosophical Logic*, 41: 1001–24.

Index